Apollo 13

The NASA Mission Reports

Compiled from the NASA archives & Edited
by Robert Godwin

"The Apollo 13 accident, which aborted man's third mission to explore the surface of the moon, is a harsh reminder of the immense difficulty of this undertaking.

The total Apollo system of ground complexes, launch vehicle, and spacecraft constitutes the most ambitious and demanding engineering development ever undertaken by man. For these missions to succeed, both men and equipment must perform to near perfection. That this system has already resulted in two successful lunar surface explorations is a tribute to those men and women who conceived, designed, built, and flew it.

Perfection is not only difficult to achieve, but difficult to maintain. The imperfection in Apollo 13 constituted a near disaster, averted only by outstanding performance on the part of the crew and the ground control team which supported them.

The Board feels that the nature of the Apollo 13 equipment failure holds important lessons which, when applied to future missions, will contribute to the safety and effectiveness of manned space flight."

Thomas Paine (Administrator NASA to the Accident Investigation Committee)

We acknowledge the financial support of the Government of Canada through the
Book Publishing Industry Development Program for our publishing activities.
Published by Apogee Books an imprint of Collector's Guide Publishing Inc., Box 62034, Burlington, Ontario, Canada, L7R 4K2
Printed and bound in Canada
Apollo 13 - The NASA Mission Reports
by Robert Godwin
ISBN 1-896522-55-6

Introduction

The previous volumes in this series of books have illustrated the enormous scope of the effort required to send men to the moon. Listing the vast resources that were marshalled to support that effort would soon exhaust this writer's supply of superlatives.

The United States had garnered unprecedented respect around the world for these feats of technological prowess while simultaneously developing an industry of air and space engineering unparalleled in the history of the world. NASA had fulfilled their mandate before the end of the 60's and yet by April 1970 a host of factors were conspiring to pull up on the reins and bring about an early end to man's exploration of the moon. By the time Jim Lovell, Fred Haise and Jack Swigert were ready to leave for the Fra Mauro highlands the budgets were already in jeopardy and the American public were showing signs of apathy towards the spectacular events transpiring a quarter of a million miles away. It would take an unforeseen adventure of epic proportions to reignite the country's interest.

No doubt if the great Homer had been alive he would have immortalised the crew of the Odyssey in a tale of danger and peril to rival Ulysses own; but it would be more than 20 years before the flight of Apollo 13 would finally ascend the lofty heights of legend with the release of Ron Howard's masterpiece motion picture. Based on Jim Lovell's book *Lost Moon* the movie *Apollo 13* went a long way towards finally clearing NASA's track record.

In 1970, even though the people who were involved knew they had accomplished an incredible achievement, it was difficult to put a positive spin on the flight of Apollo 13. President Nixon proclaimed the flight a success, but NASA's enemies were circling. The NASA team not only had to admit that the flight was a failure (and I use the word grudgingly), but they then had to endure a lengthy interrogation about what went wrong.

Most of the inquiry concentrated on the ill-fated Oxygen tank number 2 but it was clear that the elected officials of the US Congress and Senate were looking for scapegoats. The climate was still grim as the Vietnam war consumed the political agenda and many elected officials were questioning the validity of further manned lunar exploration. Little time was available to comprehend the implications of inflammable insulation and unconventional methods of emptying a cryogenic oxidizer tank.

This book is inevitably different to its predecessors. The crew of Apollo 13 did not fulfill their destiny and land in the lunar highlands, as a result of this the Press Kit and Pre-Operation Mission Reports show what *should* have happened. The Post - Operation report and the Technical Debriefing relate, in a very business-like and perfunctory fashion, what did happen. Following these are the Accident Hearings and, as you will see, little mention is made of the ingenuity, sacrifice and bravery of the crew and controllers of Apollo 13. The investigation, perhaps rightly, concentrated solely on pinpointing the problem. This was a necessary evil designed to protect the lives of future crews.

What is almost completely lost in the mass of *Findings* and *Recommendations* is the triumph of resourcefulness that was the reality of Apollo 13. Not only in the creativity utilized to bring the crew home but in the collection of brilliantly engineered and applied science that put the crew in space in the first place.

One tiny switch out of millions of components failed and NASA was forced to circle the wagons and explain just how such a thing could happen. As many of NASA's heirarchy would often plead, "Going to the moon is expensive and dangerous." The proof of the care that went into project Apollo is that *more* things didn't go wrong.

One of the amazing discoveries uncovered by the Review Board was that the explosion in Bay 4 of the service module would have created a pressure build-up of 20-25 psi before blowing off the outer panel, and yet the report says that the Command Module was attached to the Service Module with a prescribed limit of only 10 psi.* This clearly illustrates just how close to a catastrophic failure the accident might have been.

* Pages 172 and 178

It seems that the propagation of hot gasses was such that it blew the panel before it ejected the CM. If one further examines the location of oxygen tank 2 you can see that its rupture (probably in an upward direction along the central axis of the SM) would inevitably damage the three fuel cells. Fortunately for the crew, the liquid hydrogen tank is below the ill-fated O2 tank. Meanwhile the SPS propulsion system with its huge tank of hypergolic fuel is located in the adjacent section of the SM through another small bulkhead. Speculations about any slight change in events always leads to the same conclusion of what might have happened.

Fortunately for Lovell, Haise and Swigert the weak spot on the tank was at the top where there was a blow-out disc. This forced the expanding gasses away from the explosive fuels. Even then the damage was favourably spread such that there was enough power in one of the fuel cells to allow the transfer of the navigation data to the LM computer before the CM died.

The timing of the explosion was also critical. Any earlier in the flight and they probably would have run out of air or electrical power before getting home and any later and they may have already committed to lunar orbit, or even worse, lunar descent. At that point the Lunar Module Aquarius would no longer be a viable life boat. The explosion taking place, where it did, when it did and the way it did all contributed to a possible rescue.

Just as Ulysses was strapped to the mast (albeit voluntarily) the crew of Apollo 13 were fated to continue on their own path as witnesses to the terrifying beauty of their surroundings. As the moon beckoned with siren-like beauty, Odyssey and Aquarius adjusted course to take the crew homewards.

Just as Apollo 13 was unique, so do the following pages reveal Apollo in a totally unique way. Putting aside the high drama of the events, the following documents reveal a side of NASA that is often overlooked, the talents of the management and administrators. It is a tribute to their perseverance, and the expertise of the technicians, that they could deliver such a complicated and detailed report to Congress in such a timely fashion. Based on the testimony of the crew along with the telemetry and a couple of fuzzy pictures, the team of experts isolated the sequence of events leading to the accident and recreated it in such detail that the future of Apollo was never seriously jeopardised. Apollo 14 was able to go ahead with only a minimal delay and fly an almost perfect mission.

Throughout history man's great exploratory endeavours have been plagued by mishaps and very often fatal accidents. The fulfillment of great deeds is often accompanied by great losses. History will show that the flight of Apollo 13 was a seminal event in the annals of man's exploration of the universe. It may well define the moment when humanity's ability to cope with the unknown and compensate appropriately finally reached maturity. Homer would have been proud.

Robert Godwin
(Editor)

This book is dedicated to the crew of the Odyssey.
James Lovell, Fred Haise and Jack Swigert.

Special thanks to Jim Lovell and Dr Jim Busby

Apollo 13

The NASA Mission Reports

(from the archives of the National Aeronautics and Space Administration)

PRESS KIT

PRE MISSION REPORT

LIST OF FIGURES

LIST OF TABLES

POST LAUNCH MISSION OPERATION REPORT

LIST OF TABLES

DEBRIEFING

ACCIDENT HEARINGS

STATEMENTS

Abbreviations and Acronyms See Acronyms.htm on the CDROM

Apollo 13 Press Kit

NATIONAL AERONAUTICS & SPACE ADMINISTRATION
NEWS RELEASE NO: 70-50K
FOR RELEASE: THURSDAY A.M. April 2, 1970
PROJECT: APOLLO 13

APOLLO 13 — THIRD LUNAR LANDING MISSION

Apollo 13, the third U.S. manned lunar landing mission, will be launched April 11 from Kennedy Space Center Fla., to explore a hilly upland region of the Moon and bring back rocks perhaps five billion years old.

The Apollo 13 lunar module will stay on the Moon more than 33 hours and the landing crew will leave the spacecraft twice to emplace scientific experiments on the lunar surface and to continue geological investigations. The Apollo 13 landing site is in the Fra Mauro uplands the two National Aeronautics and Space Administration previous landings were in mare or "sea" areas, Apollo 11 in the Sea of Tranquility and Apollo 12 in the Ocean of Storms.

Apollo 13 crewmen are commander James A. Lovell, Jr., command module pilot, Thomas K. Mattingly II, and lunar module pilot Fred W. Haise, Jr.. Lovell is a U.S. Navy captain, Mattingly a Navy lieutenant commander and Haise a civilian.

Launch vehicle is a Saturn V. Apollo 13 objectives are:

* Perform selenological inspection, survey and sampling of materials in a pre-selected region of the Fra Mauro formation.
* Deploy and activate an Apollo Lunar Surface Experiment Package (ALSEP).
* Develop man's capability to work in the lunar environment.
* Obtain photographs of candidate exploration sites. Currently 11 television transmissions in color are scheduled: one in Earth orbit an hour and a half after launch, three on the outward voyage to the Moon; one of the landing site from about nine miles up; two from the lunar surface while the astronauts work outside the spacecraft ; one at the command service module/lunar module docking operation; one of the Moon from lunar orbit; and two on the return trip.

The Apollo 13 landing site is in the hilly uplands to the north of the crater Fra Mauro. Lunar coordinates for the landing site are 3.6 degrees South latitude by 17.5 degrees west longitude, about 95.6 nautical miles east of the Apollo 12 landing point at Surveyor III crater.

Experiments emplaced at the Fra Mauro site as part of the ALSEP III will gather and relay long-term scientific data to Earth for at least a year on the Moon's physical and environmental properties. Five experiments are contained in the ALSEP: a lunar passive seismometer will measure and relay meteoroid impacts and moonquakes; a heat flow experiment will measure the heat flux from the lunar interior to the surface and conductivity of the surface materials to a depth of about 10 feet; a charged particle lunar environment experiment will measure protons and electrons to determine the effect of the solar wind on the lunar environment; a cold cathode gauge experiment will measure density and temperature variations in the lunar atmosphere; and a dust detector experiment.

The empty third stage of the Saturn V launch vehicle will be targeted to strike the Moon before the lunar landing and its impact will be recorded by the seismometer left by the Apollo 12 astronauts last November. The spent lunar module ascent stage, as in Apollo 12, will be directed to impact the Moon after rendezvous and final LM separation to provide a signal to both seismometers.

Candidate future Apollo landing sites — Censorinus, Davy Rille, and Descartes — will be photographed with a large-format lunar topographic camera carried for the first time on Apollo 13. The lunar topographic camera will make high resolution 4.5 inch square black and white photos in overlapping sequence for mosaics or in single frames. The camera mounts in the command module crew access hatch window when in use. After lunar orbit rendezvous with the lunar module and LM jettison the command module will make a plane-change maneuver to drive the orbital track over Descartes and Davy Rille for topographic photography.

The Apollo 13 flight profile in general follows those flown by Apollos 11 and 12 with one major exception: lunar orbit insertion burn no. 2 has been combined with descent orbit insertion and the docked spacecraft will be placed into a 7x57 nautical mile lunar orbit by use of the service propulsion system. Lunar module descent propellant is conserved by combining these maneuvers to provide 15 seconds of additional hover time during the landing.

Lunar surface touchdown is scheduled to take place at 9:55 p.m. EST April 15, and two periods of extravehicular activity are planned at 2:13 a.m. EST April 16 and 9:58 P.M. EST April 16. The LM ascent stage will lift off at 7:22 a.m. April 17 to rejoin the orbiting command module after more than 33 hours on the lunar surface.

Apollo 13 will leave lunar orbit at 1:42 p.m. EST April 18 for return to Earth. Splashdown in the mid-Pacific just south of the Equator will be at 3:17 p.m. EST April 21.

After the spacecraft has landed, the crew will put on clean coveralls and filter masks passed in to them through the hatch by a swimmer, and then transfer by helicopter to a Mobile Quarantine Facility (MQF) on the USS Iwo Jima. The MQF and crew will be off-loaded in Hawaii and placed aboard a C-141 aircraft for the flight back to the Lunar Receiving Laboratory at the Manned Spacecraft Center in Houston. The crew will remain in quarantine up to 21 days from completion of the second EVA.

The crew of Apollo 13 selected the call signs Odyssey for the command/service module and Aquarius for the lunar module. When all three crewmen are aboard the command module, the call sign will be "Apollo 13." As in the two previous lunar landing missions, an American flag will be emplaced on the lunar surface. A plaque bearing the date of the Apollo 13 landing and the crew signatures is attached to the LM.

Apollo 13 backup crewmen are USN commander John W. Young, commander; civilian John L. Swigert, Jr., command module pilot; and USAF Major Charles M. Duke, Jr., lunar module pilot.

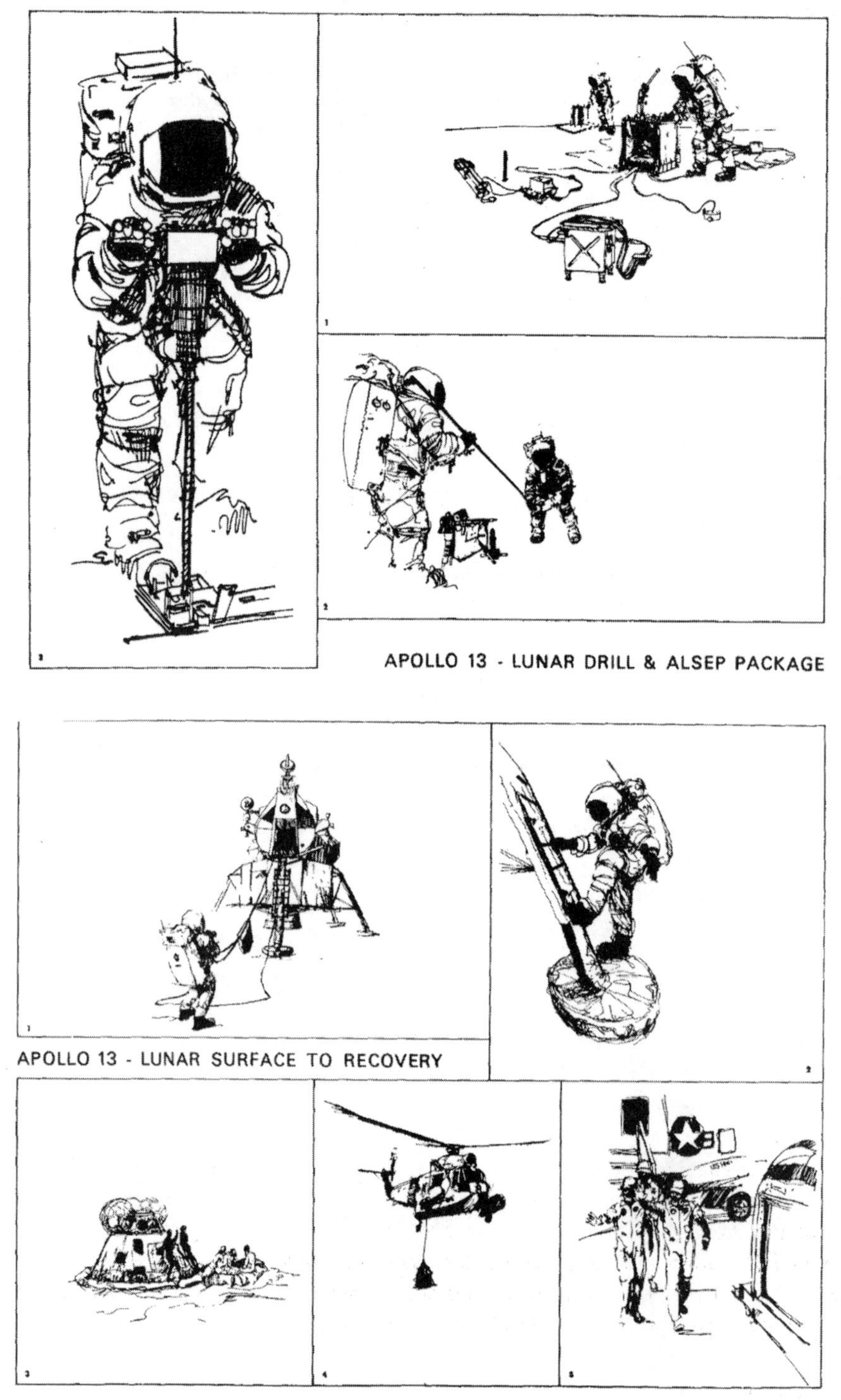

APOLLO 13 - LUNAR DRILL & ALSEP PACKAGE

APOLLO 13 - LUNAR SURFACE TO RECOVERY

APOLLO 13 COUNTDOWN

Pre-count activities for the Apollo 13 launch begin about T-4 days, when the space vehicle will be prepared for the start of the official countdown. During pre-count, final space vehicle ordnance installation and electrical connections will be accomplished. Spacecraft gaseous oxygen and gaseous helium system will be serviced, spacecraft batteries will be installed, and LM and CSM mechanical buildup will be completed. The CSM fuel cells will be activated and CSM cryogenics (liquid oxygen — liquid hydrogen) will be loaded and pressurized.

The countdown for Apollo 13 will begin at T-28 hours and will continue to T-9 hours, at which time a built-in hold is planned prior to the start of launch vehicle propellant loading.

ollowing are some of the major operations in the final count:

28 hours	Official countdown starts, LM stowage and cabin closeout (T-31:30 to T-18:00)
27 hours, 30 minutes	Install and connect LV flight batteries (to T-23 hours)
22 hours, 30 minutes	Top off of LM super critical helium (to T-20 hours, 30 minutes)
19 hours, 30 minutes	LM SHe thermal shield installation (to T-15 hours, 30 minutes) CSM crew stowage (T-19 to T-12 hours, 30 minutes
16 hours	LV range safety checks (to T-15 hours)
15 hours	Installation of ALSEP FCA (to T-14 hours, 45 minutes)
11 hours, 30 minutes	Connect LV safe and arm devices (to 10 hours, 45 minutes) CSM pre-ingress operations (to T-8 hours 45 minutes)
10 hours, 15 minutes	Start MSS move to park site
9 hours	Built-in hold for 9 hours and 13 minutes. At end of hold, pad is cleared for LV propellant loading
8 hours, 05 minutes	Launch vehicle propellant loading Three stages (LOX in first stage, LOX and LH 2 in second and third stages). Continues through T-3 hours 38 minutes
4 hours, 17 minutes	Flight crew alerted
4 hours, 02 minutes	Medical examination
3 hours, 32 minutes	Breakfast
3 hours, 30 minutes	One-hour hold
3 hours, 07 minutes	Depart Manned Spacecraft Operations Building for LC-39 via crew transfer van.
2 hours, 55 minutes	Arrive at LC-39
2 hours, 40 minutes	Start flight crew ingress
2 hours	Mission Control Center — Houston/ spacecraft command checks
1 hour, 55 minutes	Abort advisory system checks
1 hour, 51 minutes	Space Vehicle Emergency Detection System (EDS) test
43 minutes	Retract Apollo access arm to stand by position (12 degrees)
42 minutes	Arm launch escape system
40 minutes	Final launch vehicle range safety checks (to 35 minutes)
30 minutes	Launch vehicle power transfer test LM switch over to internal power
20 minutes to T-10 minutes	Shutdown LM operational instrumentation
15 minutes	Spacecraft to internal power
6 minutes	Space vehicle final status checks
5 minutes, 30 seconds	Arm destruct system
5 minutes	Apollo access arm fully retracted
3 minutes, 7 seconds	Firing command (automatic sequence)
50 seconds	Launch vehicle transfer to internal power
8.9 seconds	Ignition sequence start
2 seconds	All engines running
0	Liftoff

Jote: Some changes in the above countdown are possible as a result of experience gained in the countdown lemonstration test which occurs about 10 days before launch.

ightning Precautions

During the Apollo 12 mission the space vehicle was subjected to two distinct electrical discharge events. However, no serious damage occurred and the mission proceeded to a successful conclusion. Intensive nvestigation led to the conclusion that no hardware changes were necessary to protect the space vehicle rom similar events. For Apollo 13 the mission rules have been revised to reduce the probability that the pace vehicle will be launched into cloud formations that contain conditions conducive to initiating similar electrical discharges although flight into all clouds is not precluded.

May Launch Opportunities

The three opportunities established for May — in case the launch is postponed from April 11 — provide, in effect, the flexibility of a choice of two launch attempts. The optimum May launch window occurs on May 10. The three day window permits a choice of attempting a launch 24 hours earlier than the optimum window and if necessary a further choice of a 24 hour or 48 hour recycle. It also permits a choice of making the first aunch attempt on the optimum day with a 24 hour recycle capability. The May 9 window (T-24 hrs) requires an additional 24 hours in lunar orbit before initiating powered descent to arrive at the landing site at the same time and hence have the same Sun angle for landing as on May 10. Should the May 9 window launch attempt be scrubbed, a decision will be made at that time, based on the reason for the scrub, status of spacecraft cryogenics and weather predictions, whether to recycle for May 10 (T-0 hrs) or May 11 (T+24 hrs). If launched on May 11, the flight plan will be similar for the May 10 mission but the Sun elevation angle at lunar landing will be 18.5° instead of 7.8°.

LAUNCH, MISSION TRAJECTORY AND MANEUVER DESCRIPTION

The information presented here is based on an on-time April 11 launch and is subject to change before or during the mission to meet changing conditions.

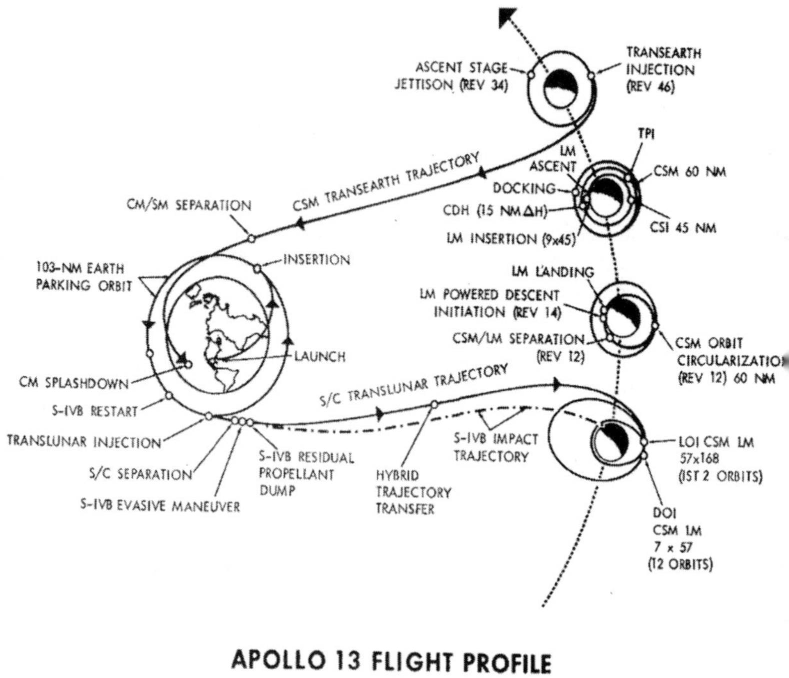

APOLLO 13 FLIGHT PROFILE

Launch

A Saturn V launch vehicle will lift the Apollo 13 spacecraft from Launch Complex 39A, NASA-Kennedy Space Center, Fla. The azimuth may vary from 72 to 96 degrees, depending on the time of launch. The azimuth changes with launch time to permit a fuel-optimum injection from Earth parking orbit to a free return circumlunar trajectory.

April 11 launch plans call for liftoff at 2:13 P.M. EST on an azimuth of 72 degrees. The vehicle will reach a altitude of 36 nautical miles before first stage cut-off 51 nm downrange. During the 2 minutes 44 seconds c powered flight, the first stage will increase vehicle velocity to 7,775 feet per second.* First stage thrust wi reach a maximum of 8,995,108 pounds before center engine cutoff. After engine shutdown and separatio from the second stage, the booster will fall into the Atlantic Ocean about 364 nm downrange from the launc site (30 degrees North latitude and 74 degrees West longitude) about 9 Minutes 4 seconds after liftoff.

The second stage (S-II) will carry the space vehicle to an altitude of 102 nm and a distance of 892 n downrange. At engine shutdown, the vehicle will be moving at a velocity Of 21,508 fps. The four outer J- engines will burn 6 minutes 32 seconds during the powered phase, but the center engine will be cut off minutes 47 seconds after S-II ignition.

At outboard engine cutoff, the S-II will separate and, following a ballistic trajectory, plunge into the Atlanti about 2,450 nm downrange from the Kennedy Space Center (31 degrees North latitude and 33.4 degree West longitude) some 20 minutes 41 seconds after liftoff.

The single engine of the Saturn V third stage (S-IVB) will ignite about 3 seconds after the S-II stage separate: The engine will fire for 143 seconds to insert the space vehicle into a circular Earth parking orbit of 103 nr beginning about 1,468 nm downrange. Velocity at Earth orbital insertion will be 24,243 fps at 11 minutes 5 second ground elapsed time (GET). Inclination will be 33 degrees to the equator.

The crew will have a backup to launch vehicle guidance during powered flight. If the Saturn instrument uni inertial platform fails, the crew can switch guidance to the command module system for first-stage powere flight automatic control. Second and third stage backup guidance is through manual takeover in whic spacecraft commander hand controller inputs are fed through the command module computer to the Satur instrument unit.

*NOTE: Multiply nautical miles by 1.1508 to obtain statute miles; multiply feet per second by 0.6818 to obtain statute miles per hou

Launch Events

Time Hrs. Min. Sec.	Event	Altitude Feet	Velocity Ft/Sec*	Range Naut. Mi.
00 00 00	First Motion	198	0	0
00 01 23	Maximum Dynamic Pressure	42,139	1,600	3
00 02 15	S-IC Center Engine Cutoff	139,856	5,120	24
00 02 44	S-IC Outboard Engines Cutoff	218,277	7,775	51
00 02 45	S-IC/S-II Separation	220,576	7,804	52
00 02 46	S-II Ignition	225,368	7,788	54
00 03 14	S-II Aft Interstage Jettison	300,222	8,173	88
00 03 20	LET Jettison	313,619	8,276	95
00 07 43	S-II Center Engine Cutoff	588,840	17,650	603
00 09 18	S-II Outboard Engines Cutoff	615,508	21,503	892
00 09 19	S-II/S-IVB Separation	615,789	21,517	895
00 09 22	S-IVB Ignition	616,616	21,518	906
00 11 45	S-IVB First Cutoff	627,996	24,239	1,429
00 11 55	Parking Orbit Insertion	628,014	24,243	1,468

* Not including velocity due to Earth's rotation, about 1,350 feet-per-second.

Apollo 13 Mission Events

Events	GET Hrs. Min.	Date/EST	Vel. Change Feet/Sec.	Purpose & Resultant Orbit
Earth orbit insertion	00:11	11 2:24 p.m.	25,593	Insertion into 103 nm circular Earth parking orbit.
Translunar injection	02:35	11 4:48 p.m.	10,437	Injection into free-return (S-IVB Engine ignition) translunar trajectory with 210 nm pericynthion.
CSM separation, docking	03:06	11 5:19 p.m.		Hard-mating of CSM and LM
Ejection from SLA	04:00	11 6:14 p.m.	1	Separates CSM-LM from S-IVB-SLA.
S-IVB evasive maneuver	04:19	11 6:32 p.m.	9.4	Provides separation prior to S-IVB propellant dump and thruster maneuver to cause lunar impact.
Midcourse correction 1	TLI+9 hrs	12 1:54 a.m.	*0	*These midcourse corrections have a nominal velocity change of 0 fps, but will be calculated in real time to correct TLI dispersions
Midcourse correction 2 (Hybrid transfer)	30:41	12 8:54 p.m.	15	MCC-2 is an SPS maneuver (15 fps) to lower pericynthion to 59 nm; trajectory then becomes non-free return.
Midcourse correction 3	LOI-22 hrs.	13 9:39 p.m.	*0	
Midcourse correction 4	LOI- 5 hrs.	14 2:38 p.m.	*0	
Lunar orbit insertion	77:25	14 7:38 p.m.	-2.815	Inserts Apollo 13 into 57x168 nm elliptical lunar orbit.
S-IVB impact	77:46	14 7:59 p.m.		Seismic event.
Descent orbit insertion	81:45	14 11:58 p.m.	-213	SPS burn places CSM/LM into 7x57 nm lunar orbit.
CSM-LM undocking	99:16	15 5:29 p.m.		Establishes equiperiod orbit for 2.5 nm separation at PDI maneuver.
CSM circularization	100:35	15 6:48 p.m.	70	Inserts CSM into 52x62 orbit.
LM powered descent initiation (PDI)	103:31	15 9:44 p.m.	-6635	Three-phase maneuver to brake LM out of transfer orbit, vertical descent and touchdown on lunar surface.
LM touchdown on lunar surface.	103:42	15 9:55 p.m.		Lunar exploration, deploy ALSEP lunar surface geological sample collection, photography.
Depressurization for first surface EVA.	108:00	16 2:13 a.m.		
CDR steps to surface	108:16	16 2:29 a.m.		
CDR collects contingency samples	108:21	16 2:34 a.m.		
LMP steps to surface	108:27	16 2:40 a.m.		
CDR unstows and erects S-band antenna	108:32	16 2:45 a.m.		
LMP mounts TV camera on tripod	108:34	16 2:47 a.m.		
LMP reenters LM to switch to S-Band antenna	108:43	16 2:56 a.m.		
LMP returns to lunar surface	108:57	16 3:10 a.m.		
CDR deploys U.S. flag	109:04	16 3:17 a.m.		
CDR and LMP begin unstowing and deployment of ALSEP	109:30	16 3:43 a.m.		
CDR and LMP return to LM collecting samples en route	111:10	16 5:23 a.m.		
CDR and LMP arrive back at LM, stow gear and samples	111:20	16 5:33 a.m.		
LMP deploys solar wind composition experiment	111:34	16 5:47 a.m.		

LMP reenters LM	111:43	16 5:56 a.m.		
CDR reenters LM	111:58	16 6:11a.m.		
LM hatch closed, repress	111:59	16 6:12 a.m.		
CSM plane change	113:43	16 7:56 a.m.		
Depress for EVA-2	127:45	16 9:58 p.m.		
CDR steps to surface	127:58	16 10:11 p.m.		
LMP steps to surface	128:07	16 10:20 p.m.		
Begin field geology traverse collect core tube and gas analysis samples, dig soil mechanics trench, magnetic sample collection.	126:18	16 10:31 p.m.		
Complete geology traverse	131:04	17 1:17 a.m.		
Return to LM area, retrieve solar wind experiment, stow gear and samples.	131:05	17 1:18 a.m.		
LMP enters LM	131:28	17 1:41 a.m.		
CDR transfers samples, LMP assists	131:35	17 1:48 a.m.		
CDR enters LM, close hatch	131:41	17 1:54 a.m.		
Repress cabin	131:44	17 1:57 a.m.		
LM ascent	137:09	17 7:22 a.m.	6,044	Boosts stage into 9x45 nm lunar orbit for rendezvous with CSM.
Insertion into lunar orbit.	137:16	17 7:29 a.m.		
LM RCS concentric sequence initiation (CSI) burn	138:19	17 8:32 a.m.	50	Raises LM perilune to 44 nm, adjusts orbital shape for rendezvous sequence.
LM RCS constant delta ht (CDH) burn	139:04	17 9:17 a.m.		Radially downward burn adjusts LM orbit to constant 15 nm below CSM.
LM RCS terminal phase	139:46	17 9:59 a.m.	24.7	LM thrusts along line of sight toward CSM, midcourse and braking maneuvers as necessary.
Rendezvous (TPF)	140:27	17 10:40 a.m.		Completes rendezvous sequence.
Docking	140:45	17 10:58 a.m.		Commander and LM pilot transfer back to CSM.
LM jettison, separation (SM RCS)	143:04	17 1:17 p.m.		Prevents recontact of CSM with LM ascent stage during remainder of lunar orbit.
LM ascent stage deorbit (RCS)	144:32	17 2:45 p.m.	-186	Seismometer records impact event.
LM ascent stage Impact	145:00	17 3:13 p.m.		Impact at about 5,508 fps, at -4° angle 35 nm from Apollo 13 ALSEP.
Plane change for photos	154:13	18 12:26 a.m.		Descartes and Davy-Rille photography.
Transearth injection (TEI) SPS	167:29	18 1:42 p.m.	3,147	Inject CSM into transearth trajectory.
Midcourse correction 5	182:31	19 4:44 a.m.	0	Transearth midcourse corrections will be computed in real time for entry corridor control and recovery area weather avoidance.
Midcourse correction 6	EI-22 hrs	20 5:03 p.m.	0	
Midcourse correction 7	EI- 3 hrs	21 12:03 p.m.	0	
CM/SM separation	240:34	21 2:47 p.m.		Command module oriented for entry.
Entry interface (400,000 feet)	240:50	21 3:03 p.m.		Command Module enters Earth's sensible atmosphere at 36,129 fps.
Splashdown	241:04	21 3:17 p.m.		Landing 1,250 nm downrange from entry 1.5° South latitude by 157.5° West longitude.

Earth Parking Orbit (EPO)

Apollo 13 will remain in Earth parking Orbit for one and one-half revolutions. The final "go" for the TLI burn will be given to the crew through the Carnarvon, Australia, Manned Space Flight Network station.

Translunar Injection (TLI)

Midway through the second revolution in Earth parking orbit, the S-IVB third-stage engine will restart at 2:35 GET over the mid-Pacific Ocean near the equator and burn for almost six minutes to inject Apollo 13 toward the Moon. The velocity will increase from 25,593 fps to 36,030 fps at TLI cutoff to a free return circumlunar trajectory from which midcourse corrections could be made with the SM RCS thrusters.

Transposition, Docking, and Ejection (TD&E)

After the TLI burn, the Apollo 13 crew will separate the command/service module from the spacecraft module adapter (SLA), thrust out away from the S-IVB, turn around and move back in for docking with the lunar module. Docking should take place at about three hours and 21 minutes GET. After the crew confirms all docking latches solidly engaged, they will connect the CSM-to-LM umbilicals and pressurize the LM with oxygen from the command module surge tank. At about 4:00 GET, the spacecraft will be ejected from the spacecraft LM adapter by spring devices at the four LM landing gear "knee" attach points. The ejection springs will impart about one fps velocity to the spacecraft. A 9.4 fps S-IVB attitude thruster evasive maneuver in plane at 4:19 GET will separate the spacecraft to a safe distance from the S-IVB.

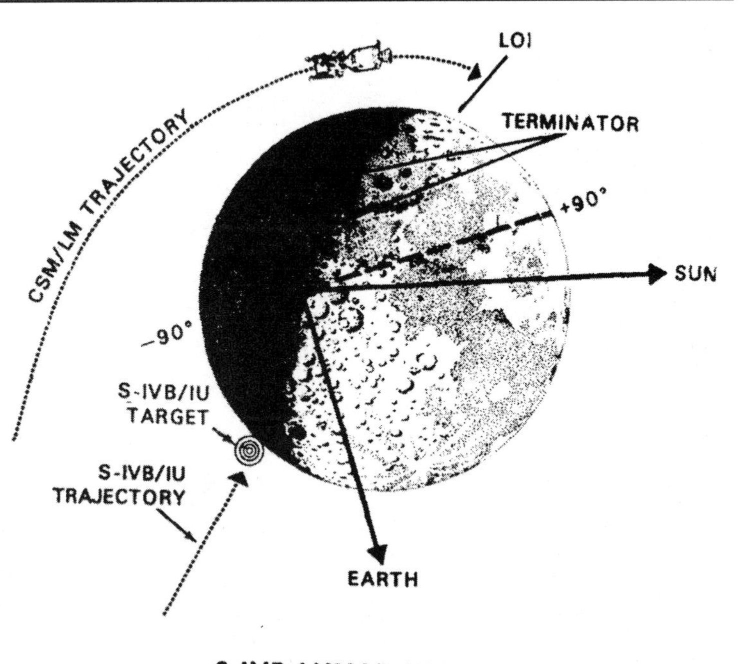

S-IVB LUNAR IMPACT

Saturn Third Stage Lunar Impact

Through a series of pre-set and ground-commanded operations, the S-IVB stage/instrument unit will be directed to hit the Moon within a target area 375 nautical miles in diameter, centered just east of Lansberg D Crater (3 degrees south latitude; 30 degrees West longitude), approximately 124 miles west of the Apollo 12 landing site.

The planned impact will provide a seismic event for the passive seismometer experiment placed on the lunar surface by the Apollo 12 astronauts in November 1969.

The residual propellants in the S-IVB will be used to attempt the lunar impact. Part of the remaining liquid oxygen (LOX) will be dumped through the engine for 48 seconds to slow the vehicle into a lunar impact trajectory. The liquid hydrogen tank's continuous venting system will vent for five minutes.

A mid-course correction will be made with the stage's auxiliary propulsion system (APS) ullage motors. A second APS burn will be used if necessary, at about 9 hours GET, to further adjust the impact point. Burn time and attitude will be determined from onboard systems and tracking data provided to ground controllers by the Manned Space Flight Network.

The LOX dump by itself would provide a lunar impact; the mid-course correction burns will place the S-IVB/IU within the desired target area for impact about 20 minutes after the command/service module enters lunar orbit.

The schedule of events concerning the lunar impact is:

Time Hrs:	Min	Event
02	42	Translunar injection (TLI) — maneuver completion
04	19	Begin S-IVB evasive maneuver (APS engines)
04	21	End evasive maneuver
04	36	LH2 tank continuous vent on
04	41	Begin LOX dump
04	41	LH2 tank continuous vent off
04	42	End LOX dump
06	00	Begin first APS burn
08	59	Begin final APS burn (if required)
09	04	APS ullage engines off
77	46	Lunar impact of S-IVB/IU

Translunar Coast

Up to four midcourse correction burns are planned during the spacecraft's translunar coast, depending upon the accuracy of the trajectory resulting from the TLI maneuver. If required, the midcourse correction burns are planned at TLI+9 hours, TLI+ 30 hours, 41 minutes, lunar orbit insertion (LOI)-22 hours and LOI-5 hours.

The MCC-2 is a 15 fps SPS hybrid transfer maneuver which lowers pericynthion from 210 nm to 59 nm an places Apollo 13 on a non-free-return trajectory.

Return to the free-return trajectory is always within the capability of the spacecraft service propulsion o descent propulsion systems. During coast periods between midcourse corrections, the spacecraft will be i the passive thermal control (PTC) or "barbecue" mode in which the spacecraft will rotate slowly about it roll axis to stabilize spacecraft thermal response to the continuous solar exposure.

Lunar Orbit Insertion (LOI)

The lunar orbit insertion burn will be made at 77:25 GET at an altitude of about 85 nm above the Moon The LOI burn will have a nominal retrograde velocity change of 2815 fps and will insert Apollo 13 into 57x168 nm elliptical lunar orbit.

Descent Orbit Insertion (DOI)

A 213 fps SPS retrograde burn at 81:45 GET will place the CSM /LM into a 7x57 nm lunar orbit from whic the LM will begin the later powered descent to landing. In Apollos 11 and 12, DOI was a separate maneuve using the LM descent engine. The Apollo 13 DOI maneuver in effect is a combination LOI-2 and DOI an produces two benefits: conserves LM descent propellant that would have been used for DOI and makes thi propellant available for additional hover time near the surface, and allows 11 lunar revolutions of spacecraf tracking in the descent orbit to enhance position/velocity (state vector) data for updating the LM guidanc computer during the descent and landing phase.

Lunar Module Separation

The lunar module will be manned and checked out for undocking and subsequent landing on the lunar surface north of the crater, Fra Mauro. Undocking during the 12th revolution will take place at 99:16 GET. A radially downward service module RCS burn of 1fps will place the CSM on an equiperiod orbit with a maximum separation of 2.5 nm.

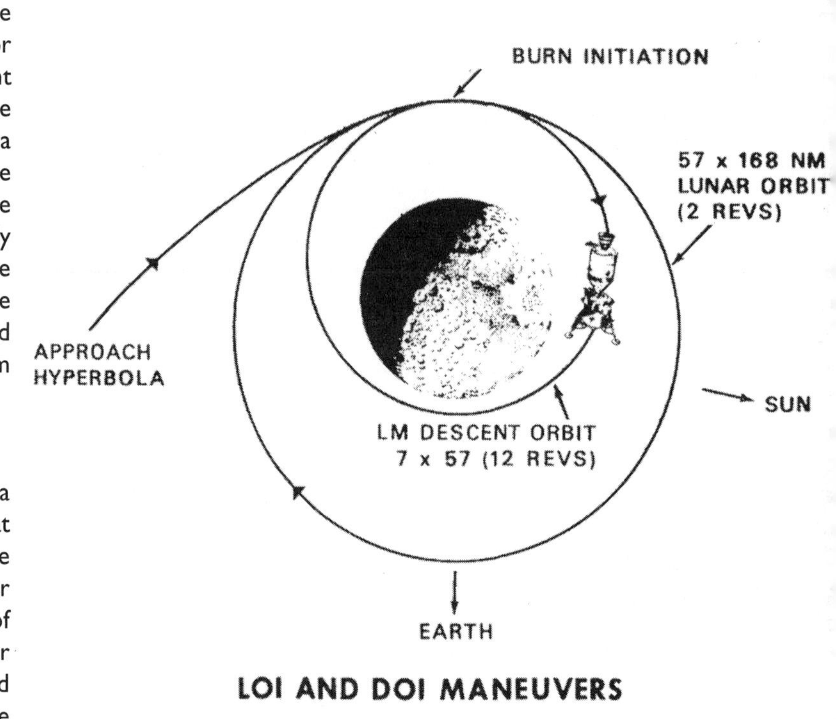

LOI AND DOI MANEUVERS

CSM Circularization

During the 12th revolution, a 70 fps posigrade SPS burn at 100:35 GET will place the CSM into 52x62 nm lunar orbit, which because of perturbations of the lunar gravitational potential, should become nearly circular at the time of rendezvous with the LM.

Power Descent Initiation (PDI), Lunar Landing

During the 14th revolution a three-phase powered descent (PD) maneuver begins at pericynthion at 103:31 GET using the LM descent engine to brake the vehicle out of the descent orbit. The guidance-controlled PD maneuver starts about 260 nm prior to touchdown, and is in retrograde attitude to reduce velocity to essentially zero at the time vertical descent begins. Spacecraft attitude will be windows up from powered descent initiation to the end of the braking phase so that the LM landing radar data can be integrated continually by the LM guidance computer and better communications can be maintained. The braking phase ıds at about 7,400 feet above the surface and the spacecraft is rotated more toward an upright ndows-forward attitude to permit a view of the landing site. The start of the approach phase is called high te, and the start of the landing phase at about 500 feet is called low gate.

oth the approach (visibility) phase and landing phase allow pilot takeover from guidance control as well as sual evaluation of the landing site. The final vertical descent to touchdown begins at about 100 feet when forward velocity is nulled out. Vertical descent rate will be 3 fps. The crew may elect to take over manual ntrol at approximately 500 feet. The crew will be able to return to automatic landing control after a period manned maneuvering if desirable. Touchdown will take place at 103:42 GET.

nar Surface Exploration

uring the 33½ hours Apollo 13 commander James Lovell and lunar module pilot Fred Haise are on the ırface, they will leave the lunar module twice for four-hour EVAs. These are extendable to five hours in real ne if the physical conditions of the astronauts and amount of remaining consumables permit.

addition to gathering more data on the lunar environment and bringing back geological samples from a ird lunar landing site, Lovell and Haise will deploy a series of experiments which will relay back to Earth ng-term scientific measurements of the Moon's physical and environmental properties.

he experiments series, called the Apollo Lunar Surface Experiment Package (ALSEP), will be left on the ırface and could transmit scientific and engineering data to the Manned Space Flight Network for at least year.

he ALSEP for Apollo 13, stowed in the LM descent stage scientific equipment bay, comprises components r the five ALSEP experiments — passive seismic, heat flow, charged particle lunar environment, cold thode gauge, and lunar dust detector.

hese experiments are aimed oward determining the structure nd state of the lunar interior, the omposition and structure of the nar surface and processes which odify the surface, and volutionary sequence leading to he Moon's present haracteristics. The Passive Seismic xperiment will become the econd point in a lunar seismic net egun with the first ALSEP at the urveyor III landing site of Apollo 2. Those two seismometers must ontinue to operate until the next eismometer is emplaced to omplete the three-station set. he heat flow experiment includes rilling two 10-foot holes with the nar surface drill.

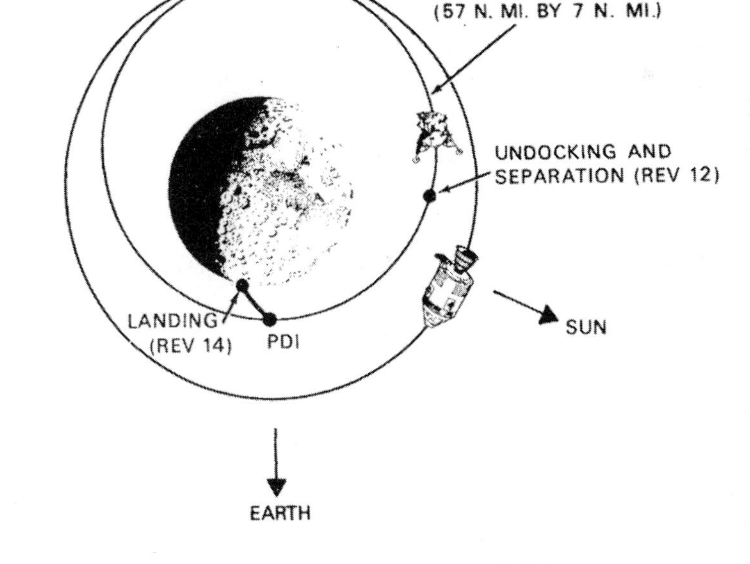

LM DESCENT ORBITAL EVENTS

While on the surface, the crew's perating radius will be limited by he range provided by the oxygen purge system (OPS), the reserve backup for each man's portable life upport system (PLSS) backpack. The OPS supplies 45 minutes of emergency breathing oxygen and suit ressure.

mong other tasks assigned to Lovell and Haise for the two EVA periods are:

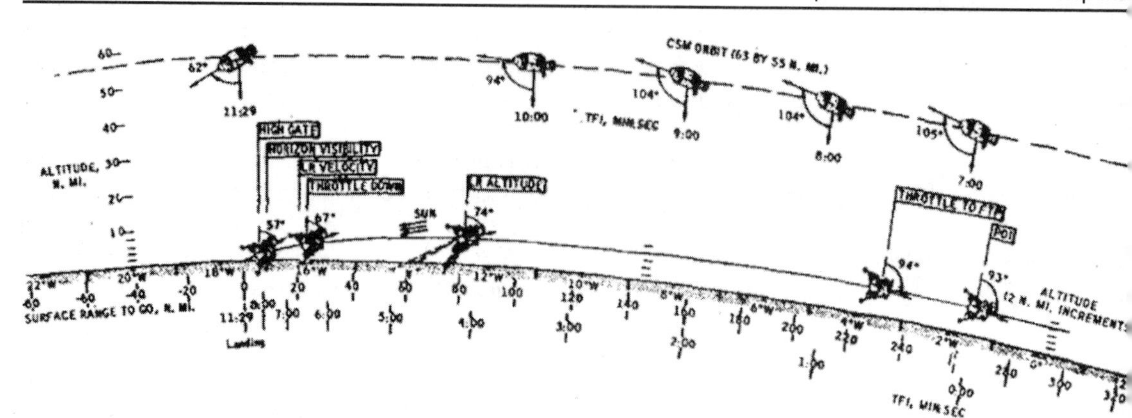

SUMMARY					
EVENT	TFI, MIN:SEC	V_I, FPS	$\dot{H}$, FPS	H, FT	ΔV, FPS
POWERED DESCENT INITIATION	0:00	5562	-4	50 700	0
THROTTLE TO MAXIMUM THRUST	0:26	5530	-3	50 611	32
LANDING RADAR ALTITUDE UPDATE	4:12	3180	-84	38 824	2427
THROTTLE RECOVERY	6:30	1466	-82	24 181	4248
LANDING RADAR VELOCITY UPDATE	6:50	1308	-110	23 105	4424
HORIZON VISIBILITY	8:14	639	-183	11 401	5205
HIGH GATE	8:34	478	-185	7 555	5406
LOW GATE	10:16	54 (67)*	-15	503	6224
LANDING	11:29	-15 (0)*	-3	5	6633

*HORIZONTAL VELOCITY RELATIVE TO SURFACE

POWERED DESCENT PROFILE

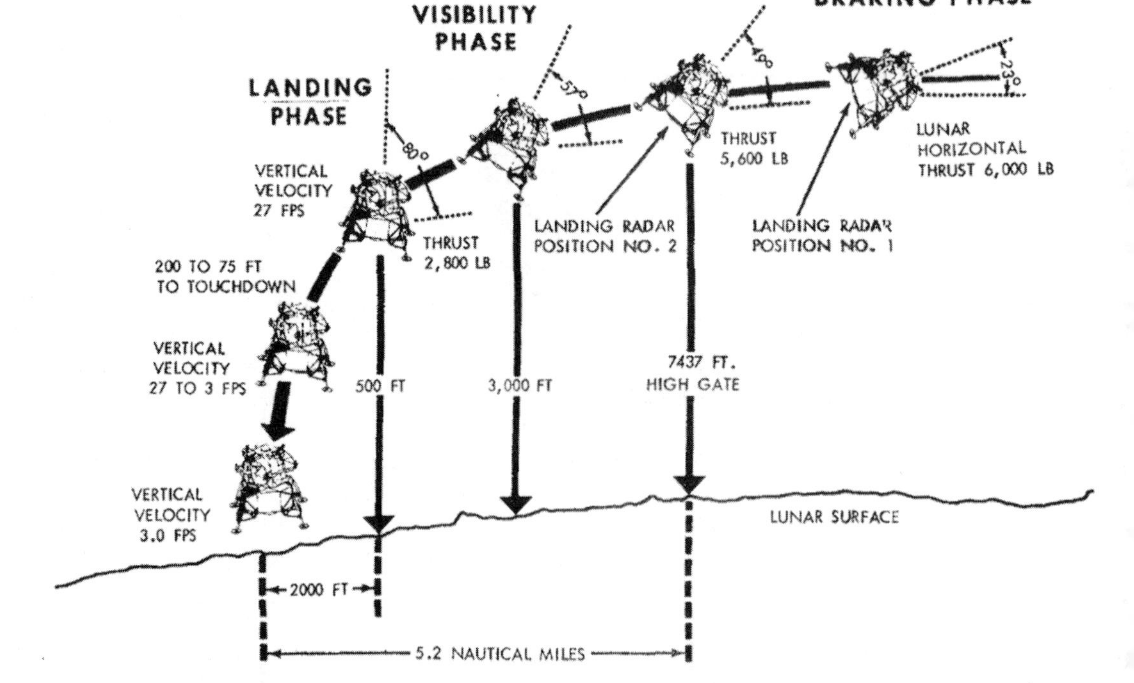

NOMINAL DESCENT TRAJECTORY FROM HIGH GATE TO TOUCHDOWN

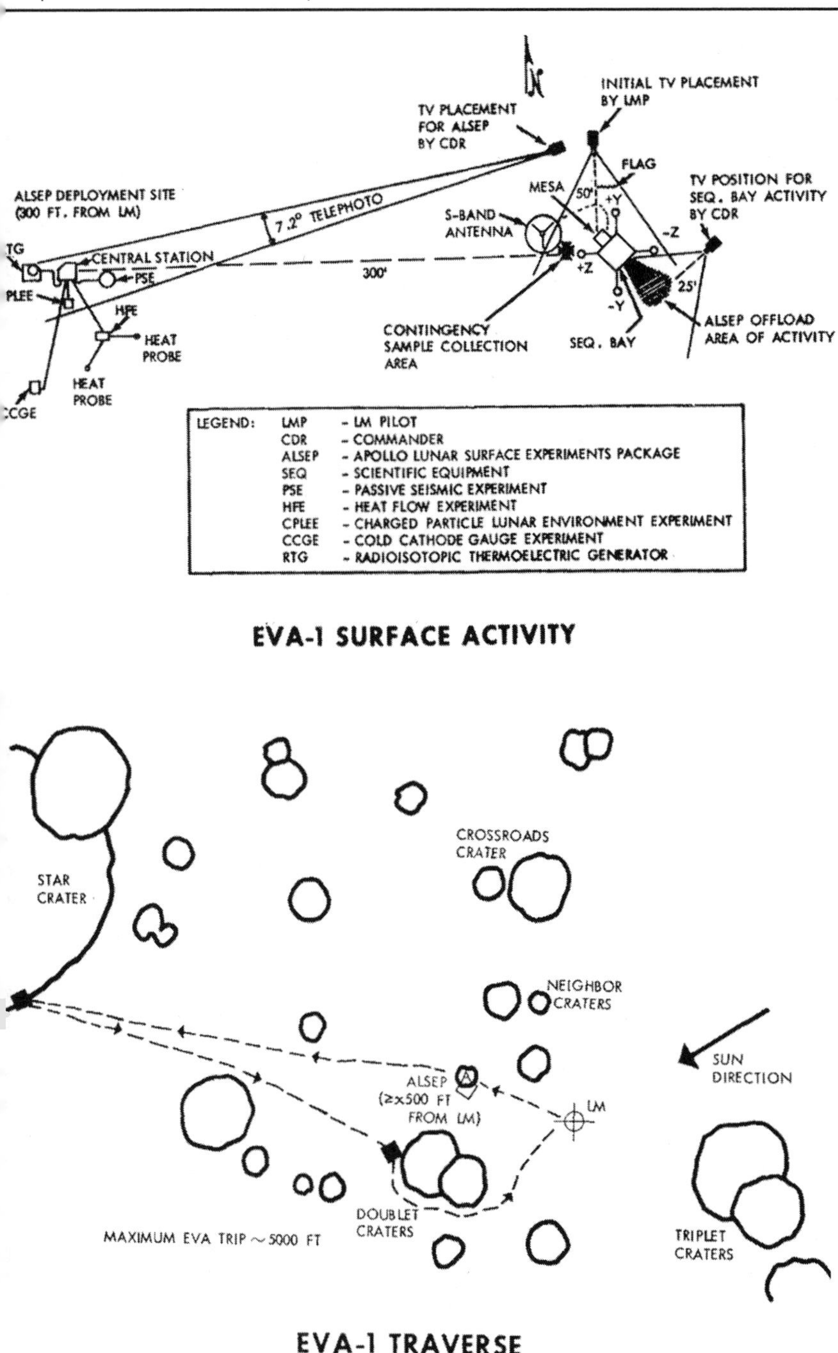

EVA-1 SURFACE ACTIVITY

EVA-1 TRAVERSE

*Collect a contingency sample of about two pounds of lunar material.

*Gather about 95 pounds of representative lunar surface material, including core samples, individual rock samples and fine-grained fragments from the Fra Mauro hilly uplands site. The crew will photograph thoroughly the areas from which samples are taken.

*Make observations and gather data on the mechanical properties and terrain characteristics of the lunar surface and conducting other lunar field geological surveys, including digging a two-foot deep trench for a soil mechanics investigation.

*Photograph with the lunar stereo close-up camera small geological features that would be destroyed in any attempts to gather them for return to Earth.

*Deploy and retrieve a window shade-like solar wind composition experiment similar to the ones used in Apollos 11 and 12.

arly in the first EVA, Lovell and Haise will set up the erectable S-Band antenna near the LM for relaying voice, V, and LM telemetry to MSFN stations. After the antenna is deployed, Haise will climb back into the LM to witch from the LM steerable S-Band antenna to the erectable antenna while Lovell makes final adjustments o the antenna's alignment. Haise will then rejoin Lovell on the lunar surface to set up a United States flag nd continue with EVA tasks.

ed stripes around the elbows and knees of Lovell's pressure suit will permit crew recognition during EVA elevision transmissions and on photographs.

scent, Lunar Orbit Rendezvous

ollowing the 33-hour lunar stay the LM ascent stage will lift off the lunar surface to begin the rendezvous equence with the orbiting CSM. Ignition of the LM ascent engine will be at 137:09 for a seven minute eight

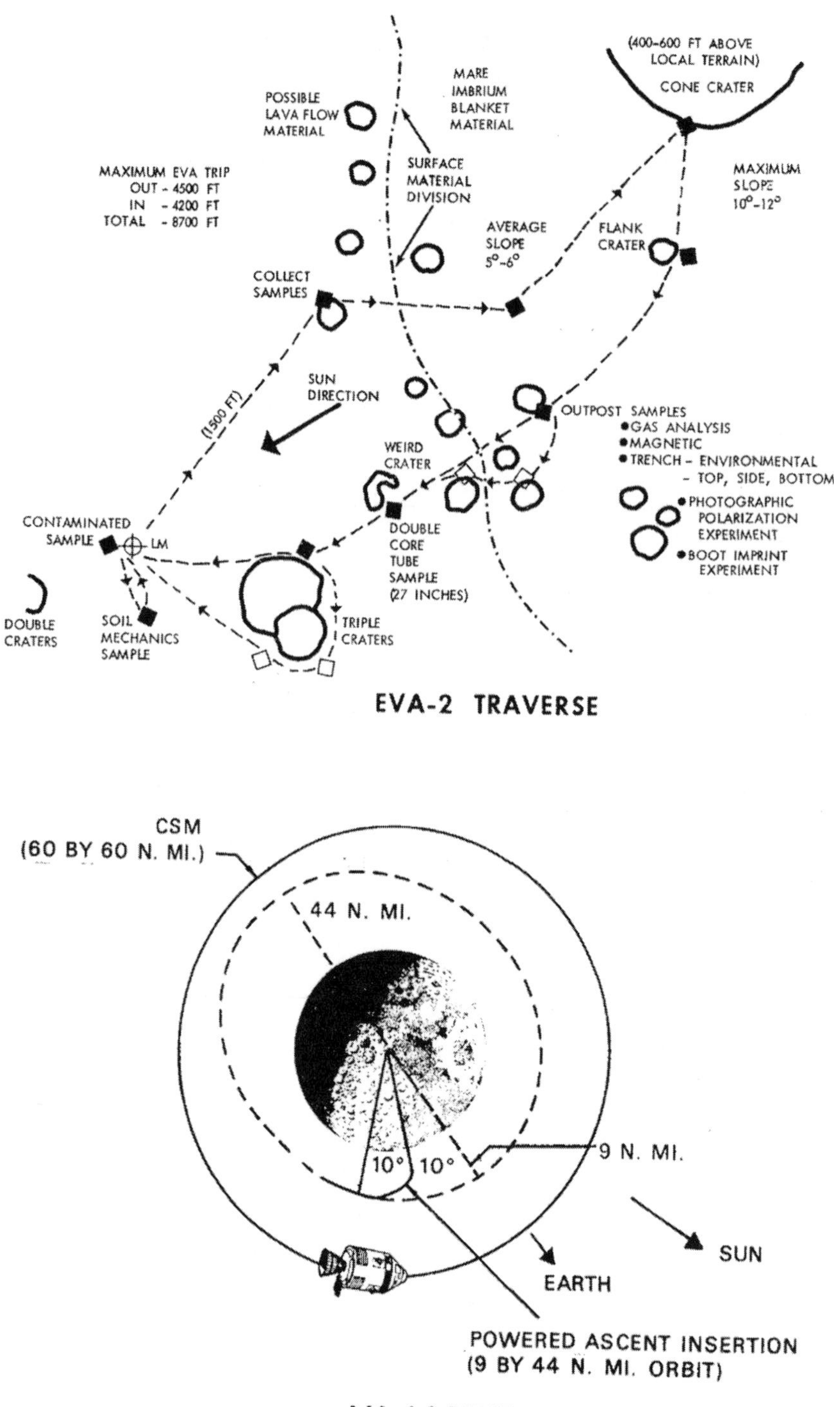

EVA-2 TRAVERSE

LM ASCENT

second burn attaining total velocity of 6,044 fp Powered ascent is in tw phases: vertical ascent fo terrain clearance and th orbital insertion phas Pitchover along the desir launch azimuth begins the vertical ascent ra reaches 50 fps about seconds after liftoff about 272 feet in altitud Insertion into a 9x44 n lunar orbit will take pla about 166 nm west of th landing site.

Following LM insertion int lunar orbit, the LM cre will compute onboard th major maneuvers fo rendezvous with the CS which is about 267 n ahead of and 51 mile above the LM at this poin All maneuvers in th sequences will be mad with the LM RCS thruster The premission rendezvo sequence maneuvers, tim and velocities, which like will differ slightly in re time, are as follows:

Concentric sequenc initiate (CSI): At first L apolune after insertio 138:19 GET, 50 fp posigrade, following som 20 minutes of L rendezvous radar trackin and CSM sextant/VH ranging navigation. CSI wi be targeted to place the L in an orbit 15 nm below th CSM at the time of th later constant delta heigh (CDH) maneuver (139:04)

The CSI burn may also initiate corrections for any out-of-plane dispersions resulting from insertion azimut errors. The resulting LM orbit after CSI will be 45x43.5 nm and will have a catch up rate to the CSM of abou 120 feet per second.

Terminal phase initiation (TPI): This maneuver occurs at 139:46 and adds 24.7 fps along the line of sigh toward the CSM when the elevation angle to the CSM reaches 26.6 degrees. The LM orbit becomes 61x4 nm and the catch up rate to the CSM decreases to a closing rate of 133 fps.

Midcourse correction maneuvers will be made if needed, followed by four braking maneuvers. Dockin nominally will take place at 140:25 GET to end the three and one-half hour rendezvous sequence.

he LM ascent stage will be jettisoned at 143:04 ET and a CSM RCS 1.0 fps maneuver will provide paration.

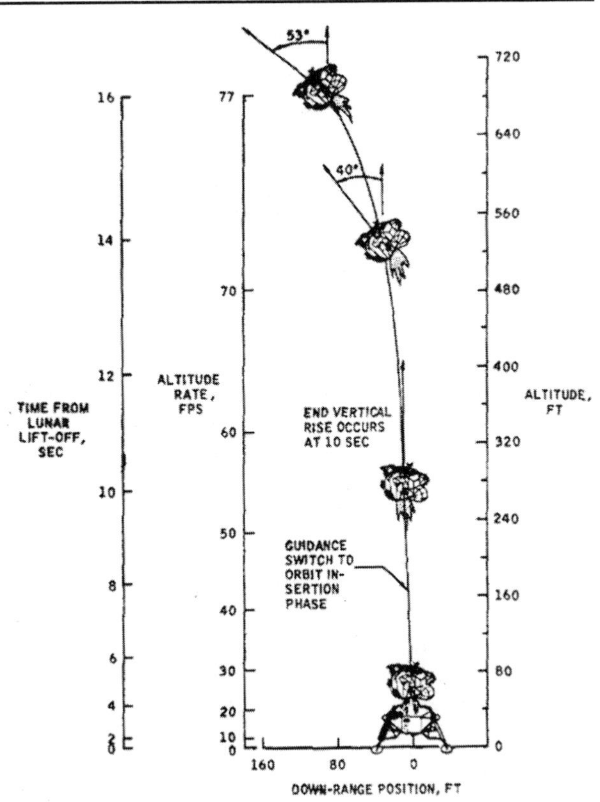

VERTICAL RISE PHASE

scent Stage Deorbit

rior to transferring to the command module, the M crew will set up the LM guidance system to aintain the ascent stage in an inertial attitude. At bout 144:32 GET the LM RCS thrusters will ignite n ground command for 186 fps retrograde burn rgeted for ascent stage impact at 145:00 about 5 miles from the landing site. The burn will have a mall out-of-plane north component so that the round track will include the original landing site. he ascent stage will impact at about 5508 fps at n angle of four degrees relative to the local orizontal. The ascent stage deorbit serves to emove debris from lunar orbit. Impacting an bject with a known velocity and mass near the nding site will provide experimenters with an vent for calibrating readouts from the ALSEP eismometer left behind.

plane change maneuver at 154:13 GET will place he CSM on an orbital track passing directly over he crater Descartes and Davy Rille eight evolutions later for photographs from orbit. The aneuver will be a 825 fps/SPS burn out of plane or a plane change of 8.8 degrees, and will result in n orbit inclination of 11.4 degrees.

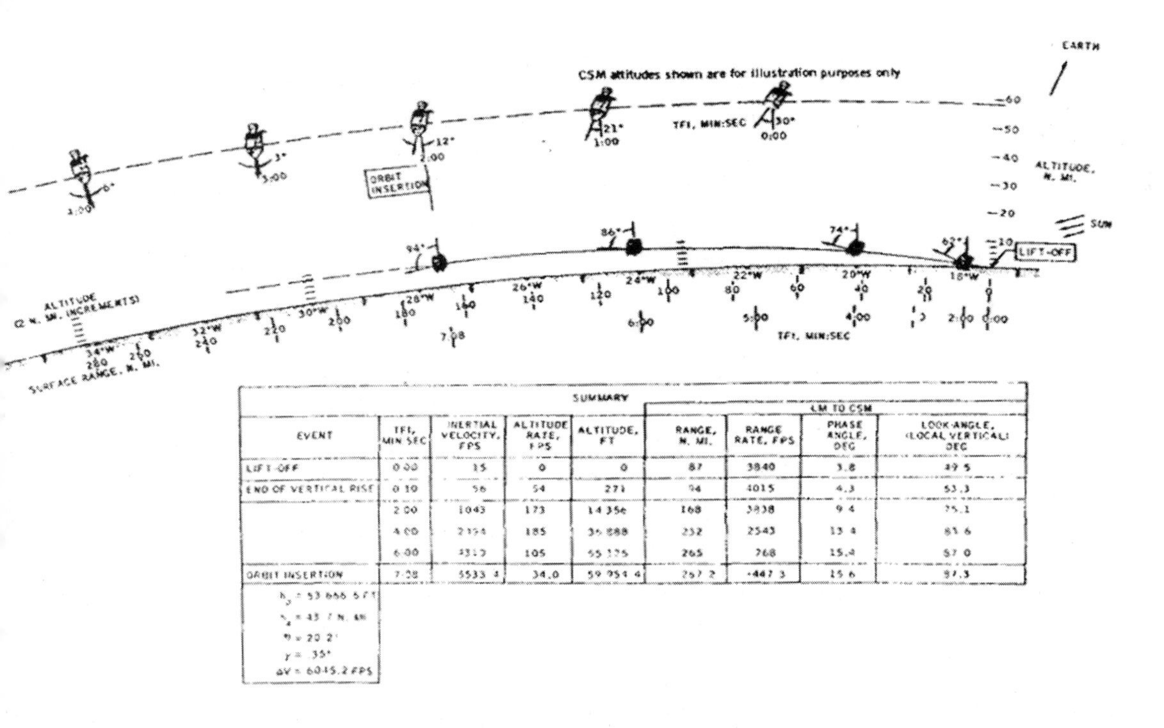

SUMMARY					LM TO CSM			
EVENT	TFI, MIN SEC	INERTIAL VELOCITY, FPS	ALTITUDE RATE, FPS	ALTITUDE, FT	RANGE, N. MI.	RANGE RATE, FPS	PHASE ANGLE, DEG	LOOK ANGLE, (LOCAL VERTICAL) DEG
LIFT-OFF	0 00	15	0	0	87	3840	3.8	49.5
END OF VERTICAL RISE	0 10	56	54	271	94	4015	4.3	53.3
	2 00	1043	173	14 356	168	3838	9.4	75.1
	4 00	2354	185	35 888	232	2543	13.4	85.6
	6 00	4313	105	55 175	265	768	15.4	57.0
ORBIT INSERTION	7 28	5533.4	34.0	59 954.4	257.2	-447.3	15.6	97.3

h_p = 53 666.5 FT
h_a = 43.7 N. MI.
θ = 20.2°
γ = .35°
ΔV = 6045.2 FPS

ORBIT INSERTION PHASE

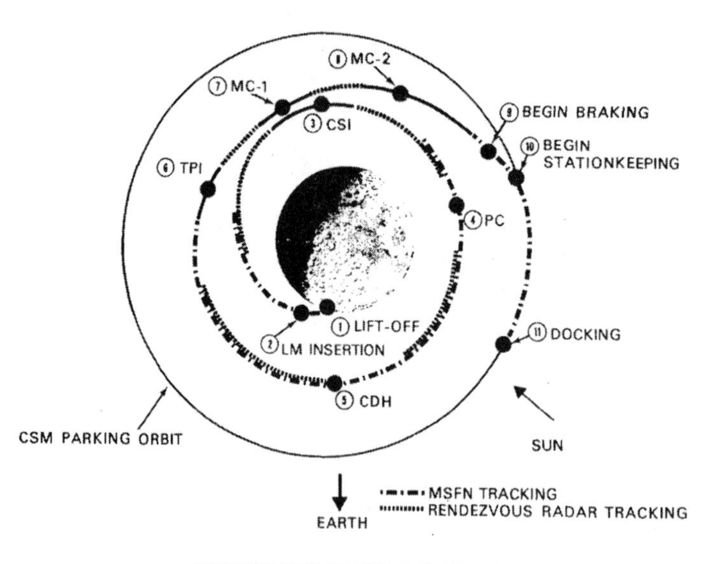

ASCENT THROUGH DOCKING

Transearth Injection (TEI)

The nominal transearth injection bu will be at 167:29 GET following hours in lunar orbit. TEI will take pla on the lunar farside, will be a 3,147 f posigrade SPS burn of two minutes seconds duration and will produce entry velocity of 36,129 fps after a hours transearth flight time.

Transearth Coast

Three entry corridor-contr transearth midcourse correctic burns will be made if needed: MCC at TEI+15 hours, MCC-6 at ent interface (EI) -22 hours and MCC-7 EI -3 hrs.

Entry, Landing

Apollo 13 will encounter the Earth atmosphere (400,000 feet) at 240:5 GET at a velocity of 36,129 fps and will land approximately 1,250 nm downrange from the entry-interfac point using the spacecraft's lifting characteristics to reach the landing point. Splashdown will be at 241:04 1.5 degrees South latitude by 157.5 degrees West longitude.

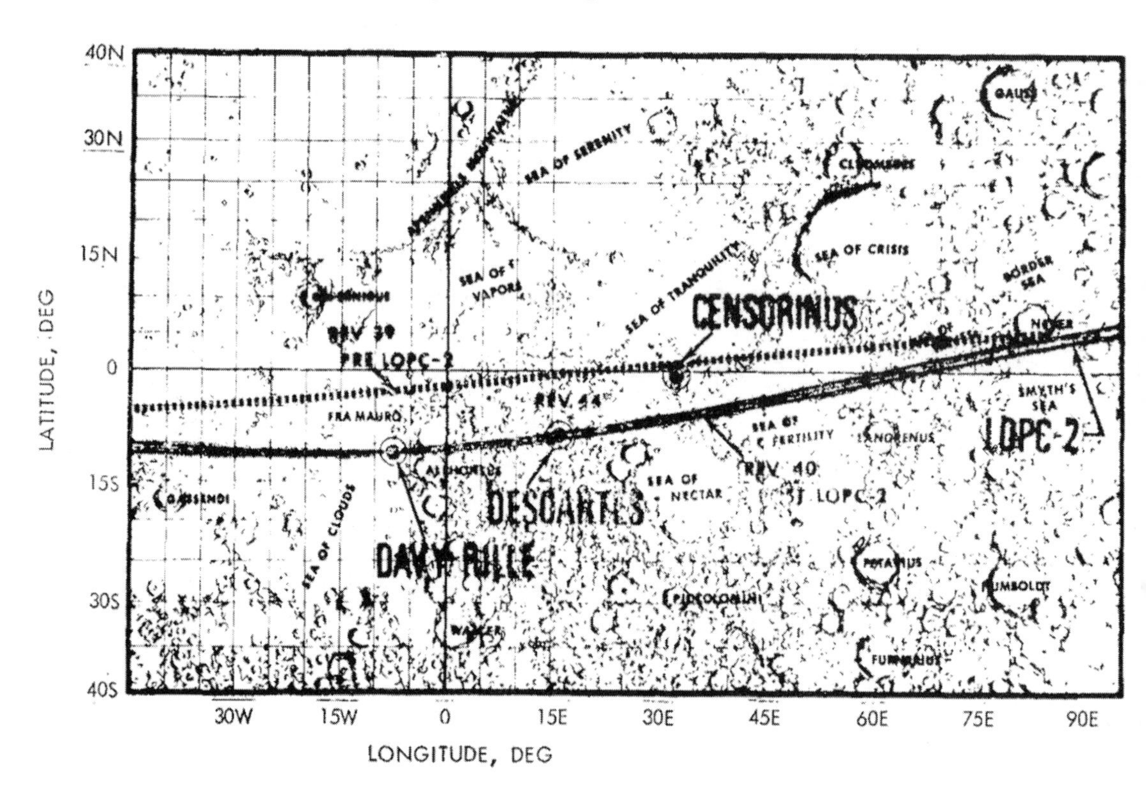

APOLLO 13 LOPC-2 TARGET LUNAR ORBIT

Recovery Operations

Launch abort landing areas extend downrange 3,400 nautical miles from Kennedy Space Center, fanwise 5 nm miles above and below the limits of the variable launch azimuth (72-96 degrees) in the Atlantic Ocean

On station in the launch abort area will be the destroyer USS New.

The landing platform-helicopter (LPH) Iwo Jima, Apollo 13 prime recovery ship, will be stationed near the Pacific Ocean end-of-mission aiming point prior entry.

Splashdown for a full-duration lunar landing mission launched on time April 11 will be at one degree 34 minutes South by 157 degrees 30 minutes West about 180 nautical miles South of Christmas Island, at 241:04 GET (3:17 p.m. EST) April 21.

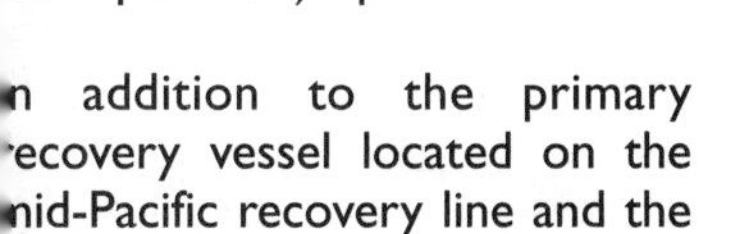

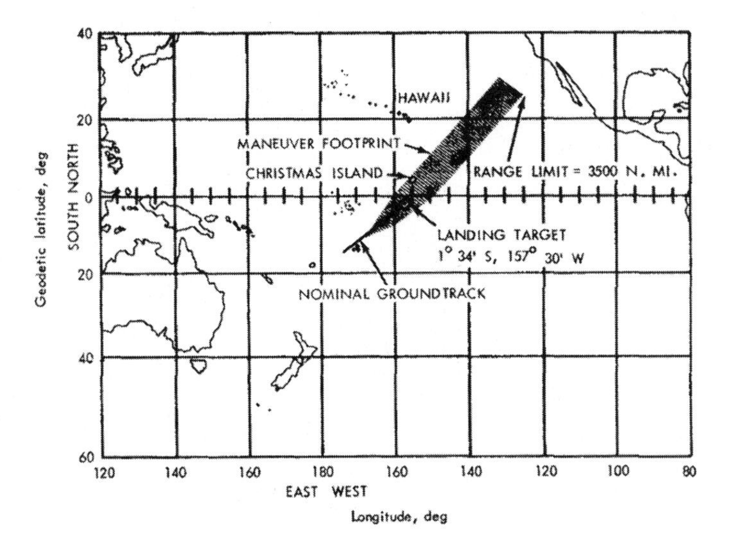

MANEUVER FOOTPRINT AND NOMINAL GROUNDTRACK

In addition to the primary recovery vessel located on the mid-Pacific recovery line and the surface vessel in the launch abort area, eight HC-130 aircraft will be on standby at five staging bases around the Earth: Guam; Hawaii; Azores; Ascension Island; and Florida.

Apollo 13 recovery operations will be directed from the Recovery Operations Control Room in the Mission Control Center, supported by the Atlantic Recovery Control Center, Norfolk, Va., and the Pacific Recovery Control Center, Kunia, Hawaii.

After splashdown, the Apollo 13 crew will don clean coveralls and filter masks passed to them through the spacecraft hatch by a recovery swimmer. The crew will be carried by helicopter to the Iwo Jima where they will enter a Mobile Quarantine Facility (MQF) about 90 minutes after landing.

APOLLO 13 ONBOARD TELEVISION

Apollo 13 will carry two color and one black-and-white television cameras. One color camera will be used for command module cabin interiors and out-the-window Earth/ Moon telecasts, and the other color camera will be stowed in the LM descent stage from where it will view the astronaut initiate egress to the lunar surface and later will be deployed on a tripod to transmit a real-time picture of the two periods of lunar surface EVA. The black-and-white camera will be carried in the LM cabin. It will only be used as a backup to the lunar surface color camera.

The two color TV cameras are essentially identical, except for additional thermal protection on the lunar surface camera. Built by Westinghouse Electric Corp., Aerospace Division, Baltimore, Md., the color cameras output a standard 525-line, 30 frame-per-second signal in color by use of a rotating color wheel system.

The color TV cameras weigh 12 pounds and are fitted with zoom lenses for wide-angle or close-up fields of view. The CM camera is fitted with a three-inch monitor for framing and focusing. The lunar surface color camera has 100 feet of cable available.

The backup black-and-white lunar surface TV camera, also built by Westinghouse, is of the same type used in the first manned lunar landing in Apollo 11. It weighs 7.25 pounds and draws 6.5 watts of 24-32 volts DC power. Scan rate is 10 frames-per-second at 325 lines-per-frame. The camera body is 10.6 inches long, 6.5 inches wide and 3.4 inches deep, and is fitted with bayonet-mount wide-angle and lunar day lenses.

During the two lunar surface EVA periods, Apollo 13 commander Lovell will be recognizable by red stripes around the elbows and knees of his pressure suit,

The following is a preliminary plan for TV transmissions based upon a 2:13 p.m. EST April 11 launch.

APOLLO 13 TV SCHEDULE

DAY	DATE	CST	GET	DURATION	ACTIVITY/SUBECT	VEH	STA
Saturday	Apr. 11	2:48 p.m.	01:35	7 Min.	Earth CSM	KSC**	
Saturday	Apr. 11	4:28 p.m.	03:15	1Hr. 8 Min.	Transposition & Docking	CSM	GDS
Sunday	Apr. 12	7:28 p.m.	30:15	30 Min.	Spacecraft Interior (MCC-2)	CSM	GDS
Monday	Apr. 13	11:13 p.m.	58:00	30 Min.	Interior & IVT to LM	CSM	GDS
Wednesday	Apr. 15	1:03 p.m.	95:50	15 Min.	Fra Mauro Landing Site	CSM	MAD
Thursday	Apr. 16	1:23 am	108:10	3 Hrs. 52 Min.	Lunar Surface (EVA-1)	LM	GDS/HSK
Thursday	Apr. 16	9:03 p.m.	127:50	6 Hrs. 35 Min.	Lunar Surface (EVA-2)	LM	GDS
Friday	Apr. 17	9:36 am	140:23	12 Min.	Docking CSM	MAD	
Saturday	Apr. 18	11:23 am	166:10	40 Min.	Lunar Surface	CSM	MAD*
Saturday	Apr. 18	1:13 p.m.	168:00	25 Min.	Lunar Surface (post TEI)	CSM	MAD*
Monday	Apr. 20	6:58 p.m.	221:45	15 Min.	Earth & Spacecraft Interior	CSM	GDS

* Recorded only, ** Tentative

APOLLO 13 SCIENCE

Lunar Orbital Photography

Science experiments and photographic tasks will be conducted from the CSM during the Apollo 13 missior During the translunar phase of the mission, photography will be taken of the Earth as well as variou operational photography.

During lunar orbit, various lunar surface features including candidate landing sites Censorinus, Descartes an Davy Rille and the Apollo 11 and 12 landing sites will be photographed with the Lunar Topographic Camera In addition, five astronomical phenomena will be photographed:

1) Photographs will be taken of the solar corona using the Moon as an occulting edge to block out the solar disk.

2) Photography will be taken of the zodiacal light which is believed to originate from reflected sunlight in the asteroid belt. Eart observation of zodiacal light is inconclusive due to atmospheric distortion.

3)Photography will be taken of lunar limb brightening, which appears as bright rim light above the horizon following lunar sunset.

4)Photographs will be taken of the Comet J.C. Bennett, 19691 which should be visible from lunar orbit during the Apollo 13 mission.

5)Photographs will be taken of the region of Gegenschein which is a faint light source covering a 20° field of view about the Earth-Su line on the opposite side of the Earth from the Sun (anti-solar axis). One of the theories for the Gegenschein source is the existenc of trapped particles of matter at the Moulton point which produce brightness due to reflected sunlight. The Moulton point is theoretical point located 940,000 statute miles from the Earth along the anti-solar axis at which the sum of all gravitational forces zero. From the vantage point of lunar orbit, the Moulton point region may be photographed from approximately 15° off the Earth/Su line. These photographs should show if Gegenschein results from the Moulton point theory or from zodiacal light or a similar source

Photographic studies will be made on Apollo 13 of the ice particle flow following a water dump and of th gaseous cloud which surrounds a manned spacecraft in a vacuum and results from liquid dumps, outgassing etc.

In addition to the photographic studies, an experiment will be conducted with the CSM VHF communication link. During this experiment, the VHF signal will be reflected from the lunar surface and received by a 150-foo antenna on Earth. By analysis of the wavelength of the received signal, certain lunar subsurface characteristic may be discernible such as the depth of the lunar regolith layer. This experiment is called VHF Bistatic Rada

Charged Particle Lunar Environment Experiment (CPLEE)

The scientific objective of the Charged Particle Lunar Environment Experiment is to measure the particl energies of protons and electrons that reach the lunar surface from the Sun. Increased knowledge on th energy distribution of these particles will help us understand how they perturb the Earth Moon system. A some point electrons and protons in the magnetospheric tail of the Earth are accelerated and plunge int the terrestrial atmosphere causing the spectacular auroras and the Van Allen radiation. When the Moon is i nterplanetary space the CPLEE measures proton and lectrons from solar flares which results in magnetic storms the Earth's atmosphere. Similar instruments have been own on Javelin rockets and on satellites. The lunar surface, owever, allows data to be gathered over a long period of me and from a relatively stable platform in space.

o study these phenomena, the CPLEE measures the energy f protons and electrons simultaneously from 50 electron olts to 50,000 electron volts (50Kev). The solar radiation henomena measured are as follows:

a. Solar wind electrons and protons 50ev- 5Kev.
b. Thermalized solar wind protons and electrons 50e v-10Kev.
c. Magnetospheric tail particles 50ev - 50Kev.
d. Low energy solar cosmic rays 40ev - 50Kev.

his experiment is distinct from the ALSEP Solar Wind pectrometer (SWS) flown on Apollo 12 which measures irection as well as energy levels. The SWS measures lectrons from 10.5ev to 1,400ev and protons from 75ev to 0,000ev.

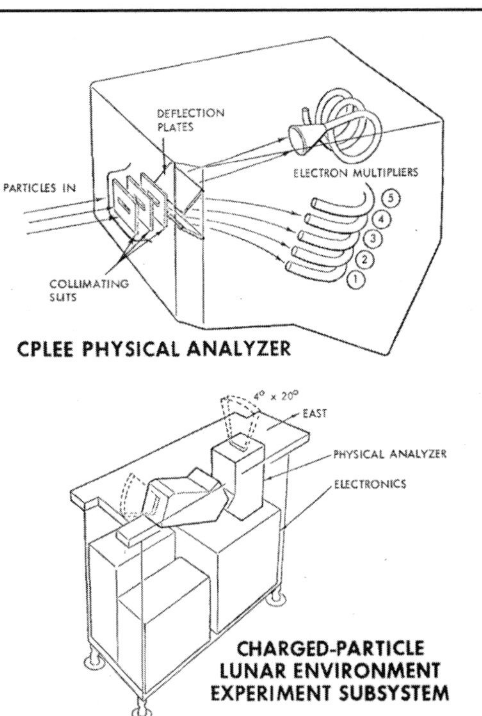

he detector package contains two spectrometers providing data on the direction of the incoming flux. ach spectrometer has six particle detectors; five C-shaped channeltron photon-multipliers and one unneltron, a helical shaped photon multiplier. Particles of a given charge and different energies on entering he spectrometer are subject to varying voltages and deflected toward the five channeltrons while particles f the opposite charge are deflected toward the funneltron. Thus electrons and protons are measured imultaneously in six different energy levels. The voltages are changed over six steps; ±35V, ±350 volts and ±3500V. In this way electrons and protons are measured from 50ev to 70Kev in a period of less than 20 econds.

he channeltron is a glass capillary tube having an inside diameter of about one millimeter and a length of 10 entimeters. The helical funneltron has an opening of 8mm. When a voltage is applied between the ends of he tube, an electric field is established down its length. Charged particles entering the tube are amplified by factor of 10^8.

he spectrometers have two ranges of sensitivity and can measure fluxes between 10^4 and 10^{10} particles/cm^2-ec-steradian.

he charged particle lunar environment experiment (CPLEE) and data analysis are the responsibility of Dr. rian O'Brien, University of Sydney (Australia) and Dr. David Reasoner, Rice University, with Dr. O'Brien ssuming the role of Principal Investigator.

unar Atmosphere Detector (LAD)

lthough the Moon is commonly described as a planetary body with no atmosphere, the existence of some tmosphere cannot be doubted. Two sources of this atmosphere are predicted: Internal, i.e., degassing from he interior of the Moon either by constant diffusion through its surface or intermittent release from active ents; external i.e., solar wind and vaporization during meteorite impacts. Telescopic observations from olarized scattered light indicate that the atmospheric pressure could not exceed one millionth of a torr (a orr is defined as 1/760 of the standard atmosphere).

easurements will be of the greatest significance if it turns out through later orbital sensors that they are of nternal origin. The Earth's atmosphere and oceans have been released from the Earth's interior by degassing. he most certain source, however, is the solar wind whose ionized particles become neutralized in the lunar

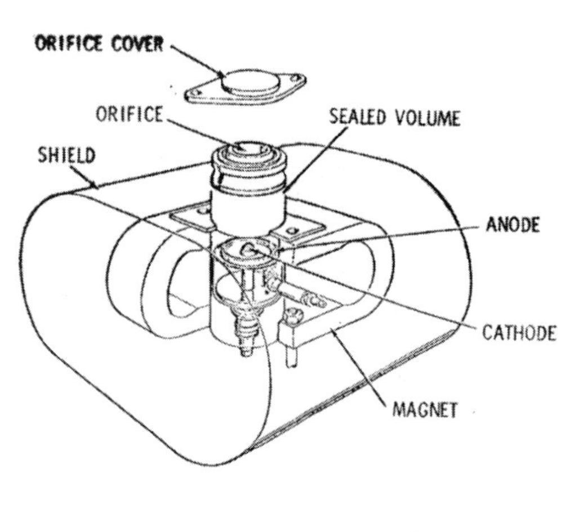

LUNAR ATMOSPHERIC DETECTOR
(COLD CATHODE ION GAUGE INSTRUMENT)

atmosphere and then are released as neutral gases. Neon is the predominant gas expected. Lighter gases such as hydrogen and helium escape and heavier ones statistically should be present in small quantities. Neutral particles are ionized in the lunar atmosphere further reducing the numbers present; others will escape as the temperature rises (and concentrate near the surface when it falls).

The LAD utilizes a cold cathode ionization gauge to measure the density of neutral particles at the lunar surface and the variations in density association with lunar phase or solar activity. The ionization gauge is basically a crossed electromagnetic field device. Electrons in the gauge are accelerated by the combined magnetic and electric fields producing a collision are collected by the cathode where they form a flow of positive ions. The positive ions current is found to be proportional to the density of the gas molecules entering the gauge. In addition, the gauge temperature is read over the range of -90° to 125°C with ±5°C accuracy.

From the density and temperature data the pressure of the ambient lunar atmosphere can then be calculated. Chemical composition of the atmosphere however is not directly measured but the gauge has been calibrated for each gas it is expected to encounter on the lunar surface and some estimates can be made on the chemical composition. Any one of seven different dynamic ranges may be selected permitting detection of neutral particles from 10^{-6} Torr (highest pressure predicted) to 10^{-12} Torr (maximum capability of gauge). For pressure greater than 10^{-10} Torr accuracy of ±30% will be obtained; for pressures less then 10^{-10} Torr accuracy ± 50% will be obtained. The experiment, therefore, will reduce the present uncertainty from a magnitude to a factor.

The Lunar Atmosphere Detector (LAD) and data are the responsibility of Francis Johnson, University of Texas (Dallas) and Dallas Evans, Manned Spacecraft Center, with Dr. Johnson serving as Principal Investigator.

Lunar Heat Flow Experiment (HFE)

The scientific objective of the Heat Flow experiment is to measure the steady-state heat flow from the lunar interior. Two predicted sources of heat are 1) original heat at the time of the Moon's formation and 2) radioactivity. Scientists believe that heat could have been generated by the in-falling of material and its subsequent compaction as the Moon was formed. Moreover, varying amounts of the radioactive elements uranium, thorium and potassium were found present in the Apollo 11 and 12 lunar samples which if present at depth, would supply significant amounts of heat. No simple way has been devised for relating the contribution of each of these sources to the present rate of heat loss. In addition to temperature, the experiment is capable of measuring the thermal conductivity of the lunar rock material.

The combined measurement of temperature and thermal conductivity gives the net heat flux from the lunar interior through the lunar surface. Similar measurements on Earth have contributed basic information to our understanding of volcanoes, earthquakes and mountain building processes. In conjunction with the seismic and magnetic data obtained on other lunar experiments the values derived from the heat flow measurements will help scientists to build more exact models of the Moon and thereby give us a better understanding of its origin and history.

The Heat Flow experiment consists of instrument probes, electronics and emplacement tool and the lunar surface drill. Each of two probes is connected by a cable to an electronics box which rests on the lunar surface. The electronics, which provide control, monitoring and data processing for the experiment, is connected to the ALSEP central station.

Each probe consists of two identical 20-inch (50 cm) long sections each of which contains a "gradient" sensor bridge, a "ring" sensor bridge and two heaters. Each bridge consists of four platinum resistors mounted in a thin-walled fiberglass cylindrical shell. Adjacent areas of the bridge are located in sensors at opposite ends of the 20-inch fiberglass probe sheath. Gradient bridges consequently measure the temperature difference between two sensor locations.

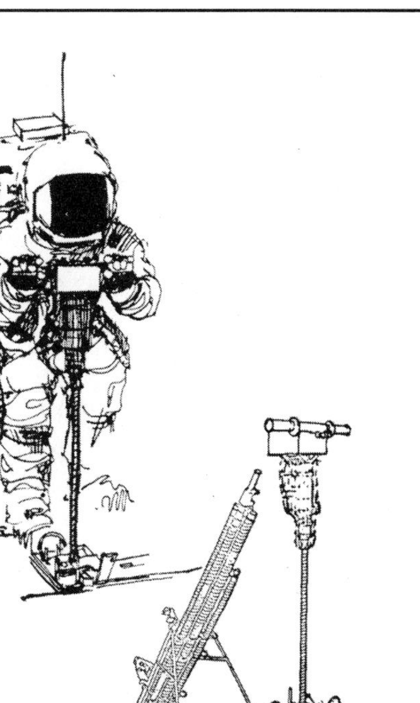

APOLLO LUNAR SURFACE DRILL

In thermal conductivity measurements at very low values a heater surrounding the gradient sensor is energized with 0.002 watts and the gradient sensor values monitored. The rise in temperature of the gradient sensor is a function of the thermal conductivity of the surrounding lunar material. For higher range of values, the heater is energized at 0.5 watts of heat and monitored by a ring sensor. The rate of temperature rise, monitored by the ring sensor is a function of the thermal conductivity of the surrounding lunar material. The ring sensor, approximately four inches from the heater, is also a platinum resistor. A total of eight thermal conductivity measurements can be made. The thermal conductivity mode of the experiment will be implemented about twenty days (500 hours) after-deployment. This is to allow sufficient time for the perturbing effects of drilling and emplacing the probe in the bore-hole to decay; i.e. for the probe and casings to come to equilibrium with the lunar subsurface.

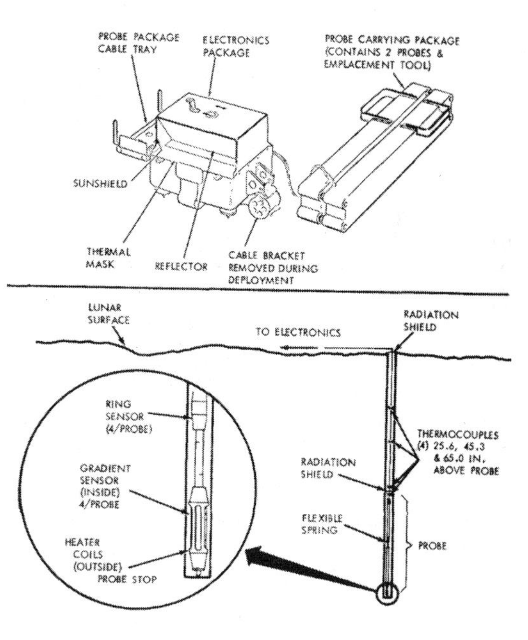

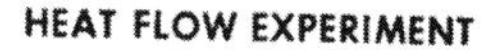

HEAT FLOW EXPERIMENT

A 30-foot (10 meter) cable connects each probe to the electronics box. In the upper six feet of the bore-hole the cable contains four evenly spaced thermocouples; at the top of the probe; at 26" (65 cm), 45" (115 cm), and 66" (165 cm). The thermocouples will measure temperature transients propagating downward from the Lunar surface. The reference junction temperature for each thermocouple is located in the electronics box. In fact, the feasibility of making a heat flow measurement depends to a large degree on the low thermal conductivity of the lunar surface layer, the regolith. Measurement of lunar surface temperature variations by Earth-based telescopes as well as the Surveyor and Apollo missions show a remarkably rapid rate of cooling. The wide fluctuations in temperature of the lunar surface (from -250°F to +250°) are expected to influence only the upper six feet and not the bottom 3 feet of the bore-hole.

The astronauts will use the Apollo Lunar Surface Drill (ALSD) to make a lined bore-hole in the lunar surface for the probes. The drilling energy will be provided by a battery-powered rotary percussive power head. The drill rod consists of fiberglass tubular sections reinforced with boron filaments (each about 20 inches or 50 cm long). A closed drill bit, placed on the first drill rod, is capable of penetrating the variety of rock including three feet of vesicular basalt (40 per cent porosity). As lunar surface penetration progresses, additional drill rod sections will be connected to the drill string. The drill string is left in place to serve as a hole casing.

An emplacement tool is used by the astronaut to insert the probe to full depth. Alignment springs position the probe within the casing and assure a well-defined radiative thermal coupling between the probe and the bore-hole. Radiation shields on the hole prevent direct sunlight from reaching the bottom of the hole.

The astronaut will drill a third hole near the HFE and obtain cores of lunar material for subsequent analysis of thermal properties.

Heat flow experiment, design and data analysis are the responsibility of Dr. Marcus Langseth of the Lamont-Doherty Geological Observatory; Dr. Sydney Clark, Jr., Yale University, and Dr. M.G. Simmons, MIT, with Dr. Langseth assuming the role of Principal Investigator.

Passive Seismic Experiment (PSE)

The ALSEP Passive Seismic Experiment (PSE) will measure seismic activity or the Moon and obtain information on the physical properties of the lunar crust and interior. The PSE will detect surface tilt produced by tidal deformations, moonquakes and meteorite impacts. The passive seismometer design and subsequent experiment analysis are the responsibility of Dr. Gary Lathan of the Lamont-Doherty Geological Observatory.

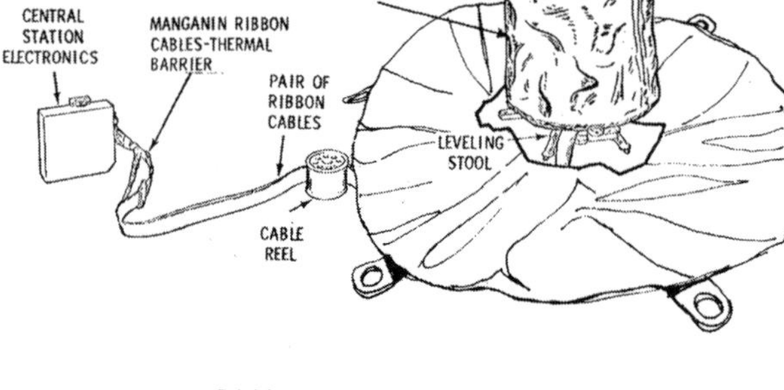

PASSIVE SEISMIC EXPERIMENT

A similar passive seismic experiment was deployed as part of the Apollo 12 ALSEP station at Surveyor crater last November and has transmitted Earthward lunar surface seismic activities since that time. The Apollo 12 and 13 seismometers differ from the seismometer left at Tranquility Base in July 1969 by the Apollo 11 crew in that they are continuously powered by a SNAP-27 radioisotope electric generator, while the Apollo 11 seismometer was powered by solar energy and could output data only during the lunar day at its location.

After Lovell and Haise ascend from the lunar surface and rendezvous with the command module in lunar orbit, the lunar module ascent stage will be jettisoned and later ground-commanded to impact on the lunar surface about 42 statute miles from the Apollo 13 landing site at Fra Mauro. Impact of an object of known mass and velocity will assist in calibrating the Apollo 13 seismometer readouts as well as providing comparative readings between the Apollo 12 and 13 seismometers forming the first two stations of a lunar surface seismic network.

There are three major physical components of the PSE:

The sensor assembly consists of three, long-period seismometers with orthogonally-oriented, capacitance type seismic sensors, measuring along two horizontal axes and one vertical axis. This is mounted on a gimbal platform assembly. There is one short period seismometer which has magnet-type sensors. It is located directly on the base of the sensor assembly.

The leveling stool allows manual leveling of the sensor assembly by the astronaut to within ±5°, and final leveling to within 3 arc seconds by control motors. The thermal shroud covers and helps stabilize the temperature of the sensor assembly. Also, two radioisotope heaters will protect the instrument from the extreme cold of the lunar night.

Solar Wind Composition Experiment (SWCE)

The scientific objective of the solar wind composition experiment is to determine the elemental and isotopic composition of the noble gases in the solar wind. (This is not an ALSEP experiment). The solar wind composition detector experiment design and subsequent data analysis are the responsibility of J. Geiss and P. Eberhardt, University of Bern (Switzerland) and P. Signer, Swiss Federal Institute of Technology, with Professor Geiss assuming the responsibility of Principal Investigator.

As in Apollo 11 and 12 the SWC detector will be deployed on the Moon and brought back to Earth by the astronauts. The detector, however, will be exposed to the solar wind flux for 20 hours instead of two hours as in Apollo 11 and 18 hours 42 minutes on Apollo 12.

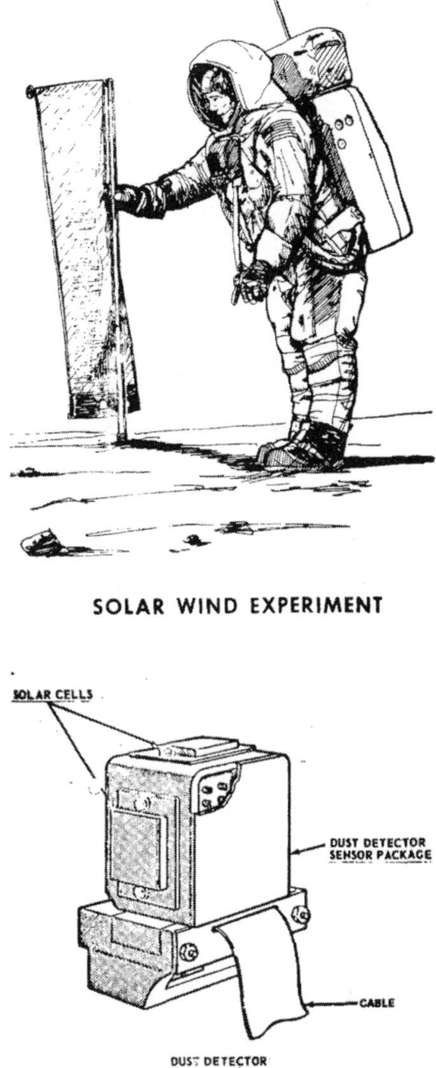

SOLAR WIND EXPERIMENT

DUST DETECTOR

The solar wind composition detector consists of an aluminum foil four square feet in area and about 0.5 mils thick rimmed by Teflon for resistance to tear during deployment. A staff and yard arrangement will be used to deploy the foil and to maintain the foil approximately perpendicular to the solar wind flux. Solar wind particles will penetrate into the foil while cosmic rays will pass right through. The solar wind particles will be firmly trapped at a depth of several hundred atomic layers. After exposure on the lunar surface, the foil is reeled and returned to Earth.

Dust Detector

The ALSEP Dust Detector is an engineering measurement designed to detect the presence of dust or debris that may impinge on the ALSEP or accumulate during its operating life. The measurement apparatus consists of three calibrated solar cells, one pointing in east, west and vertical to face the elliptic path of the Sun. The detector is located on the central station. Dust accumulation on the surface of the three solar cells will reduce the solar illumination detected by the cells. The temperature of each cell will be measured and compared with predicted values.

Field Geology Investigations

The scientific objectives of the Apollo Field Geology Investigations are to determine the composition of the Moon and the processes which shape its surfaces. This information will help to determine the history of the Moon and its relationship to the Earth. Apollo 11 visited the Sea of Tranquility (Mare Tranquillitatis) and Apollo 12 studied the Ocean of Storms (Oceanus Procellarum). The results of these studies should help establish the nature of Mare-type areas. Apollo 13 will investigate a hilly upland area.

Geology investigation of the Moon actually began with the telescope. Systematic geology mapping began 10 years ago with a team of scientists at the U.S. Geological Survey. Ranger, Surveyor, and especially Lunar Orbiter data enormously increased the detail and accuracy of these studies. The Apollo 11 and 12 investigations represent another enormous advancement in providing new evidence on the Moon's great age, its curious chemistry, the surprisingly high density of the lunar surface material.

On Apollo 13, almost the entire second EVA will be devoted to the Field Geology Investigations and the collection of documented samples. The sample locations will be carefully photographed before and after sampling. The astronauts will carefully describe the setting from which the sample is collected. In addition to specific tasks, the astronauts will be free to photograph and sample phenomena which they judge to be unusual, significant, and interesting. The astronauts are provided with a package of detailed photo maps which they will use for planning traverses. Photographs will be taken from the LM window. Each feature or family of features will be described, relating to features on the photo maps. Areas and features where photographs should be taken and representative samples collected will be marked on the maps. The crew and their ground support personnel will consider real-time deviation from the nominal plan based upon an on-the-spot analysis of the actual situation. A trench will be dug for soil mechanics investigations.

The Earth-based geologists will be available to advise the astronauts in real-time and will work with the data returned, the photos, the samples of rock and the astronauts' observations to reconstruct here on Earth the astronauts traverse on the Moon.

Each astronaut will carry a Lunar Surface Camera (a modified 70 mm electric Hasselblad). The camera has a 60 mm lens and a Reseau plate. Lens apertures range from f/5.6 to f/45. Its focus range is from three feet to infinity. A removable polarizing filter is attached to the lens of one of the cameras and can be rotated in 45-degree increments for light polarizing studies.

A gnomon, used for metric control of near field (less than 10 feet) stereoscopic photography, will provide angular orientation relative to the local vertical. Information on the distances to objects and on the pitch, roll, and azimuth of the camera's optic axis are thereby included in each photograph. The gnomon is a weighted tube suspended vertically on a tripod supported gimbal. The tube extends one foot above the gimbal and is painted with a gray scale in bands one centimeter wide. Photogrammetric techniques will be used to produce three-dimensional models and maps of the lunar surface from the angular and distance relationship between specific objects recorded on the film.

The 16 mm Data Acquisition Camera will provide times: sequence coverage from within the LM. It can be operated in several automatic modes, ranging from one frame/second to 24 frames/second. Shutter speeds, which are independent of the frame rates, range from 1/1000 second to 1/60 second. Time exposures are also possible. While a variety of lenses is provided, the 18 mm lens will be used to record most of the geological activities in the one frame/second mode. A similar battery powered 16 mm camera will be carried in EVA.

The Lunar Surface Close-up Camera will be used to obtain very high resolution close-up stereoscopic photographs of the lunar surface to provide fine scale information on lunar soil and rock textures. Up to 100 stereo pairs can be exposed on the preloaded roll of 35 mm color film. The handle grip enables the astronaut to operate the camera from a standing position. The film drive and electronic flash are battery-operated. The camera photographs a 3"x3" area of the lunar surface.

Geological sampling equipment includes tongs, scoop, hammer, and core tubes. A 24-inch extension handle is provided for several of the tools to aid the astronaut in using them without kneeling.

Sample return containers (SRC) have been provided for return of up to 40 pounds each of lunar material for Earth-based analysis. The SRC's are identical to the ones used on the Apollo 11 and 12 missions. They are machined from aluminum forging and are designed to maintain an internal vacuum during the outbound and return flights. The SRC's will be filled with representative samples of lunar surface material, collected and separately bagged by the astronauts on their traverse and documented by verbal descriptions and photography. Subsurface samples will be obtained by using drive tubes 16 inches long and one inch in diameter. A few grams of material will be preserved under lunar vacuum conditions in a special environmental sample container. This container will be opened for analysis under vacuum conditions equivalent to that at the lunar surface. Special containers are provided for a magnetic sample and a gas analysis sample.

<u>SNAP-27</u>

SNAP-27 is one of a series of radioisotope thermoelectric generators, or atomic batteries, developed by the U.S. Atomic Energy Commission under its SNAP program. The SNAP (Systems for Nuclear Auxiliary Power) Program is directed at development of generators and reactors for use in space, on land, and in the sea.

SNAP-27 was first used in the Apollo 12 mission to provide electricity for the first Apollo Lunar Surface Experiments Package (ALSEP). A duplicate of the Apollo 12 SNAP-27 will power the Apollo 13 ALSEP.

The basic SNAP-27 unit is designed to produce at least 63 electrical watts of power. It is a cylindrical generator fueled with the radioisotope plutonium 238. It is about 18 inches high and 16 inches in diameter including the heat radiating fins. The generator, making maximum use of the lightweight material beryllium, weighs about 28 pounds unfueled.

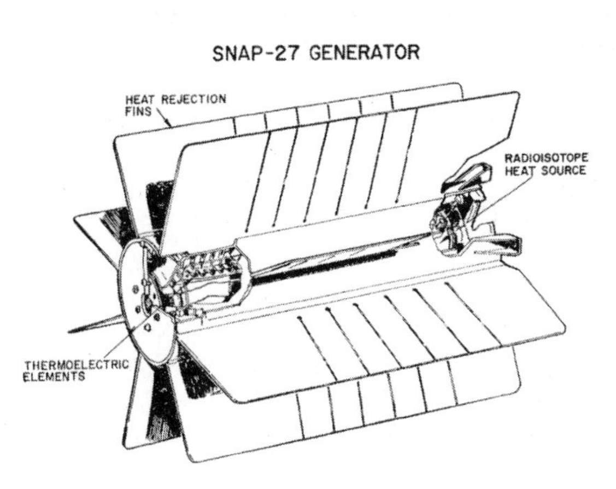

'he fuel capsule, made of a super alloy material, is 16.5 ıches long and 2.5 inches in diameter. It weighs about 5.5 pounds, of which 8.36 pounds represent fuel. The lutonium 238 fuel is fully oxidized and is chemically nd biologically inert.

'he rugged fuel capsule is contained within a graphite ɹel cask from launch through lunar landing. The cask ; designed to provide reentry heating protection and dded containment for the fuel capsule in the unlikely vent of an aborted mission. The cylindrical cask with emispherical ends includes a primary graphite heat hield, a secondary beryllium thermal shield, and a fuel apsule support structure made of titanium and ıconel materials. The cask is 23 inches long and eight ıches in diameter and weighs about 24.5 pounds. With the fuel capsule installed, it weighs about 40 pounds. is mounted on the lunar module descent stage by a titanium support structure.

)nce the lunar module is on the Moon, the lunar module pilot will remove the fuel capsule from the cask nd insert it into the SNAP-27 generator which will have been placed on the lunar surface near the module.

'he spontaneous radioactive decay of the plutonium 238 within the fuel capsule generates heat in the enerator. An assembly of 442 lead telluride thermoelectric elements converts this heat — 1480 thermal vatts — directly into electrical energy — at least 63 watts. There are no moving parts.

lutonium 238 is an excellent isotope for use in space nuclear generators. At the end of almost 90 years, lutonium 238 will still supply half of its original heat. In the decay process, plutonium 238 emits mainly the uclei of helium (alpha radiation), a very mild type of radiation with a short emission range.

efore the use of the SNAP-27 system in the Apollo program was authorized, a thorough review was onducted to assure the health and safety of personnel involved in the mission and the general public. xtensive safety analyses and tests were conducted which demonstrated that the fuel would be safely ontained under almost all credible accident conditions.

Contractors for SNAP-27

General Electric Co., Missile and Space Division, Philadelphia, Pa., designed, developed, and fabricated the NAP-27 generator for the ALSEP.

'he 3M Co., St. Paul, Minn., fabricated the thermoelectric elements and assembled the SNAP-27 generator.

olar Division of International Harvester, San Diego, Calif., fabricated the generator's beryllium structure.

litco, Gardena, Calif., fabricated the graphite structure for the SNAP-27 Graphite LM Fuel Cask.

andia Corp., a subsidiary of Western Electric, operator of AEC's Sandia Laboratory, Albuquerque, N.M., rovided technical direction for the SNAP-27 program.

avannah River Laboratory, Aiken, S.C., operated by the DuPont Co. for the AEC, prepared the raw plutonium ɹel.

Mound Laboratory, Miamisburg Ohio, operated by Monsanto Research Corp., for the AEC, fabricated the raw ɹel into the final fuel form and encapsulated the fuel.

PHOTOGRAPHIC EQUIPMENT

Still and motion pictures will be made of most spacecraft maneuvers and crew lunar surface activities, an mapping photos from orbital altitude to aid in planning future landing missions. During lunar surface activitie emphasis will be on photographic documentation of lunar surface features and lunar material sampl collection.

Camera equipment stowed in the Apollo 13 command module consists of two 70mm Hasselblad electri cameras, a 16mm motion picture camera and the Hycon lunar topographic camera (LTC). The LTC, to b flown on Apollos, 13, 14 and 15, is stowed beneath the commander's couch. In use, the camera mounts in-th crew access hatch window. The LTC with 18-inch focal length f/4.0 lens provides resolution of objects as sma as 15-25 feet from a 60-nm altitude and as small as 3 to 5 feet from the 8 nm pericynthion. Film format 4.5-inch square frames on 100 foot long rolls, with a frame rate variable from 4 to 75 frames a minute. Shutte speeds are 1/50, 1/100, and 1/200 second. Spacecraft forward motion during exposures is compensated fo by a servo-controlled rocking mount. The film is held flat in the focal plane by a vacuum platen connected t the auxiliary dump valve.

The camera weighs 65 pounds without film, is 28 inches long, 10.5 inches wide, and 12.25 inches high. It is modification of an aerial reconnaissance camera.

Future lunar landing sites and targets of scientific interest will be photographed with the lunar topographi camera in overlapping sequence of single frame modes. A candidate landing site northwest of the crate Censorinus will be photographed from the 8-mile pericynthion during the period between descent orbi insertion and CSM/LM separation. Additional topographic photos of the Censorinus site and sites near Dav Rille and Descartes will be made later in the mission from the 60-nm circular orbit. The camera again will b unstowed and mounted for 20 minutes of photography of the lunar disc at 5 minute intervals starting at hours after transearth injection.

Cameras stowed in the lunar module are two 70mm Hasselblad data cameras fitted with 60mm Zeiss Metri lenses, a 16mm motion picture camera fitted with a 10mm lens, and a Kodak close-up stereo camera for hig resolution photos on the lunar surface. The LM Hasselblads have crew chest mounts that leave both hand free.

One of the command module Hasselblad electric cameras is normally fitted with an 80mm f/2.8 Zeiss Plana lens, but bayonet mount 250mm and 500mm lenses my be substituted for special tasks. The secon Hasselblad camera is fitted with an 80mm lens and a Reseau plate which allows greater dimensional contro on photographs of the lunar surface. The 500mm lens will be used only as a backup to the lunar topographi camera.

The 80mm lens has a focussing range from 3 feet to infinity and has a field of view of 38 degrees vertical an horizontal on the square-format film frame. Accessories for the command module Hasselblads include spotmeter, intervalometer, remote control cable, and film magazines. Hasselblad shutter speeds range from time exposure and one second to one 1/500 second.

The Maurer 16mm motion picture camera in the command module has lenses of 5, 18, and 75mm available The camera weighs 2.8 pounds with a 130-foot film magazine attached. Accessories include a right-angl mirror, a power cable, and a sextant adapter which allows the camera to use the navigation sextant optica system. The LM motion picture camera will be mounted in the right-hand window to record descent an landing and the two EVA periods and later will be taken to the surface.

The 35 mm stereo close-up camera stowed in the LM MESA shoots 24mm square stereo pairs with an imag scale of one half actual size. The camera is fixed focus and is equipped with a stand-off hood to position th camera at the proper focus distance. A long handle permits an EVA crewman to position the camera withou stooping for surface object photography. Detail as small as 40 microns can be recorded. The camera allow photography of significant surface structure which would remain intact only in the lunar environment, suc as fine powdery deposits, cracks or holes, and adhesion of particles. A battery-powered electronic flas provides illumination, and film capacity is a minimum of 100 stereo pairs.

LUNAR DESCRIPTION

Terrain — Mountainous and crater-pitted, the mountains rising as high as 29 thousand feet and the crater ranging from a few inches to 180 miles in diameter. The craters are thought to be formed primarily by th mpact of meteorites. The surface is covered with a layer of fine grained material resembling silt or sand, as vell as small rocks and boulders.

Environment — No air no wind and no moisture. The temperature ranges from 243 degrees F. in the wo-week lunar day to 279 degrees below zero in the two-week lunar night. Gravity is one-sixth that of Earth. Micrometeoroids pelt the Moon since there is no atmosphere to burn them up. Radiation might present a problem during periods of unusual solar activity.

Far Side — The far or hidden side of the Moon no longer is a complete mystery. It was first photographed by a Russian craft and since then has been photographed many times, particularly from NASA's Lunar Orbiter and Apollo spacecraft.

Origin — There is still no agreement among scientists on the origin of the Moon. The three theories: (1) the Moon once was part of Earth and split off into its own orbit (2) it evolved as a separate body at the same time as Earth, and (3) it formed elsewhere in space and wandered until it was captured by Earth's gravitational field.

Physical Facts

Diameter	2,160 miles (about 1/4 that of Earth)
Circumference	6,790 miles (about 1/4 that of Earth)
Distance from Earth	238,857 miles (mean; 221,463 min. to 252,710 max.)
Surface temperature	+243°F (Sun at zenith) -279°F (night)
Surface gravity	1/6 that of Earth
Mass	1/100th that of Earth
Volume	1/50th that of Earth
Lunar day and night	14 Earth days each
Mean velocity in orbit	2,287 miles-per-hour
Escape velocity	1.48 miles-per-second
Month (period of rotation around Earth)	27 days, 7 hours, 43 minutes

Landing Site
The landing site selected for Apollo 13 is located at 3° 40' 7"S, 17° 27' 3"W, about 30 miles north of the Fra Mauro crater. The site is in a hilly, upland region. This will be the first Apollo landing to other than a lunar mare, the flat dark areas of the Moon once thought to be lunar seas. This hilly region has been designated as

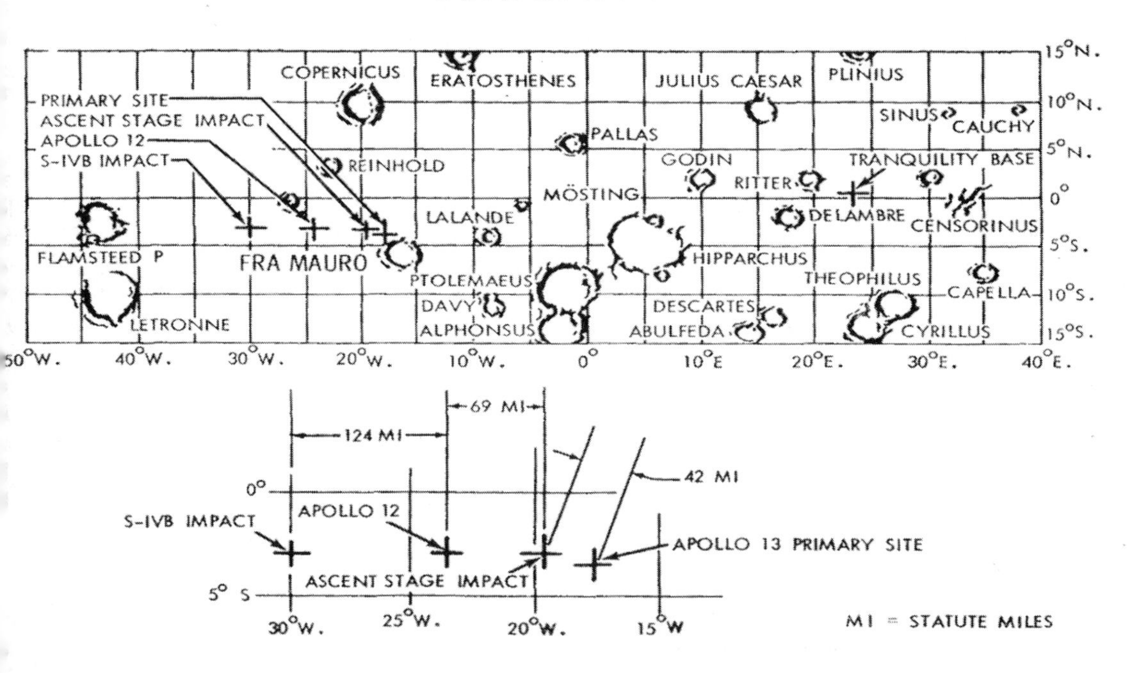

the Fra Mauro formation, a widespread geological unit covering large portions of the lunar surface around Mare Imbrium (Sea of Rains). The Fra Mauro formation is interpreted by lunar geologists to be an ejecta blanket of material thrown out by the event which created the circular Mare Imbrium basin.

The interpretation of the Fra Mauro formation as ejecta from Mare Imbrium gives rise to the expectation that surface material originated from deep within the Moon, perhaps from a hundred miles below the Moon's surface. If the interpretation proves correct, it will also be possible to date the Mare Imbrium event, believed to be a major impact, perhaps the in-fall of a smaller Moon, which was swept up in the primordial accretionary evolution of the Moon. Based on this theory, rocks from the Fra Mauro formation should predate the rocks returned from either Apollo 11 (4.6 billion years) or Apollo 12 (3.5 billion years) and be close to the original age of the Moon.

APOLLO 13 FLAGS, LUNAR MODULE PLAQUE

The United States flag to be erected on the lunar surface measures 30 by 48 inches and will be deployed on a two-piece aluminum tube eight feet long. The folding horizontal bar which keeps the flag standing out from the staff on the airless Moon has been improved over the mechanisms used on Apollo 11 and 12. The flag, made of nylon, will be stowed in the lunar module descent stage modularized equipment stowage assembly (MESA) instead of in a thermal-protective tube on the LM front leg, as in Apollo 11 and 12. Also carried on the mission and returned to Earth will be 25 United States and 50 individual state flags, each 4 by 6 inches. A 7 by 9 inch stainless steel plaque, similar to those flown on Apollos 11 and 12, will be fixed to the LM front leg. The plaque has on it the words "Apollo 13" with "Aquarius" beneath, the date, and the signatures or the three crewmen.

SATURN V LAUNCH VEHICLE

The Saturn V launch vehicle (SA-508) assigned to the APOLLO 13 mission was developed at the Marshall Space Flight Center, Huntsville, Ala. The vehicle is almost identical to those used in the missions of Apollo 8 through 12.

First Stage

The first stage (S-IC) of the Saturn V is built by the Boeing Company at NASA's Michoud Assembly Facility, Now Orleans, La. The stage's five F-1 engines develop a total of about 7.6 million pounds of thrust at launch. Major components of the stage are the forward skirt, oxidizer tank, intertank structure, fuel tank, and thrust structure. Propellant to the five engines normally flows at a rate of 29,364.5 pounds (3,400 gallons) each second. One engine is rigidly mounted on the stage's centerline; the other four engines are mounted on a ring at 90° angles around the center engine. These four outer engines are gimbaled to control the vehicle's attitude during flight.

Second Stage

The second stage (S-II) is built by the Space Division of the North American Rockwell Corporation at Seal Beach, Calif. Five J-2 engines develop a total of about 1.16 million pounds of thrust during flight. Major structural components are the forward skirt, liquid hydrogen and liquid oxygen tanks (separated by an insulated common bulkhead), a thrust structure, and an interstage section that connects the first and second stages. The five engines are mounted and used in the same way as the first stage's F-1 engines: four outer engines can be gimbaled, the center one is rigid.

Third Stage

The third stage (S-IVB) is built by the McDonnell Douglas Astronautics Company at Huntington Beach, Calif. Major components are the aft interstage and skirt, thrust structure, two propellant tanks with a common bulkhead, a forward skirt, and a single J-2 engine. The gimbaled engine has a maximum thrust of 230,000 pounds and can be shut off and restarted.

Instrument Unit

The instrument unit (IU), built by the International Business Machines Corp., at Huntsville, Ala., contains avigation guidance and control quipment to steer the launch ehicle into its Earth orbit and into ranslunar trajectory. The six major ystems are structural, thermal ontrol, guidance and control, neasuring and telemetry, radio requency, and electric.

'he instrument unit provides a ath-adaptive guidance scheme vherein a programmed trajectory s used during first stage boost with uidance beginning during second tage burn. This scheme prevents novements that could cause the ehicle to break up while ttempting to compensate for vinds or jet streams in the tmosphere.

'he instrument unit's inertial latform (heart of the navigation, uidance and control system) rovides space-fixed reference oordinates and measures cceleration along three mutually erpendicular axes of a coordinate ystem. If the platform fails during oost, systems in the Apollo pacecraft are programmed to rovide guidance for the launch vehicle. After second stage ignition, the spacecraft commander could nanually steer the vehicle in the event of loss of the launch vehicle inertial platform.

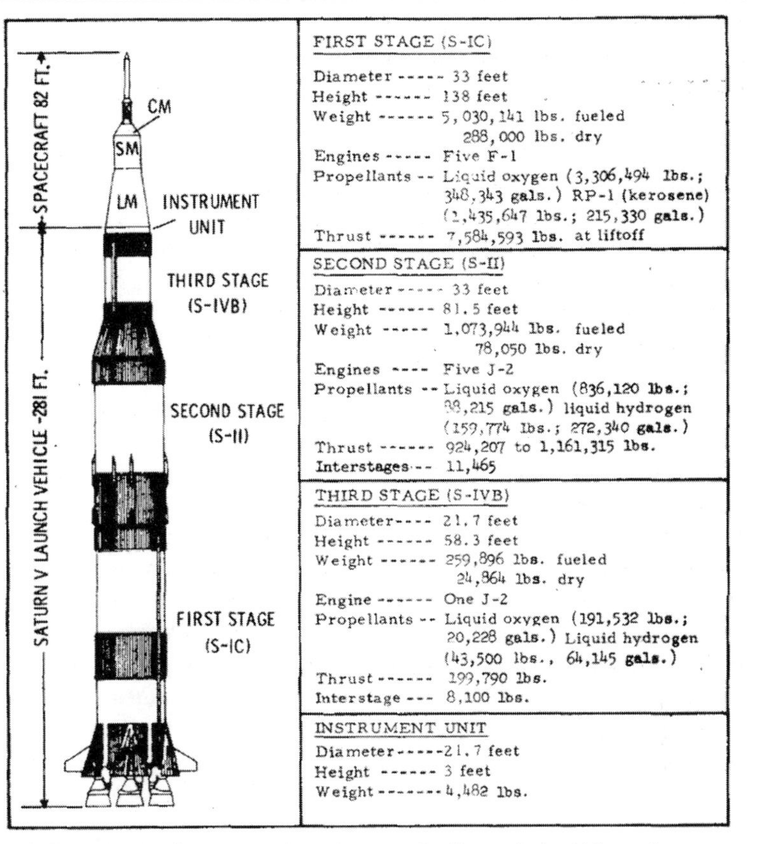

SPACECRAFT 82 FT.
CM
SM
LM
INSTRUMENT UNIT
THIRD STAGE (S-IVB)
SECOND STAGE (S-II)
FIRST STAGE (S-IC)
SATURN V LAUNCH VEHICLE -281 FT.
FIRST STAGE (S-IC)
Diameter ----- 33 feet
Height ------ 138 feet
Weight ------ 5,030,141 lbs. fueled
288,000 lbs. dry
Engines ----- Five F-1
Propellants -- Liquid oxygen (3,306,494 lbs.; 348,343 gals.) RP-1 (kerosene) (1,435,647 lbs.; 215,330 gals.)
Thrust ------ 7,584,593 lbs. at liftoff
SECOND STAGE (S-II)
Diameter ----- 33 feet
Height ------ 81.5 feet
Weight ----- 1,073,944 lbs. fueled
78,050 lbs. dry
Engines ---- Five J-2
Propellants -- Liquid oxygen (836,120 lbs.; 88,215 gals.) liquid hydrogen (159,774 lbs.; 272,340 gals.)
Thrust ------ 924,207 to 1,161,315 lbs.
Interstages -- 11,465
THIRD STAGE (S-IVB)
Diameter---- 21.7 feet
Height ------ 58.3 feet
Weight ------ 259,896 lbs. fueled
24,864 lbs. dry
Engine ------ One J-2
Propellants -- Liquid oxygen (191,532 lbs.; 20,228 gals.) Liquid hydrogen (43,500 lbs., 64,145 gals.)
Thrust ------ 199,790 lbs.
Interstage --- 8,100 lbs.
INSTRUMENT UNIT
Diameter-----21.7 feet
Height ------ 3 feet
Weight-------4,482 lbs.

NOTE: Weights and measures given above are for the nominal vehicle configuration for Apollo 12. The figures may vary slightly due to changes before launch to meet changing conditions. Weights of dry stages and propellants do not equal total weight because frost and miscellaneous smaller items are not included in chart.

Editor's note: The original document mistakenly refers to Apollo 12.

'ropulsion

'he Saturn V has 37 propulsive units, with thrust ratings ranging from 70 pounds to more than 1.5 million ounds. The large main engines burn liquid propellants; the smaller units use solid or hypergolic propellants.

'he five F-1 engines on the first stage burn a combination of RP-1 (kerosene) as fuel and liquid oxygen as xidizer. Each engine develops approximately 1,516,918 pounds of thrust at liftoff, building to about 1,799,022 ounds before cutoff. The five-engine cluster gives the first stage a thrust range of from 7,584,593 pounds at ftoff to 8,995,108 pounds just before center engine cutoff. The F-1 engine weighs almost 10 tons, is more han 18 feet long and has a nozzle exit diameter of nearly 14 feet. The engine consumes almost three tons f propellant every second.

'he first stage also has eight solid-fuel retrorockets that fire to separate the first and second stages. Each etrorocket produces a thrust of 87,900 pounds for 0.6 seconds.

'he second and third stages are powered by J-2 engines that burn liquid hydrogen (fuel) and liquid oxygen oxidizer). J-2 engine thrust varies from 184,841 to 232,263 pounds during flight. The 3,500-pound J-2 engine s considered more efficient than the F-1 engine because the J-2 burns high energy liquid hydrogen. F-1 and -2 engines are built by the Rocketdyne Division of the North American Rockwell Corp. The second stage lso has four 21,000 pound thrust solid fuel ullage rockets that settle liquid propellant in the bottom of the nain tanks and help attain a "clean" separation from the first stage. Four retrorockets, located in the S-IVB's ft Interstage (which never separates from the S-II), separate the S-II from the S-IVB. There are two jettisonable ullage rockets for propellant settling before engine ignition. Eight smaller engines in the two auxiliary propulsion system modules on the S-IVB stage provide three-axis attitude control.

COMMAND AND SERVICE MODULE STRUCTURE, SYSTEMS

The Apollo spacecraft for the Apollo 13 mission is comprised of Command Module 109, Service Module 109, Lunar Module 7, a spacecraft-lunar module adapter (SLA) and a launch escape system. The SLA houses the lunar module and serves as a mating structure between the Saturn V instrument unit and the SM.

Launch Escape System (LES) — Would propel command module to safety in an aborted launch. It has three solid-propellant rocket motors: a 147,000 pound-thrust launch escape system motor, a 2,400-pound-thrust pitch control motor, and a 31,500 pound-thrust tower jettison motor. Two canard vanes deploy to turn the command module aerodynamically to an attitude with the heat-shield forward. The system is 33 feet tall and 4 feet in diameter at the base, and weighs 8,945 pounds.

Command Module (CM) Structure — The command module is a pressure vessel encased in heat shields, cone-shaped, weighing 12,365 pounds at launch. The command module consists of a forward compartment which contains two reaction control engines and components of the Earth landing system; the crew compartment or inner pressure vessel containing crew accommodations, controls and displays, and many of the spacecraft systems; and the aft compartment housing ten reaction control engines, propellant tankage, helium tanks, water tanks, and the CSM umbilical cable. The crew compartment contains 210 cubic feet of habitable volume. Heat-shields around the three compartments are made of brazed stainless steel honeycomb with an outer layer of phenolic epoxy resin as an ablative material. CSM and LM are equipped with the probe-and-drogue docking hardware. The probe assembly is a powered folding coupling and impact

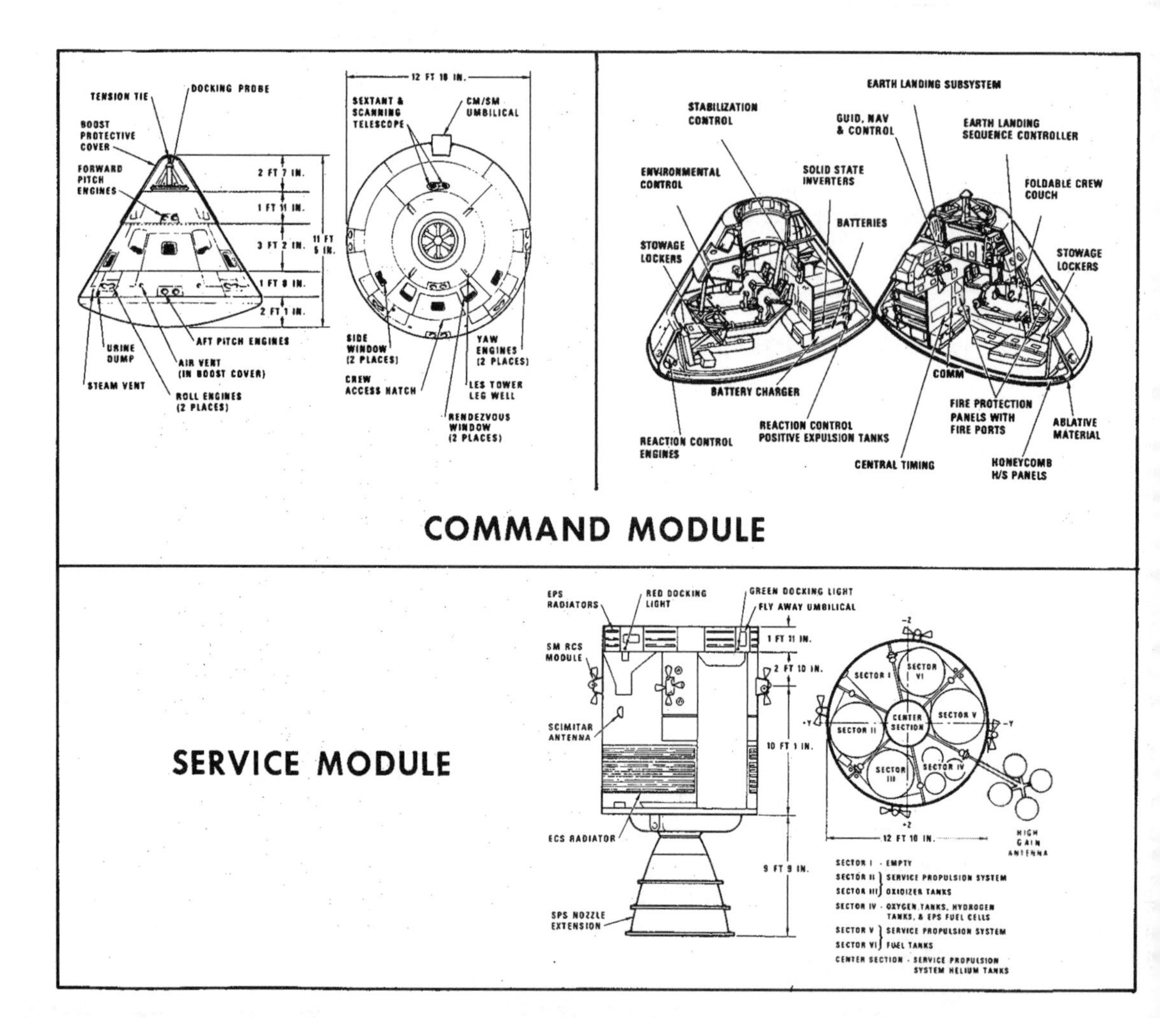

ttenuating device mounted in the CM tunnel ıat mates with a conical drogue mounted in ıe LM docking tunnel. After the 12 automatic ocking latches are checked following a ocking maneuver, both the probe and drogue re removed to allow crew transfer between ıe CSM and LM.

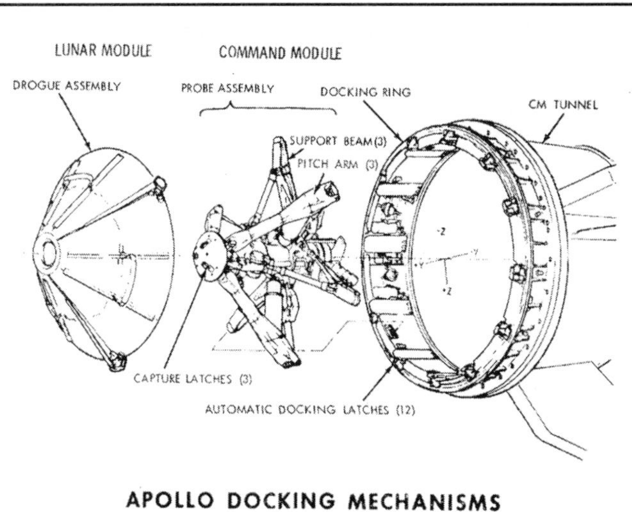

APOLLO DOCKING MECHANISMS

ervice Module (SM) Structure — At launch, ıe service module for the Apollo 13 mission ill weigh 51,105 pounds. Aluminum oneycomb panels one inch thick form the uter skin, and milled aluminum radial beam eparate the interior into six sections around a entral cylinder containing two helium spheres, our sections containing service propulsion ystem fuel-oxidizer tankage, another ontaining fuel cells, cryogenic oxygen and hydrogen, and one sector essentially empty.

pacecraft-LM Adapter (SLA) Structure — The spacecraft LM adapter is a truncated cone 28 feet long apering from 260 inches diameter at the base to 154 inches at the forward end at the service module mating ne. The SLA weighs 4,000 pounds and houses the LM during launch and Earth orbital flight.

SM Systems

uidance, Navigation and Control Systems (GNCS) — Measures and controls spacecraft position attitude, and velocity, calculates ajectory, controls spacecraft propulsion system thrust vector, and displays abort data. The guidance system consists of three ıbsystems: Inertial, made up of an inertial measurement unit and associated power and data components; computer which processes formation to or from other components; and optics consisting of scanning telescope and sextant for celestial and/or landmark sighting r spacecraft navigation. VHF ranging device serves as a backup to the LM rendezvous radar.

abilization and Control Systems (SCS) — Controls spacecraft rotation, translation, and thrust vector and provides displays for ew-initiated maneuvers; backs up the guidance system for control functions. It has three subsystems ; attitude reference, attitude ontrol, and thrust vector control.

ervice Propulsion System (SPS) — Provides thrust for large spacecraft velocity changes through a gimbal-mounted 20,500-pound-thrust pergolic engine using a nitrogen tetroxide oxidizer and a 50-50 mixture of unsymmetrical dimethyl hydrazine and hydrazine fuel. This stem is in the service module. The system responds to automatic firing commands from the guidance and navigation system or to anual commands from the crew. The engine thrust level is not throttleable. The stabilization and control system gimbals the engine to rect the thrust vector through the spacecraft center of gravity.

elecommunications System — Provides voice, television, telemetry, and command data and tracking and ranging between the spacecraft nd Earth, between the command module and the lunar module and between the spacecraft and astronauts during EVA. It also provides tercommunications between astronauts.

he high-gain steerable S-Band antenna consists of four, 31-inchdiameter parabolic dishes mounted on a folding boom at the aft end of e service module. Signals from the ground stations can be tracked either automatically or manually with the antenna's gimballing stem. Normal S-Band voice and uplink/downlink communications will be handled by the omni and high-gain antennae.

equential System — Interfaces with other spacecraft systems and subsystems to initiate time critical functions during launch, docking aneuvers, sub-orbital aborts, and entry portions of a mission. The system also controls routine spacecraft sequencing such as service odule separation and deployment of the Earth landing system.

nergency Detection System (EDS) — Detects and displays to the crew launch vehicle emergency conditions, such as excessive pitch r roll rates or two engines out, and automatically or manually shuts down the booster and activates the launch escape system; functions ntil the spacecraft is in orbit.

arth Landing System (ELS) — Includes the drogue and main parachute system as well as postlanding recovery aids. In a normal entry escent, the command module forward heat shield is jettisoned at 24,000 feet, permitting mortar deployment of two reefed 16.5-foot ameter drogue parachutes for orienting and decelerating the spacecraft. After disreef and drogue release, three mortar deployed pilot utes pull out the three main 83.3 foot diameter parachutes with two-stage reefing to provide gradual inflation in three steps. Two main arachutes out of three can provide a safe landing.

Reaction Control System (RCS) — The SM RCS has four identical RCS "quads" mounted around the SM 90 degrees apart. Each qu has four 100 pound-thrust engines, two fuel and two oxidizer tanks and a helium pressurization sphere. Attitude control and sm velocity maneuvers are made with the SM RCS. The CM RCS consists of two independent six-engine subsystems of six 93 pound-thru engines each used for spacecraft attitude control during entry. Propellants for both CM and SM ECS are monomethyl hydrazine fuel a nitrogen tetroxide oxidizer with helium pressurization. These propellants burn spontaneously when combined (without an igniter).

Electrical Power System (EPS) — Provides electrical energy sources, power generation and control, power conversion, conditioning, a distribution to the spacecraft. The primary source of electrical power is the fuel cells mounted in the SM. The fuel cell also furnish drinking water to the astronauts as a by-product. Three silver-zinc oxide storage batteries supply power to the CM during entry a after landing, provide power for sequence controllers, and supplement the fuel cells during periods of peak power demand. A batte charger assures a full charge prior to entry. Two other silver-zinc oxide batteries supply power for explosive devices for CM/S separation, parachute deployment and separation, third-stage separation, launch escape tower separation, and other pyrotechnic uses

Environmental Control System (ECS) — Controls spacecraft atmosphere, pressure and temperature and manages water. In addition regulating cabin and suit gas pressure, temperature and humidity, the system removes carbon dioxide, odors and particles and ventilat the cabin after landing. It collects and stores fuel cell potable water for crew use, supplies water to the glycol evaporators for coolir and dumps surplus water overboard through the waste H2O dump nozzle. Proper operating temperature of electronics and electric equipment is maintained by this system through the use of the cabin heat exchangers, the space radiators, and the glycol evaporator

Recovery Aids — Recovery aids include the uprighting system, swimmer interphone connections, sea dye marker, flashing beacon, VH recovery beacon, and VHF transceiver. The uprighting system consists of three compressor-inflated bags to upright the spacecraft if should land in the water apex down (stable II position).

Caution and Warning System — Monitors spacecraft systems for out-of-tolerance conditions and alerts crew by visual and audib alarms.

Controls and Displays — Provide status readouts and control functions or spacecraft systems in the command and service modules. A controls are designed to be operated by crewmen in pressurized suits. Displays are grouped by system and located according to th frequency of use and crew responsibility.

LUNAR MODULE STRUCTURES, WEIGHT

The lunar module is a two-stage vehicle designed for space operations near and on the Moon. The luna module stands 22 feet 11 inches high and is 31 feet wide (diagonally across landing gear). The ascent an descent stages of the LM operate as a unit until staging, when the ascent stage functions as a single spacecra for rendezvous and docking with the CM.

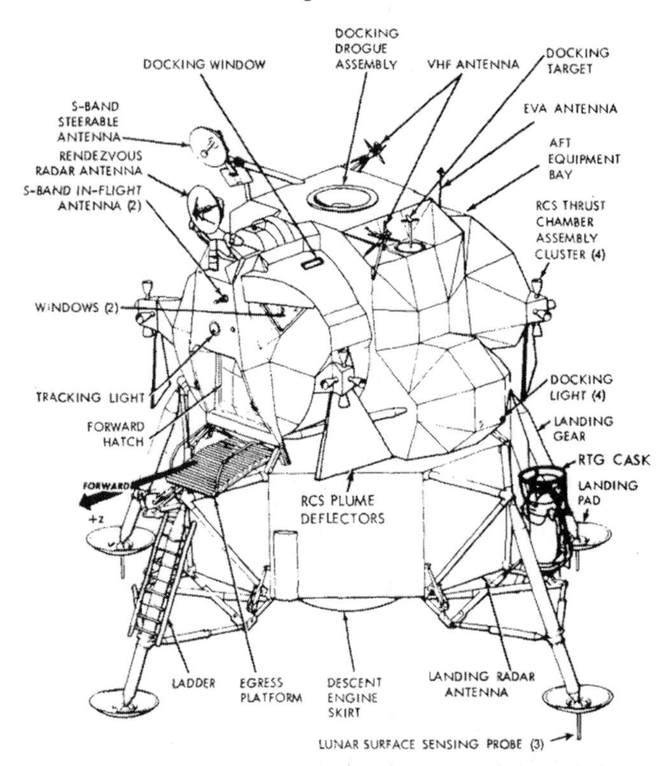

APOLLO LUNAR MODULE

Ascent Stage

Three main sections make up the ascer stage: the crew compartment, midsection, an aft equipment bay. Only the cre compartment and midsection are pressurize (4.8 psig). The cabin volume is 235 cubic fe (6.7 cubic meters). The stage measures 12 fe 4 inches high by 14 feet 1 inch in diamete The ascent stage has six substructural area crew compartment, midsection, aft equipme bay, thrust chamber assembly cluste supports, antenna supports and thermal an micrometeoroid shield.

The cylindrical crew compartment is 9 inches (2.35 m) in diameter and 42 inche (1.07 m) deep. Two flight stations ar equipped with control and display panel armrests, body restraints, landing aids, tw front windows, an overhead docking windo and an alignment optical telescope in th center between the two flight stations. Th habitable volume is 160 cubic feet.

tunnel ring atop the ascent stage meshes with the command module docking latch assemblies. During locking, the CM docking ring and latches are aligned by the LM drogue and the CSM probe.

'he docking tunnel extends downward into the midsection 16 inches (40 am). The tunnel is 32 inches (81 m) in diameter and is used for crew transfer between the CSM and LM. The upper hatch on the inboard nd of the docking tunnel opens inward and cannot be opened without equalizing pressure on both hatch urfaces.

thermal and micrometeoroid shield of multiple layers of Mylar and a single thickness of thin aluminum skin ncases the entire ascent stage structure.

)escent Stage

'he descent stage center compartment houses the descent engine, and descent propellant tanks are housed n the four square bays around the engine. Quadrant II (Seq bay) contains ALSEP, and Radioisotope 'hermoelectric Generator (RTG) externally. Quadrant IV contains the MESA. The descent stage measures 0 feet 7 inches high by 14 feet 1 inch in diameter and is encased in the Mylar and aluminum alloy thermal nd micrometeoroid shield.

'he LM egress platform, or "porch", is mounted on the forward outrigger just below the forward hatch. A adder extends down the forward landing gear strut from the porch for crew lunar surface operations.

'he landing gear struts are explosively extended and provide lunar surface landing impact attenuation. The nain struts are filled with crushable aluminum honeycomb for absorbing compression loads. Footpads 37 nches (0.95 m) in diameter at the end of each landing gear provide vehicle support on the lunar surface.

:ach pad (except forward pad) is fitted with a 66 inch long lunar surface sensing probe which signals the crew o shut down the descent engine upon contact with the lunar surface.

.M-7 flown on the Apollo 13 mission has a launch weight of 33,476 pounds. The weight breakdown is as ollows:

Ascent stage, dry	4,668 lbs.	Includes water and oxygen; no crew
Descent stage, dry	4,650 lbs.	
RCS propellants (loaded)	590 lbs.	
DPS propellants (loaded)	18,339 lbs.	
APS propellants (loaded)	5,229 lbs.	
Total	33,476 lbs.	

.unar Module Systems

:lectrical Power System — The LM DC electrical system consists of six silver-zinc batteries in the descent stage and two in the ascent tage. Twenty-eight-volt DC power is distributed to all LM systems. AC power (117v 400 Hz) is supplied by two inverters.

:nvironmental Control System — Consists of the atmosphere revitalization section, oxygen supply and cabin pressure control section, vater management, heat transport section, and outlets for oxygen and water reservicing of the portable life support system (PLSS). Components of the atmosphere revitalization section are the suit circuit assembly which cools and ventilates the pressure garments, educes carbon dioxide levels, removes odors, noxious gases and excessive moisture; the cabin recirculation assembly which ventilates nd controls cabin atmosphere temperatures; and the steam flex duct which vents to space steam from the suit circuit water evaporator. 'he oxygen supply and cabin pressure section supplies gaseous oxygen to the atmosphere revitalization section for maintaining suit and abin pressure. The descent stage oxygen supply provides descent flight phase and lunar stay oxygen needs, and the ascent stage oxygen upply provides oxygen needs for the ascent and rendezvous flight phase.

Vater for drinking, cooling, fire fighting, food preparation, and refilling the PLSS cooling water servicing tank is supplied by the water nanagement section. The water is contained in three nitrogen pressurized bladder-type tanks one of 367-pound capacity in the descent tage and two of 47.5-pound capacity in the ascent stage. The heat transport section has primary and secondary water-glycol solution oolant loops. The primary coolant loop circulates waterglycol for temperature control of cabin and suit circuit oxygen and for thermal ontrol of batteries and electronic components mounted on cold plates and rails. If the primary loop becomes inoperative, the econdary loop circulates coolant through the rails and cold plates only. Suit circuit cooling during secondary coolant loop operation is rovided by the suit loop water boiler. Waste heat from both loops is vented overboard by water evaporation or sublimators.

Communications System — Two S-band transmitter-receivers, two VHF transmitter-receivers a signal processing assembly, ar associated spacecraft antenna make up the LM communications system. The system transmits and receives voice and tracking and rangir data, and transmits telemetry data on about 270 measurements and TV signals to the ground. Voice communications between the L and ground stations is by S-band, and between the LM and CSM voice is on VHF. Although no real-time commands can be sent to th LM, the digital uplink processes guidance officer commands, such as state vector updates, transmitted from Mission Control Center t the LM guidance computer. The data storage electronics assembly (DSEA) is a four-channel voice recorder with timing signals, with 10-hour recording capacity, which will be brought back into the CSM for return to Earth. DSEA recordings cannot be "dumped" t ground stations. LM antennas are one 26-inch-diameter parabolic S-band steerable antenna, two S-band inflight antennas, two VHF infligh antennas, EVA antenna, and an erectable S-band antenna (optional) for lunar surface.

Guidance, Navigation and Control System — Comprised of six sections: primary guidance and navigation section (PGNS), abort guidanc section (AGS), radar section, control electronics section (CES), and orbit rate display Earth and lunar (ORDEAL).

* The PGNS is an aided inertial guidance system updated by the alignment optical telescope, an inertial measurement unit, and th rendezvous and landing radars. The system provides inertial reference data for computations, produces inertial alignment reference b feeding optical sighting data into the LM guidance computer, displays position and velocity data, computes LM-CSM rendezvous dat from radar inputs, controls attitude and thrust to maintain desired LM trajectory, and controls descent engine throttling and gimbalin The LM-7 primary guidance computer has the Luminary 1C Software program, which is an improved version over that in LM-6.
* The AGS is an independent backup system for the PGNS, having its own inertial sensors and computer.
* The radar section is made up of the rendezvous radar which provides CSM range and range rate, and line-of-sight angles for maneuve computation to the LM guidance computer; and the landing radar which provides altitude and velocity data to the LM guidanc computer during lunar landing. The rendezvous radar has an operating range from 80 feet to 400 nautical miles. The ranging ton transfer assembly, utilizing VHF electronics, is a passive responder to the CSM VHF ranging device and is a backup to the rendezvou radar.
* The CES controls LM attitude and translation about all axes. It also controls by PGNS command the automatic operation of th ascent and descent engine and the reaction control thrusters. Manual attitude controller and thrust-translation controller command are also handled by the CES.
* ORDEAL, displayed on the flight director attitude indicator, is the computed local vertical in the pitch axis during circular Earth o lunar orbits.

Reaction Control System — The LM has four RCS engine clusters of four 100 pound (45.4kg) thrust engines each, which use heliur pressurized hypergolic propellants. The oxidizer is nitrogen tetroxide, fuel is Aerozine 50 (50/50 blend of hydrazine and unsymmetric dimethyl hydrazine). Interconnect valves permit the RCS system to draw from ascent engine propellant tanks. The RCS provides sma stabilizing impulses during ascent and descent burns, controls LM attitude during maneuvers, and produces thrust for separation, and fo ascent/descent engine tank ullage. The system may be operated in either the pulse or steady-state modes.

Descent Propulsion System — Maximum rated thrust of the descent engine is 9,870 pounds (4,380.9 kg) and is throttleable betwee 1,050 (476.7 kg) 6,300 pounds (2,860.2 kg). The engine can be gimbaled six degrees in any direction in response to attitude command and to compensate for center of gravity offsets. Propellants are helium-pressurized Aerozine 50 and nitrogen tetroxide.

Ascent Propulsion System — The 3,500 pound (1,589 kg) thrust ascent engine is not gimbaled and performs at full thrust. The engin remains dormant until after the ascent stage is separated from the descent stage. Propellants are the same as are burned by the RC engines and the descent engine.

Caution and Warning, Controls and Displays — These two systems have the same function aboard the lunar module as they do aboar the command module (See CSM systems section.)

Tracking & Docking Lights — A flashing tracking light (once per second 20 millisecond duration on the front face of the lunar modul is an aid for contingency CSM-active rendezvous LM rescue. Visibility ranges from 400 nautical miles through the CSM Sextant to 13 miles with the naked eye. Five docking lights analogous to aircraft running lights are mounted on the LM for CSM-active rendezvou two forward yellow lights, aft white light, port red light and starboard green light. All docking lights have about a 1,000-foot visibility.

APOLLO 13 CREW AND CREW EQUIPMENT

Life Support Equipment — Space Suits

Apollo 13 crewmen will wear two versions of the Apollo space suit; an intravehicular pressure garmen assembly worn by the command module pilot and the extravehicular pressure garment assembly worn b the commander and the lunar module pilot. Both versions are basically identical except that th extravehicular version has an integral thermal/ meteoroid garment over the basic suit.

From the skin out, the basic pressure garment consists of a nomex comfort layer, a neoprene-coated nylo pressure bladder and a nylon restraint layer. The outer layers of the intravehicular suit are, from the insid out, nomex and two layers of Teflon-coated Beta cloth. The extravehicular integral thermal/meteoroid cove consists of a liner of two layers of neoprene-coated nylon, seven layers of Beta/Kapton spacer laminate, an an outer layer of Teflon-coated Beta fabric.

The extravehicular suit, together with a liquid cooling garment, portable life support system (PLSS), oxygen purge system, lunar extravehicular visor assembly and other components make up the extravehicular mobility unit (EMU). The EMU provides an extravehicular crewman with life support for a four hour mission outside the lunar module without replenishing expendables. EMU total weight is 183 pounds. The intravehicular suit weighs 35.6 pounds.

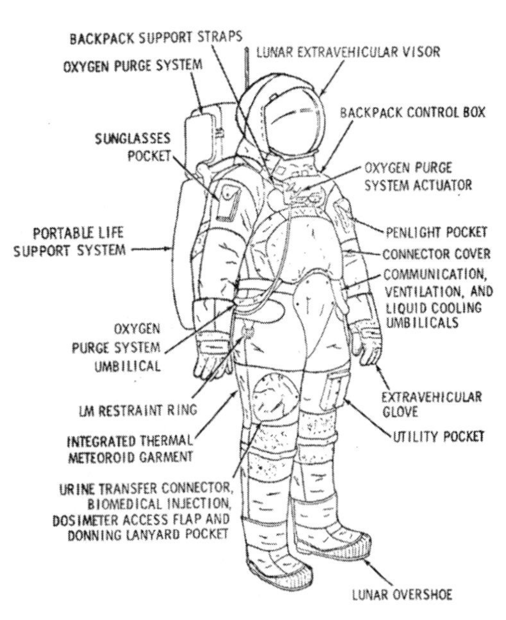

EXTRAVEHICULAR MOBILITY UNIT

Liquid Cooling Garment — A knitted nylon-spandex garment with a network of plastic tubing through which cooling water from the PLSS is circulated. It is worn next to the skin and replaces the constant wear-garment during EVA only.

Portable life Support System — A backpack supplying oxygen at 3.9 PSI and cooling water to the liquid cooling garment. Return oxygen is cleansed of solid and gas contaminants by a lithium hydroxide canister. The PLSS includes communications and telemetry equipment, displays and controls, and a main power supply. The PLSS is covered by a thermal insulation jacket. (Two stowed in LM).

Oxygen purge system — Mounted atop the PLSS, the oxygen purge system provides a contingency 45-minute supply of gaseous oxygen in two two-pound bottles pressurized to 5,880 psia. The system may also be worn separately on the front of the pressure garment assembly torso. It serves as a mount for the VHF antenna for the PLSS. (Two stowed in LM).

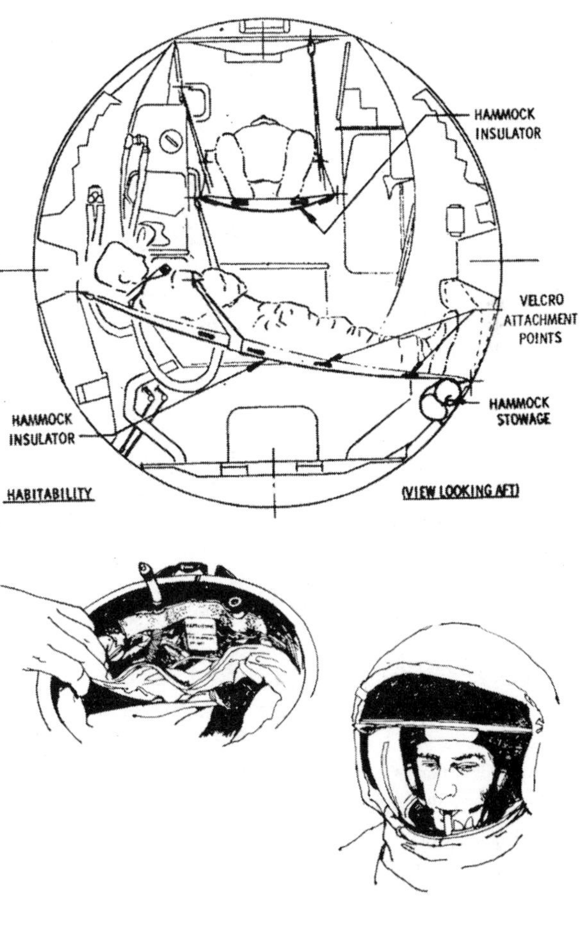

WATER BAG

APOLLO 13 SPACESUIT WATER BAG FOR USE DURING EVA.

Lunar extravehicular visor assembly — A polycarbonate shell and two visors with thermal control and optical coatings on them. The EVA visor is attached over the pressure helmet to provide impact, micrometeoroid, thermal and ultraviolet/infrared light protection to the EVA crewman. Since Apollo 12, a sunshade has been added to the outer portion of the LEVVA in the middle portion of the helmet rim.

Extravehicular gloves — Built of an outer shell of Chromel - R fabric and thermal insulation to provide protection when handling extremely hot and cold objects. The finger tips are made of silicone rubber to provide more sensitivity. A one-piece constant-wear garment, similar to "long Johns," is worn as an undergarment for the space suit in intravehicular operations and for the inflight coveralls. The garment is porous-knit cotton with a waist-to-neck zipper for donning. Biomedical harness attach points are provided.

During periods out of the space suits, crewmen wear two piece Teflon fabric inflight coveralls for warmt and for pocket stowage of personal items. Communications carriers ("Snoopy Hats") with redundar microphones and earphones are worn with the pressure helmet; a lightweight headset is worn with th inflight coveralls. Another modification since Apollo 12 has been the addition of eight-ounce drinking wate bags ("Gunga Dins") attached to the inside neck rings of the EVA suits. The crewmen can take a sip of wate from the 6 X 8 inch bag through a 1/8-inch diameter tube within reach of his mouth. The bags are filled fror the lunar module potable water dispenser.

Apollo Lunar Hand Tools

Special Environmental Container — The special environmental sample is collected in a carefully selected are and sealed in a special container which will retain a high vacuum. The container is opened in the Luna Receiving Laboratory where it will provide scientists the opportunity to study lunar material in its origin environment.

Extension handle — This tool is of aluminum alloy tubing with a malleable stainless steel cap designed to b used as an anvil surface. The handle is designed to be used as an extension for several other tools and t permit their use without requiring the astronaut to kneel or bend down. The handle is approximately 2 inches long and 1 inch in diameter. The handle contains the female half of a quick disconnect fitting designe to resist compression, tension, torsion, or a combination of these loads.

Three core tubes — These tubes are designed to be driven or augured into loose gravel, sandy material o into soft rock such as feather rock or pumice. They are about 15 inches in length and an inch in diamete and are made of aluminum tubing. Each tube is supplied with a removable non-serrated cutting edge and screw-on cap incorporating a metal-to-metal crush seal which replaces the cutting edge. The upper end of each tube is sealed and designed to be used with the extension handle or as an anvil. Incorporated into eac tube is a spring device to retain loose materials in the tube.

Scoops (large and small) — These tools are designed for use as trowel and as a chisel. The scoop is fabricate primarily of aluminum with a hardened-steel cutting edge riveted on and a nine-inch handle. A malleabl stainless steel anvil is on the end of the handle. The angle between the scoop pan and the handle allows a compromise for the dual use. The scoop is used either by itself or with the extension handle. The large scoop has a sieve which permits particles smaller than 1/2 cm to pass through.

Sampling hammer — This tool serves three functions, as a sampling hammer, as a pick or mattock, and as a hammer to drive the core tubes or scoop. The head has a small hammer face on one end, a broad horizontal blade on the other, and large hammering flats on the sides. The handle is 14 inches long and is made of formed tubular aluminum. The hammer has on its lower end a quick-disconnect to allow attachment to the extension handle for use as a hoe. The head weight has been increased to provide more impact force.

Tongs — The tongs are designed to allow the astronaut to retrieve small samples from the lunar surface while in a standing position. The tines are of such angles, length, and number to allow samples of from 3/8 up to 2-1/2 inch diameter to be picked up. This tool is 24 inches in overall length.

Brush/Scriber/Hand Lens — A composite tool

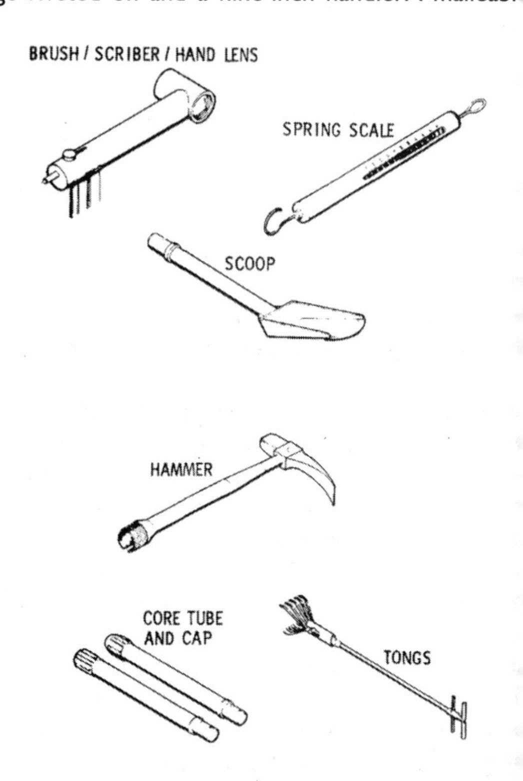

GEOLOGIC SAMPLING TOOLS

) Brush — To clean samples prior to selection
.) Scriber — To scratch samples for selection and to mark for identification
) Hand lens — Magnifying glass to facilitate sample selection

ring Scale — To weigh two rock boxes and other bags containing lunar material samples, to maintain weight idget for return to Earth.

strument staff — The staff holds the Hasselblad camera. The staff breaks down into sections. The upper ction telescopes to allow generation of a vertical stereoscopic base of one foot for photography. Positive ops are provided at the extreme of travel. A shaped hand grip aids in aiming and carrying. The bottom ction is available in several lengths to suit the staff to astronauts of varying sizes. The device is fabricated om tubular aluminum.

nomon — This tool consists of a weighted staff suspended on a two-ring gimbal and supported by a tripod. he staff extends 12 inches above the gimbal and is painted with a gray scale. The gnomon is used as a notographic reference to indicate local vertical, sun angle, and scale. The gnomon has a required accuracy f vertical indication of 20 minutes of arc. Magnetic damping is incorporated to reduce oscillations.

olor Chart — The color chart is painted with three primary colors and a gray scale. It is used as a alibration for lunar photography. The scale is mounted on the tool carrier but may easily be removed and eturned to Earth for reference. The color chart is 6 inches in size.

ool Carrier — The carrier is the stowage container for the tools during the lunar flight. After the landing ne carrier serves as support for the astronaut when he kneels down, as a support for the sample bags and amples, and as a tripod base for the instrument staff. The carrier folds flat for stowage. For field use it opens nto a triangular configuration. The carrier is constructed of formed sheet metal and approximates a truss ructure. Six-inch legs extend from the carrier to elevate the carrying handle sufficiently to be easily grasped y the astronaut.

ield Sample Bags — Approximately 80 bags four inches by five inches are included in the Apollo lunar hand ools for the packaging of samples. These bags are fabricated from Teflon FEP.

Collection Bag — This is a large bag (4 X 8 inches) attached to the astronaut's side of the tool carrier. Field ample bags are stowed in this bag after they have been filled. It can also be used for general storage or to old items temporarily. (Two in each SRC).

renching Tool — A trenching tool with a pivoting scoop has been provided for digging the two-foot deep oil mechanics investigation trench. The two-piece handle is five feet long. The scoop is eight inches long and ve inches wide and pivots from in-line with the handle to 90° — similar to the trenching tool carried on nfantry backpacks. The trenching tool is stowed in the MESA rather than in the tool carrier.

unar Surface Drill — The 29.4 pound Apollo Lunar Surface Drill (ALSD) is stowed in the ALSEP Subpackage lo. 2 and will be used for boring two ten-foot deep 1.25-inch diameter holes for ALSEP heat flow xperiment probes, and one approximately eight-foot-deep, one-inch-diameter core sample. The silver-zinc attery-powered rotary percussive drill has a clutch to limit torque to 20 foot-pounds. A treadle assembly erves as a drilling platform and as a core stem lock during the drill string decoupling operation as the string withdrawn from the lunar soil. Bore stems for the heat flow experiment holes are of boron/fiberglass, and ne core sample core stems are titanium. Cutting bits are tungsten carbide.

<u>pollo 13 Crew Menu</u>

lore than 70 items comprise the food selection list of freeze-dried rehydratable, wet-pack and spoon-bowl oods. Balanced meals for five days have been packed in man/day overwraps. Items similar to those in the daily nenus have been packed in a snack pantry. The snack pantry permits the crew to locate easily a food item a smorgasbord mode without having to "rob" a regular meal somewhere down deep in a storage box.

Water for drinking and rehydrating food is obtained from two sources in the command module — dispenser for drinking water and a water spigot at the food preparation station supplying water at about I or 55°F. The potable water dispenser squirts water continuously as long as the trigger is held down, and t food preparation spigot dispenses water in one-ounce increments. A continuous-feed hand water dispens similar to the one in the command module is used aboard the lunar module for cold water rehydration food packets stowed aboard the LM. After water has been injected into a food bag, it is kneaded for abo three minutes. The bag neck is then cut off and the food squeezed into the crewman's mouth. After a m germicide pills attached to the outside of the food bags are placed in the bags to prevent fermentation a gas formation. The bags are then rolled and stowed in waste disposal compartments.

The day-by-day, meal-by-meal Apollo 13 Menu for Commander Lovell is on the following page as a typical fi day menu for each crewman.

TYPICAL CREW MENU IS THAT OF APOLLO 13 COMMANDER LOVELL:

MEAL	Day 1*,5**,9	Day 2,6,10	Day 3,7,11	Day 4,8
A.	Peaches RSB Canadian Bacon Applesauce RSB Bacon Squares (8) IMB Cocoa R Orange Drink R	Pears IMB Bacon Squares (8) IMB Scrambled Eggs RSB Grapefruit Drink R Coffee (b) R	Peaches, IMB Canadian Bacon Applesauce RSB Sugar Coated Corn Flakes RSB Cocoa R Grape Drink R	Apricots IMB Bacon Squares (8) IMB Scrambled Eggs RSB Orange-G.F. Drink R Coffee (B) R
B.	Salmon Salad RSB Beef A Gravy WP Jellied Candy IMB Grape Drink R	Frankfurters WP Cranberry-Orange RSB Chocolate Pudding RSB Orange-G.F. Drink R	Cream of Chicken Soup RSB Bread Slice ***Sandwich Spread WP Chocolate Bar IMB P.A.-G.F. Drink R	Chicken & Rice Soup RSB Meatballs with Sauce WP Caramel Candy IMB Orange Drink R
C.	Pea soup RSB Chicken & Rice RSB Date Fruitcake (4) DB P.A.-G.F. Drink R	Shrimp Cocktail RSB Pork & Scalloped Potatoes RSB Apricots IMB Orange Drink R	Chicken Stew RSB Turkey & Gravy WP Butterscotch Pudding RSB Grapefruit Drink R	Tuna Salad RSB Beef Stew RSB Banana Pudding RSB Grape Punch R
TOTAL CALORIES	2106	2073	2183	2043

DB — Dry Bits, IMB — Intermediate Moisture Bits, R — rehydratable, RSB — Rehydratable Spoon Bowl, WP — Wet Pack

* Day 1 consists of Meal B and C only; extra meal consists of: Ham & Cheese Sandwich (frozen), Caramel Candy, Orange, G.F. Drink
** Day 3 consists of Meal A only
*** Bread: Choose, Rye, White Sandwich Spreads: Chicken, Ham, Tuna Salad, Cheddar Cheese Spread, Peanut Butter, Jelly.

Personal Hygiene

Crew personal hygiene equipment aboard Apollo 13 includes body cleanliness item, the waste manageme system and one medical kit. Packaged with the food are a toothbrush and a two-ounce tube of toothpas for each crewman. Each man-meal package contains a 3.5-by-4-inch wet-wipe cleansing towel. Additional three packages of 12-by-12-inch dry towels are stowed beneath the command module pilot's couch. Eac package contains seven towels. Also stowed under the command module pilot's couch are seven tissu dispensers containing 53 three ply tissues each. Solid body wastes are collected in plastic defecation ba which contain a germicide to prevent bacteria and gas formation. The bags are sealed after use and stowe in empty food containers for post-flight analysis. Urine collection devices are provided for use while weari either the pressure suit or the inflight coveralls. The urine is dumped overboard through the spacecraft urir dump valve in the CM and stored in the LM.

Medical Kit

The 5x5x8-inch medical accessory kit is stowed in a compartment on the spacecraft right side wall besid the lunar module pilot couch. The medical kit contains three motion sickness injectors, three pa suppression injectors, one two ounce bottle first aid ointment, two one-ounce bottles eye drops, three nas sprays, two compress bandages, 12 adhesive bandages, one oral thermometer, and four spare crew biomedic ırnesses. Pills in the medical kit are 60 antibiotic, 12 nausea, 12 stimulant, 18 pain killer, 60 decongestant, 24 arrhea, 72 aspirin and 21 sleeping. Additionally, a small medical kit containing four stimulant, eight diarrhea, vo sleeping and four pain killer pills, 12 aspirin, one bottle eye drops, two compress bandages, 8 econgestant pills, one pain injector, one bottle nasal spray is stowed in the lunar module flight data file ompartment.

urvival Gear

he survival kit is stowed in two rucksacks in the right-hand forward equipment bay above the lunar module ilot. Contents of rucksack No. 1 are: two combination survival lights, one desalter kit, three pair sunglasses, ne radio beacon, one spare radio beacon battery and spacecraft connector cable, one knife in sheath, three ater containers, and two containers of Sun lotion, two utility knives, three survival blankets and one utility etting. Rucksack No. 2: one three-man life raft with CO_2 inflater, one sea Anchor, two sea dye markers, three unbonnets, one mooring lanyard, three mainlines and two attach brackets. The survival kit is designed to rovide a 48-hour postlanding (water or land) survival capability for three crewmen between 40° North and outh latitudes.

iomedical Inflight Monitoring

he Apollo 13 crew biomedical telemetry data received by the Manned Space Flight Network will be relayed or instantaneous display at Mission Control Center where heart rate and breathing rate data will be isplayed on the flight surgeon's console. Heart rate and respiration rate average, range and deviation are omputed and displayed on digital TV screens.

ı addition, the instantaneous heart rate, real-time and delayed EKG and respiration are recorded on strip harts for each man. Biomedical telemetry will be simultaneous from all crewmen while in the CSM, but electable by a manual onboard switch in the LM. Biomedical data observed by the flight surgeon and his team ı the Life Support System Staff Support Room will be correlated with spacecraft and space suit nvironmental data displays. Blood pressures are no longer telemetered as they were in the Mercury and emini programs. Oral temperatures, however, can be measured onboard for diagnostic purposes and voiced own by the crew in case of inflight illness. Energy expended by the crewman during EVA will be determined ıdirectly using a metabolic computation program based on three separate measurements:

1) Heart rate portion — Heart rate will be determined from telemetered EKG and converted to oxygen consumption (litre/min) and heat production (BTU/hour) based on pre-flight calibration curves. These curves are determined from exercise response tests utilizing a bicycle ergometer.

2) Oxygen usage portion — Oxygen usage will be determined from the telemetered measurement of PLSS oxygen supply pressure. Suit leak determined preflight is taken into account. Heat production will be calculated from oxygen usage.

3) Liquid cooled garment temperature portion — The amount of heat taken up by the liquid cooled garment will be determined from telemetered measurements of the LCG water temperature inlet and change in/out. This measurement (the amount of heat taken up by the water) plus an allowance made for sensible and latent heat loss, radiant heat load, and possible heat storage will provide an indication of heat production by the crewman.

raining

he crewman of Apollo 13 have spent more than five hours of formal crew training for each hour of the unar-launching mission's ten-day duration. More than 1,000 hours of training were in Apollo 13 crew training yllabus over and above the normal preparations for the mission technical briefings and reviews, pilot neetings and study.

he Apollo 13 crewmen also took part in prelaunch testing at Kennedy Space Center, such as altitude hamber tests and the countdown demonstration tests (CDDT) which provided the crew with thorough perational knowledge of the complex vehicle.

Highlights of specialized Apollo 13 crew training topics are:

* Detailed series of briefings on spacecraft systems, operation and modifications.
* Saturn launch vehicle briefings on countdown, range safety, flight dynamics, failure modes and abort conditions. The launch vehi briefings were updated periodically.
* Apollo Guidance and Navigation system briefings at the Massachusetts Institute of Technology Instrumentation Laboratory.
* Briefings and continuous training on mission photographic objectives and use of camera equipment.
* Extensive pilot participation in reviews of all flight procedures for normal as well as emergency situations.
* Stowage reviews and practice in training sessions in the spacecraft's mockups and command module simulators allowed t crewmen to evaluate spacecraft stowage of crew-associated equipment.
* More than 400 hours of training per man in command module and lunar module simulators at MSC and KSC, including closed-lo simulations with flight controllers in the Mission Control Center. Other Apollo simulators at various locations were used extensive for specialized crew training.
* Lunar surface briefings and some 20 suited 1-g walkthroughs of lunar surface EVA operations covering lunar geology a microbiology and deployment of experiments in the Apollo Lunar Surface Experiment Package (ALSEP). Training in lunar surface E included practice sessions with lunar surface sample gathering tools and return containers, cameras, the erectable S-band anten and the modular equipment stowage assembly (MESA) housed in the LM descent stage.
* Proficiency flights in the lunar landing training vehicle (LLTV) for the commander.
* Zero-g and one-sixth 9 aircraft flights using command module and lunar module mockups for EVA and pressure suit doffing/domi practice and training.
* Underwater zero-g training in the MSC Water Immersion Facility using spacecraft mockups to further familiarize crew with aspects of CSM-LM docking tunnel intravehicular transfer and EVA in pressurized suits.
* Water egress training conducted in indoor tanks as well as in the Gulf of Mexico, included uprighting from the Stable II Positi (apex down) to the Stable I position (apex up), egress onto rafts donning Biological Isolation Garments (BIGs), decontaminati procedures and helicopter pickup.
* Launch pad egress training from mockups and from the actual spacecraft on the launch pad for possible emergencies such as fir contaminants and power failures.
* The training covered use of Apollo spacecraft fire suppression equipment in the cockpit.
* Planetarium reviews at Morehead Planetarium, Chapel Hill, N.C., and at Griffith Planetarium, Los Angeles, Calif., of the celesti sphere with special emphasis on the 37 navigational stars used by the Apollo guidance computer.

NATIONAL AERONAUTICS AND SPACE ADMINISTRATION WASHINGTON D.C. 20546
BIOGRAPHICAL DATA

NAME: James Arthur Lovell, Jr. (Captain. USN) NASA Astronaut

BIRTHPLACE AND DATE; Born March 25, 1928, in Cleveland Ohio. His mother, Mrs. Blanche Lovell, resides at Edgewater Beac Florida,

PHYSICAL DESCRIPTION: Blond hair; blue eyes; height: 5 feet 11 Inches; weight: 170 pounds.

EDUCATION: Graduated from Juneau High School, Milwaukee, Wisconsin; attended the University of Wisconsin for 2 years, the received a Bachelor of Science degree from the United States Naval Academy in 1952; presented an Honorary Doctorate from Illino Wesleyan University in 1969.

MARITAL STATUS: Married to the former Marilyn Gerlach of Milwaukee, Wisconsin. Her parents, Mr. and Mrs. Carl Gerlach, a residents of Milwaukee.

CHILDREN: Barbara L., October 13 1953 James A., February 15, 1955; Susan K., July 14, 1958; Jeffrey C., January 14, 1966.

RECREATIONAL INTERESTS: His hobbies are golf, swimming, handball, and tennis.

ORGANIZATIONS: Member of the Society of Experimental Test Pilots and the Explorers Club.

SPECIAL HONORS: Awarded the NASA Distinguished Service Medal, two NASA Exceptional Service Medals, the Navy Astrona Wings, the Navy Distinguished Service Medal, and two Navy Distinguished Flying Crosses; recipient of the 1967 FAI De Laval and Go Space Medals (Athens, Greece), the American Academy of Achievement Golden Plate Award, the City of New York Gold Medal in 196 the City of Houston Medal for Valor in 1969, the National Geographic Society's Hubbard Medal in 1969, the National Academy Television Arts and Sciences Special Trustees Award in 1969, and the Institute of Navigation Award In 1969.

Co-recipient of the American Astronautical Society Flight Achievement Awards in 1966 and 1968, the Harmon International Trophy 1966 and 1967, the Robert H. Goddard Memorial Trophy in 1969, the H. H. Arnold Trophy for 1969, the General Thomas D. White USA Space Trophy for 1968, the Robert J. Collier Trophy for 1968, and the 1969 Henry G. Bennett Distinguished Service Award.

EXPERIENCE: Lovell, a Navy Captain, received flight training following graduation from Annapolis In 1952.

He has had numerous naval aviator assignments including a 4-year tour as a test pilot at the Naval Air Test Center, Patuxent Rive aryland. While there he served as program manager for the F4H weapon system evaluation. A graduate of the Aviation Safety School the University of Southern California, he also served as a flight Instructor and safety engineer with Fighter Squadron 101 at the Naval r Station, Oceana, Virginia.

e has logged more than 4,407 hours flying time — more than 3,000 hours in jet aircraft.

URRENT ASSIGNMENT: Captain Lovell was selected as an astronaut by NASA in September 1962. He has since served as backup ot for the Gemini 4 flight and backup command pilot for the Gemini 9 flight.

n December 4, 1965, he and Command Pilot Frank Borman were launched into space on the history-making Gemini 7 mission. The ht lasted 330 hours and 35 minutes, during which the following space firsts were accomplished: longest manned space flight; first ndezvous of two manned maneuverable spacecraft, as Gemini 7 was joined in orbit by Gemini 6; and longest multi-manned space flight. was also on this flight that numerous technical and medical experiments were completed successfully.

e Gemini 12 mission, with Lovell and pilot Edwin Aldrin, began on November 11, 1966. This 4-day, 59-revolution flight brought the emini Program to a successful close. Major accomplishments of the 94-hour 35-minute flight included a third-revolution rendezvous th the previously launched Agena (using for the first time backup onboard computations due to radar failure); a tethered station-eping exercise; retrieval of a micrometeorite experiment package from the spacecraft exterior; an evaluation of the use of body straints specially designed for completing work tasks outside of the spacecraft; and completion of numerous photographic periments, highlights of which are the first pictures taken from space of an eclipse of the sun.

emini 12 ended with retrofire at the beginning of the 60th revolution, followed by the second consecutive fully automatic controlled entry of a spacecraft, and a lending in the Atlantic within 2½ miles of the USS WASP.

s a result of his participation in the Gemini 7 and 12 flights, Lovell logged 425 hours and 10 minutes in space. Aldrin established a new 'A record by completing 5½ hours outside the spacecraft during two standup EVAs and one umbilical EVA.

ovell served as command module pilot for the epic six day journey of Apollo 8 — man's maiden voyage to the moon — December -27, 1968. Apollo 8 was the first manned spacecraft to be lifted into near-earth orbit by a 7½ million pound thrust Saturn V launch hicle, and all events in the flight plan occurred as scheduled with unbelievable accuracy.

"go" for the translunar injection burn was given midway through the second near-earth orbit, and the restart of the S-IVB third stage effect this maneuver increased the spacecraft's velocity to place it on an intercept course with the moon. Lovell and fellow crew embers, Frank Borman (spacecraft commander) and William A. Anders (lunar module pilot), piloted their spacecraft some 223,000 iles to become the first humans to leave the earth's influence; and upon reaching the moon on December 24, they performed the first itical maneuver to place Apollo 8 into a 60 by 168 nautical miles lunar orbit.

vo revolutions later, the crew executed a second maneuver using the spacecraft's 20,500-pound thrust service module propulsion stem to achieve a circular lunar orbit of 60 nautical miles. During their ten revolutions of the moon, the crew conducted live television ansmissions of the lunar surface and performed such tasks as landmark and Apollo landing site tracking, vertical stereo photography d stereo navigation photography, and sextant navigation using lunar landmarks and stars. At the end of the tenth lunar orbit, they ecuted a transearth injection burn which placed Apollo 8 on a proper trajectory for the return to earth.

he final leg of the trip required only 58 hours, as compared to the 69 hours used to travel to the moon, and Apollo 8 came to a ccessful conclusion on December 27, 1968. Splashdown occurred at an estimated 5,000 yards from the USS YORKTOWN, following e successful negotiation of a critical 28-mile high reentry corridor at speeds close to 25,000 miles per hour.

aptain Lovell has since served as the backup spacecraft commander for the Apollo 11 lunar landing mission. He has completed three ace flights and holds the U.S. Astronaut record for time in space with a total of 572 hours and 10 minutes.

PECIAL ASSIGNMENT: In addition to his regular duties as an astronaut, Captain Lovell continues to serve as Special Consultant the President's Council on Physical Fitness and Sports — an assignment he has held since June 1967.

URRENT SALARY: $1,717.28 per month.

AME: Thomas Kenneth Mattingly II (Lieutenant Commander, USN) NASA Astronaut

IRTHPLACE AND DATE: Born in Chicago, Ill., March 17, 1936. His parents, Mr. and Mrs. Thomas K. Mattingly, now reside in Hialeah, a.

HYSICAL DESCRIPTION: Brown hair; blue eyes; height: 5 feet 10 inches; weight: 140 pounds.

DUCATION: Attended Florida elementary and secondary schools and is a graduate of Mimi Edison High School, Miami, Fla.; received Bachelor of Science degree in Aeronautical Engineering from Auburn University In 1958.

ARITAL STATUS: Single

ECREATIONAL INTERESTS: Enjoys water skiing and playing handball and tennis.

ORGANIZATIONS: Member of the American Institute of Aeronautics and Astronautics and the U.S. Naval Institute.

EXPERIENCE: Prior to reporting for duty at the Manned Spacecraft Center, he was a student at the Air Force Aerospace Resea Pilot School.

He began his Naval career as an Ensign in 1958 and received his wings in 1960. He was then assigned to VA-35 and flew A1H aircr aboard the USS SARATOGA from 1960 to 1963. In July 1963, he served in VAH-11 deployed aboard the USS FRANKLIN D. ROOSEVE where he flew the A3B aircraft for two years.

He has logged 3,700 hours of flight time — 1,946 hours in jet aircraft.

CURRENT ASSIGNMENT: Lt. Commander Mattingly is one of the 19 astronauts selected by NASA in April 1966. He served a member of the astronaut support crews for the Apollo 8 and 11 missions.

CURRENT SALARY: $1,293.33 per month.

BIOGRAPHICAL DATA

NAME: Fred Wallace Haise, Jr. (Mr.) NASA Astronaut

BIRTHPLACE AND DATE: Born in Biloxi, Miss., on Nov. 14, 1933; his mother, Mrs. Fred W. Haise, Sr., resides in Biloxi.

PHYSICAL DESCRIPTION: Brown hair; brown eyes; height; 5 feet 9½ inches; weight: 150 pounds.

EDUCATION: Graduated from Biloxi High School, Biloxi, Miss.; attended Perkinston Junior College (Association of Arts); receivec Bachelor of Science degree with honors in Aeronautical Engineering from the University of Oklahoma In 1959.

MARITAL STATUS: Married to the former Mary Griffin Grant of Biloxi, Miss. Her parents, Mr. and Mrs. William J. Grant, Jr., reside Biloxi.

CHILDREN: Mary M., January 25, 1956; Frederick T., May 13, 1958; Stephen W., June 30, 1961.

ORGANIZATIONS: Member of the Society of Experimental Test Pilots, Tau Beta Pi, Sigma Gamma Tau, and Phi Theta Kappa.

SPECIAL HONORS: Recipient of the A. B. Honts Trophy as the outstanding graduate of class 64A from the Aerospace Research Pil School in 1964; awarded the American Defense Ribbon and the Society of Experimental Test Pilots Ray E. Tenhoff Award for 1966.

EXPERIENCE: Haise was a research pilot at the NASA Flight Research Center at Edwards, Calif., before coming to Houston and t Manned Spacecraft Center; and from September 1959 to March 1963, he was a research pilot at the NASA Lewis Research Center Cleveland, Ohio. During this time, he authored the following papers which have been published: a NASA TND, entitled "An Evaluati of the Flying Qualities of Seven General Aviation Aircraft;" NASA TND 3380, "Use of Aircraft for Zero Gravity Environment, May 196 SAE Business Aircraft Conference Paper, entitled "An Evaluation of General-Aviation Aircraft Flying Qualities," March 30-April 1, 196 and a paper delivered at the tenth symposium of the Society of Experimental Test Pilots, entitled "A Quantitative/ Qualitative Handli Qualities Evaluation of Seven General Aviation Aircraft," 1966.

He was the Aerospace Research Pilots School's outstanding graduate of Class 64A and served with the U.S. Air Force from Octob 1961 to August 1962 as a tactical fighter pilot and as Chief of the 164th Standardization-Evaluation Flight of the 164th Tactical Fight Squadron at Mansfield, Ohio. From March 1957 to September 1959, he was a fighter-interceptor pilot with the 185th Fighter Intercept Squadron in the Oklahoma Air National Guard.

He also served as a tactics and all weather flight instructor in the U.S. Navy Advanced Training Command at NAAS Kingsville, Texas, a was assigned as a U.S. Marine Corps fighter pilot to VMF-533 and 114 at MCAS Cherry Point, N.C., from March 1954 to Septemb 1956.

His military career began in October 1952 as a Naval Aviation Cadet at the Naval Air Station in Pensacola, Fla.

He has accumulated 5,800 hours flying time, including 3,000 hours in jets.

CURRENT ASSIGNMENT: Mr. Haise is one of the 19 astronauts selected by NASA in April 1966. He served as backup lunar modu pilot for the Apollo 8 and 11 missions.

CURRENT SALARY: $1,698.00 per month.

LAUNCH COMPLEX 39

aunch Complex 39 facilities at the Kennedy Space Center were planned and built specifically for the Apollo aturn V the space vehicle being used in the United States manned lunar exploration program.

omplex 39 introduced the mobile concept of launch operations in which the space vehicle is thoroughly necked out in an enclosed building before it is moved to the launch pad for final preparations. This affords reater protection from the elements and permits a high launch rate since pad time is minimal.

aturn V stages are shipped to the Kennedy Space Center by oceangoing vessels and specially designed ircraft. Apollo spacecraft modules are transported by air and first taken to the Manned Spacecraft Operations Building in the Industrial Area south of Complex 39 for preliminary checkout, altitude chamber esting, and assembly.

pollo 13 is the sixth Saturn V/ Apollo space vehicle to be launched from Complex 39's Pad A, one of two ctagonal launch pads which are 3,000 feet across. The major components of Complex 39 include:

1. The Vehicle Assembly Building heart of the complex, is where the 363 foot tall space vehicle is assembled and tested. It contains 129.5 million cubic feet of space, covers eight acres, is 716 feet long and 518 feet wide. Its high bay area, 525 feet high, contains four assembly and checkout bays and its low bay area — 210 feet high, 442 feet wide and 274 feet long — contains eight stage-preparation and checkout cells. There are 141 lifting devices in the building, ranging from one-ton hoists to two 250-ton high lift bridge cranes.

2. The Launch Control Center, a four-story structure adjacent and to the south of the Vehicle Assembly Building is a radical departure from the dome-shaped, "hardened" blockhouse at older launch sites. The Launch Control Center is the electronic "brain" of Complex 39 and was used for checkout and test operations while Apollo 13 was being assembled inside the Vehicle Assembly Building high bay. Three of the four firing rooms contain identical sets of control and monitoring equipment so that launch of one vehicle and checkout of others may continue simultaneously. Each firing room is associated with a ground computer facility to provide data links with the launch vehicle on its mobile launcher at the pad or inside the Vehicle Assembly Building.

3. The Mobile Launcher, 445 feet tall and weighing 12 million pounds, in a transportable launch base and umbilical tower for the space vehicle.4. The Transporters used to move mobile launchers into the Vehicle Assembly Building and then with their space vehicles to the launch pad, weigh six million pounds and are among the largest tracked vehicles known. The Transporters — there are two — are 131 feet long and 114 feet wide. Powered by electric motors driven by two 2,750 - horsepower diesel engines, the vehicles move on four double-tracked crawlers, each 10 feet high and 40 feet long. Maximum speed is about one-mile-per-hour loaded and two miles-per-hour unloaded. The three and one-half mile trip to Pad A with a mobile launcher and space vehicle takes approximately seven hours. Apollo 13 rollout to the pad occurred on December 15, 1969.

5. The Crawlerway is the roadway for the transporter and is 131 feet wide divided by a median strip. This is the approximate width of an eight-lane turnpike and the roadbed is designed to accommodate a combined weight of more than 18 million pounds.

6. The Mobile Service Structure is a 402-foot-tall, 9.8 million pound tower used to service the Apollo space vehicle at the pad. Moved into place about the Saturn V/ Apollo space vehicle and its mobile launcher by a transporter, it contains five work platforms and provides 360-degree platform access to the vehicle being prepared for launch. It is removed to a parking area about 11 hours before launch.

7. A Water Deluge System will provide about a million gallons of industrial water for cooling and fire prevention during the launch of Apollo 13. The water is used to cool the mobile launcher, the flame trench and the flame deflector above which the mobile launcher is positioned.

8. The Flame Deflector is an "A" shaped 1.3 million pound structure moved into the flame trench benea the launcher prior to launch. It is covered with a refractory material designed to withstand the launc environment. The flame trench itself is 58 feet wide and approximately six feet above mean sea level at th base.

9. The Pad Areas, A and B, are octagonal in shape and have center handstands constructed of heavi reinforced concrete. The top of Pad A stands about 48 feet above sea level. Saturn V propellants, liqu oxygen, liquid hydrogen and RP-1, the latter a high grade kerosene, are stored in large tanks spaced ne the pad perimeter and carried by pipelines from the tanks to the pad, up the mobile launcher and into th launch vehicle propellant tanks. Also located in the pad area are pneumatic, high pressure gas, electric and industrial water support facilities. Pad B, used for the launch of Apollo 10, is located 8,700 feet nort of Pad A.

MISSION CONTROL CENTER

The Mission Control Center at the Manned Spacecraft Center, Houston, is the focal point for Apollo fligh control activities. The center receives tracking and telemetry data from the Manned Space Flight Netwo which in turn is processed by the MCC Real-Time Computer Complex for display to flight controllers in th Mission Operations Control Room (MOCR) and adjacent staff support rooms.

Console positions in the two identical MOCRs in Mission Control Center fall into three basic operatio groups: mission command and control, systems operations, and flight dynamics.

Positions in the command and control group are:

Mission Director — Responsible for overall mission conduct.
Flight Operations Director — Represents MSC management.
Flight Director — Responsible for operational decisions and actions in the MOCR.
Assistant Flight Director — Assists flight director and acts in his absence.
Flight Activities Officer — Develops and coordinates flight plan.
Department of DefenseRepresentative — Coordinates and directs DOD mission support.
Network Controller — Responsible to FD for MSFN status and troubleshooting; MCC equipment operation.
Surgeon — Monitors crew medical condition and informs FD of my medical situation affecting missio
Spacecraft Communicator (CapCom) — Serves as voice contact with flight crew.
Experiments Officer — Coordinates operation and control of onboard flight experiments.
Public Affairs Officer — Reports mission progress to public through commentary & relay of live air-to-ground transmissions.

Systems Operations Group:

Environmental, Electrical and Instrumentation Engineer (EECOM) — monitors and troubleshoots command/service modul environmental, electrical, and sequential systems.

Guidance, Navigation and Control Engineer (GNC) — monitors and troubleshoots CSM guidance, navigation, control, and propulsio systems.

LM Environmental and Electrical Engineer (TELCOM) — LM counterpart to EECOM.

LM Guidance, Navigation and Control Engineer (Control) — LM counterpart to GNC.

Booster Systems Engineer (BSE) (three positions) — responsible for monitoring launch vehicle performance and for sending functio commands.

Communications Systems Engineer (CSE) (call sign INCO) and Operations and Procedures Officer (O&P) — share responsibility fo monitoring and troubleshooting spacecraft and lunar surface communication systems and for coordinating MCC procedures with othe NASA centers and the network.

Flight Dynamics Group:

Flight Dynamics Officer (FIDO) — monitors powered flight events and plans spacecraft maneuvers.

Retrofire Officer (Retro) — responsible for planning deorbit maneuvers in Earth orbit and entry calculations on lunar return trajectories.

Guidance Officer (Guido) — responsible for monitoring and updating CSM and LM guidance systems and for monitoring systems performance during powered flight.Each MOCR operations group has a staff support room on the same floor in which detailed monitoring and analysis is conducted. Other supporting MCC areas include the spaceflight Meteorological Room, the Space Environment (radiation) Console, Spacecraft Planning and Analysis (SPAN) ROOM for detailed spacecraft performance analysis, Recovery Operations Control Room and the Apollo Lunar Surface Experiment Package Support Room.

Located on the first floor of the MCC are the communications, command, and telemetry system (CCATS) for processing incoming data from the tracking network, and the real-time computer complex (RTCC) which converts flight data into displays usable to MOCR flight controllers.

MANNED SPACE FLIGHT NETWORK

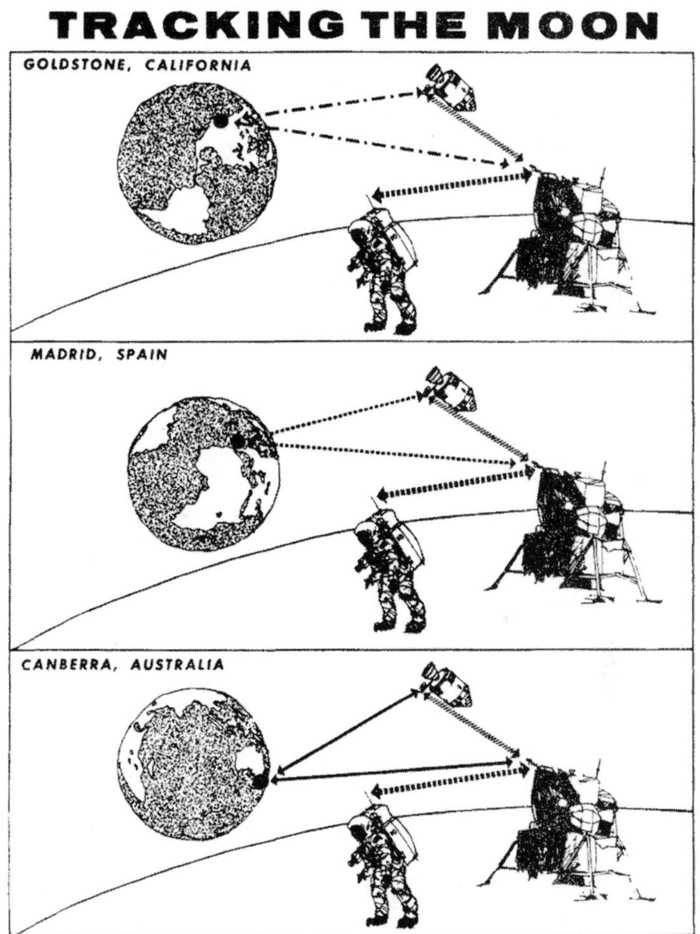

The worldwide Manned Space Flight Network (MSFN) provides reliable, continuous, and instantaneous communications with the astronauts, launch vehicle, and spacecraft from liftoff to splashdown. Following the flight, the network will continue in support of the link between Earth and the Apollo experiments left on the lunar surface by the Apollo crew.

The MSFN is maintained and operated by the NASA Goddard Space Flight Center, Greenbelt, Md., under the direction of NASA's Office of Tracking and Data Acquisition. In the MSFN Operations Center (MSFNOC) at Goddard, the Network Director and his team of Operations Managers, with the assistance or a Network Support Team, keep the entire complex tuned for the mission support. Should Houston's mission control center be seriously impaired for an extended time, the Goddard Center becomes an emergency mission control center.

The MSFN employs 12 ground tracking stations equipped with 30- and 85-foot antennas, an instrumented tracking ship, and four instrumented aircraft. For Apollo 13, the network will be augmented by the 210-foot antenna system at Goldstone Calif. and at Parkes, Australia, (Australian Commonwealth Scientific and Industrial Research Organization).

NASA Communications Network (NASCOM). The tracking network is linked together by the NASA Communications Network. All information flows to and from MCC Houston and the Apollo spacecraft over this communications system.

The NASCOM consists of almost three million circuit miles of diversely routed communications channels. It uses satellites, submarine cables, land lines, microwave systems, and high frequency radio facilities for access links.

NASCOM control center is located at Goddard. Regional communication switching centers are in Londo Madrid, Canberra, Australia; Honolulu and Guam.

Three Intelsat communications satellites will be used for Apollo 13. One satellite over the Atlantic will lin Goddard with stations at Madrid, Canary Islands, Ascension and the Vanguard tracking ship. Another Atlanti satellite will provide a direct link between Madrid and Goddard for TV signals received from the spacecraf The third satellite over the mid Pacific will link Carnarvon, Canberra, and Hawaii with Goddard through ground station at Brewster Flats, Wash.

At Goddard, NASCOM switching computers simultaneously send the voice signals directly to the Housto flight controllers and the tracking and telemetry data to computer processing complexes at; Houston an Goddard. The Goddard Real Time Computing Complex verifies performance of the tracking network an uses the collected tracking data to drive displays in the Goddard Operations Control Center.

Establishing the Link — The Merritt Island tracking station monitors prelaunch test, the terminal countdow and the first minutes of launch.

An Apollo instrumentation ship (USNS VANGUARD) fills the gaps beyond the range of land tracking station For Apollo 13 this ship will be stationed in the Atlantic to cover the insertion into Earth orbit. Apoll instrumented aircraft provide communications support to the land tracking stations during transluna injection and reentry and cover a selected abort area in the event of "no-go" decision after insertion int Earth orbit.

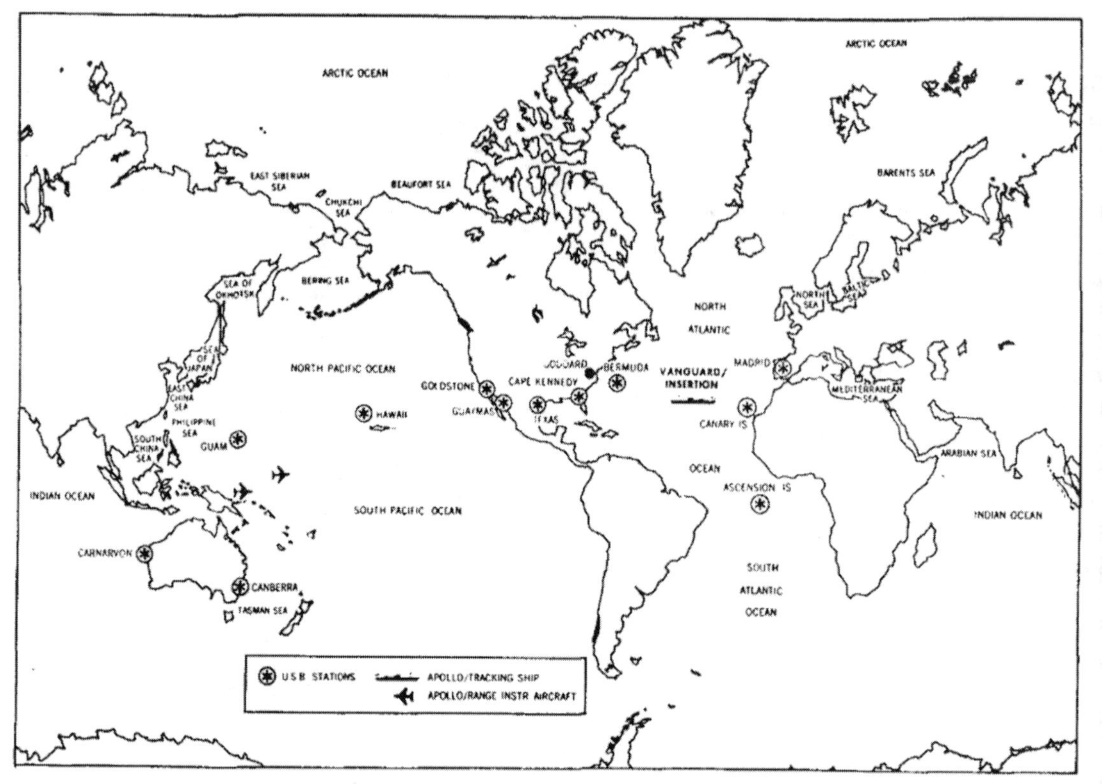

MANNED SPACE FLIGHT TRACKING NETWORK

Lunar Bound — Approximately one hour after the spacecraft has been injected into its translunar trajector (some 10,000 miles from the Earth), three prime tracking stations spaced nearly equidistant around the Eart will take over tracking and communicating with Apollo.

Each of the prime stations, located at Goldstone, Madrid and Canberra, has a dual system for use whe tracking the command module in lunar orbit and the lunar module in separate flight paths or at rest on th

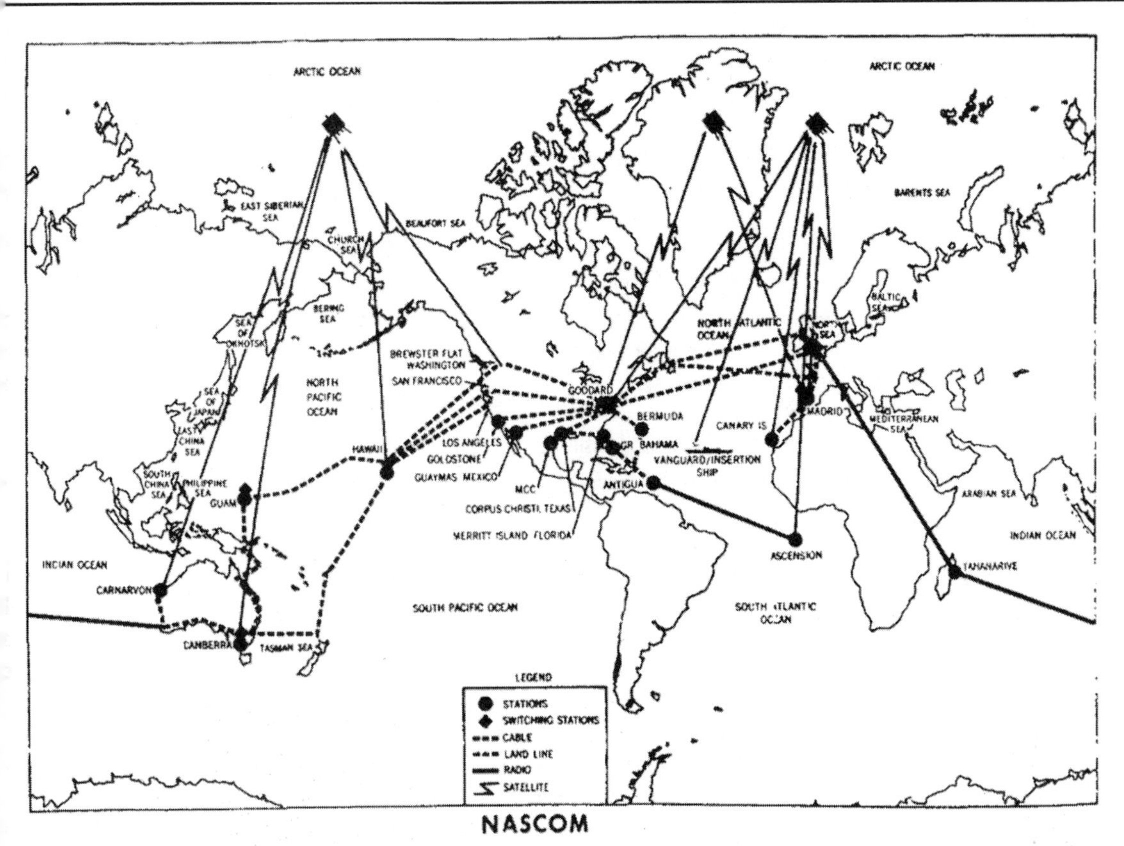

NASCOM

Moon. These stations are equipped with 85-foot antennas.

The Return Trip — To make an accurate reentry, data from the tracking stations are fed into the MCC computers to develop necessary information for the Apollo 13 crew.

Appropriate MSFN stations, including the aircraft in the Pacific, provide support during the reentry. Through the journey to the Moon and return, television will be received from the spacecraft at the three prime stations. In addition, a 210-foot antenna at Goldstone (an antenna of NASA's Deep Space Network) will augment the television coverage while Apollo 13 is near and on the Moon. For black and white TV, scan converters at the stations permit immediate transmission of commercial quality TV via NASCOM to Houston, where it will be released to U.S. TV networks.

Black and white TV can be released simultaneously in Europe and the Far East through the MSFN stations in Spain and Australia.

For color TV, the signal will be converted to commercial quality at the MSC Houston. A black and white version of the color signal can be released locally simultaneously through the stations in Spain and Australia.

Network Computers

At fraction-of-a-second intervals, the network's digital data processing system with NASA's Manned Spacecraft Center as the focal point, "talk" to each other or to the spacecraft. High speed computers at the remote sites (tracking ship included) relay commands or "up-link" data on such matters as control of cabin pressure, orbital guidance commands, or "go/no-go" indications to perform certain functions. When information originates from Houston, the computers refer to their pre-programmed information for validity before transmitting the required data to the spacecraft. Such "up-link" information is communicated at a rate of about 1,200 bits-per-second. Communication of spacecraft data between remote ground sites and the Mission Control Center, via high speed communications links occurs at twice the rate. Houston reads information from these ground sites at 8,800 bits-per second.

The computer systems perform many other functions, including:

* Assuring the quality of the transmission lines by continually testing data paths.
* Verifying accuracy of the messages.
* Constantly updating the flight status.
* For "down-link" data, sensors built into the spacecraft continually sample cabin temperature, pressure and physical information on the astronauts such as heartbeat and respiration. These data are transmitted to the ground stations at 51.2 kilobits (12,800 decimal digits) per second.

At MCC the computers:

* Detect and select changes or deviations, compare with their stored programs, and indicate the problem areas or pertinent data to the flight controllers;
* Provide displays to mission personnel;
* Assemble output data in proper formats;
* Log data on magnetic tape for the flight controllers.

The Apollo Ship Vanguard

The USNS vanguard will perform tracking, telemetry, and communication functions for the launch phase and Earth orbit insertion. Vanguard will be stationed about 1,000 miles southeast of Bermuda (28 degrees N., 49 degrees W.).

Apollo Range Instrumentation Aircraft (ARIA)

During the Apollo 13 TLI maneuver, two ARIA will record telemetry data from Apollo and relay voice communication between the astronauts and the Mission Control Center at Houston. The ARIA will be located between Australia and Hawaii. For reentry, two ARIA will be deployed to the landing area to relay communications between Apollo and Mission Control at Houston and provide position information on the spacecraft after the blackout phase of reentry has passed. The total ARIA fleet for Apollo Missions consists of four EC135A (Boeing 707) jets with 7-foot parabolic antennas installed in the nose section.

CONTAMINATION CONTROL PROGRAM

In 1966 an Interagency Committee on Back Contamination (ICBC) was established to assist NASA in developing a program to prevent contamination of the Earth from lunar materials following manned lunar exploration and to review and approve plans and procedures to prevent back contamination. Committee membership includes representatives from Public Health Service, Department of Agriculture, Department of the Interior, NASA, and the National Academy of Sciences.

The Apollo Back Contamination Program can be divided into three phases. The first phase covers procedures which are followed by the crew while in flight to reduce and, if possible, eliminate the return of lunar surface contamination in the command module. The second phase includes recovery, isolation, and transport of the crew, spacecraft, and lunar samples to the Manned Spacecraft Center. The third phase encompasses quarantine operations and preliminary sample analysis in the Lunar Receiving Laboratory.

A primary step in preventing back contamination is careful attention to spacecraft cleanliness following lunar surface operations. This includes use of special cleaning equipment, stowage provisions for lunar-exposed equipment, and crew procedures for proper "housekeeping." Prior to reentering the LM after lunar surface exploration, the crewmen brush lunar surface dust or dirt from the space suit using special brushes. They will scrape their overboots on the LM footpad and while ascending the LM ladder, dislodge any clinging particles by a kicking action. After entering and pressurizing the LM cabin, the crew doff their portable life support system, oxygen purge system, lunar boots, EVA gloves, etc.

Following LM rendezvous and docking with the CM, the CM tunnel will be pressurized and checks made to insure that an adequate pressurized seal has been made. During the period, some of the equipment may be vacuumed. The lunar module cabin atmosphere will be circulated through the environmental control system uit circuit lithium hydroxide (LiOH) canister to filter particles from the atmosphere. A minimum of five ours weightless operation and filtering will essentially eliminate the original airborne particles.

'he CM pilot will transfer lunar surface equipment stowage bags into the LM one at a time. The equipment ransferred will be bagged before being transferred. The only equipment which will not be bagged at this time re the crewmen's space suits and flight logs.

Command Module Operations — Through the use of operational and housekeeping procedures the ommand module cabin will be purged of lunar surface and/or other particulate contamination prior to Earth eentry. These procedures start while the LM is docked with the CM and continue through reentry into the arth's atmosphere.

During subsequent lunar orbital flight and the transearth phase, the command module atmosphere will be ontinually filtered through the environmental control system lithium hydroxide canister. This will remove ssentially all airborne dust particles. After about 96 hours operation essentially none of the original ontaminates will remain.

unar Mission Recovery Operations

ollowing landing and the attachment of the flotation collar to the command module, a swimmer will open he spacecraft hatch, pass in three clean flight coveralls and three filter masks and close the hatch. Crew etrieval will be accomplished by helicopter to the carrier and subsequent crew transfer to the Mobile Quarantine Facility. The spacecraft will be retrieved by the aircraft carrier and isolated.

LUNAR RECEIVING LABORATORY (LRL)

The final phase of the back contamination program is completed in the MSC Lunar Receiving Laboratory. The rew and spacecraft are quarantined for a minimum of 21 days after completion of lunar EVA operations and re released based upon the completion of prescribed test requirements and results. The lunar sample will e quarantined for a period of 50 to 80 days depending upon results of extensive biological tests.

The LRL serves four basic purposes:

* Quarantine of crew and spacecraft, the containment of lunar and lunar-exposed materials, and quarantine testing to search for adverse effects of lunar material upon terrestrial life.
* The preservation and protection of the lunar samples.
* The performance of time critical investigations.
* The preliminary examination of returned samples to assist in an intelligent distribution of samples to principal investigators.

The LRL has the only vacuum system in the world with space gloves operated by a man leading directly into a vacuum chamber at pressures of about 10 billionth of an atmosphere. It has a low level counting facility, the ackground count is an order of magnitude better than other known counters. Additionally, it is a facility that can handle a large variety of biological specimens inside Class III biological cabinets designed to contain extremely hazardous pathogenic material.

The LRL covers 83,000 square feet of floor space and includes a Crew Reception Area (CRA), vacuum aboratory, Sample Laboratories (Physical and Bio-Science) and an administrative and support area. Special uilding systems are employed to maintain air flow into sample handling areas and the CRA, to sterilize liquid vastes and to incinerate contaminated air from the primary containment systems.

The biomedical laboratories provide for quarantine tests to determine the effect of lunar samples on errestrial life. These tests are designed to provide data upon which to base the decision to release lunar material from quarantine.

Among the tests:

a). Lunar material will be applied to 12 different culture media and maintained under several environment conditions. The media will be observed for bacterial or fungal growth. Detailed inventories of the microbi flora of the spacecraft and crew have been maintained so that any living material found in the sample testin can be compared against this list of potential contaminants taken to the Moon by the crew or spacecraf

b). Six types of human and animal tissue culture cell lines will be maintained in the laboratory and togethe with embryonated eggs are exposed to the lunar material. Based on cellular and/or other changes, th presence of viral material can be established so that special tests can be conducted to identify and isolat the type of virus present.

c). Thirty-three species of plants and seedlings will be exposed to lunar material. Seed germination, growt of plant cells or the health of seedlings are then observed, and histological, microbiological and biochemic techniques are used to determine the cause of any suspected abnormality.

d). A number of lower animals will be exposed to lunar material, including germ-free mice, fish, bird oysters, shrimp, cockroaches houseflies, planaria, paramecia and euglena. If abnormalities are noted, furthe tests will be conducted to determine if the condition is transmissible from one group to another.

The crew reception area provides biological containment for the flight crew and 12 support personnel. Th nominal occupancy is about 14 days but the facility is designed and equipped to operate for considerab longer.

Sterilization and Release of the Spacecraft

Post-flight testing and inspection of the spacecraft is presently limited to investigation of anomalies whic happened during the flight. Generally, this entails some specific testing of the spacecraft and removal c certain components Of systems for further analysis. The timing of post-flight testing is important so tha corrective action may be taken for subsequent flights.

The schedule calls for the spacecraft to be returned to port where a team will deactivate pyrotechnics, an flush and drain fluid system (except water). This operation will be confined to the exterior of the spacecraf The spacecraft will then be flown to the LRL and placed in a special room for storage, sterilization, and post flight checkout.

LUNAR RECEIVING LABORATORY TENTATIVE SCHEDULE

April 20	Activate secondary barrier; support people enter Crew Reception Area and Central Status Station manned; LRL on mission status.
April 21	Command module landing, recovery.
April 22	First sample return container (SRC) arrives.
April 23	First SRC opened in vacuum lab, second SRC arrives; film, tapes, LM tape recorder begin decontamination; second SRC opened in Bioprep lab.
April 24	First sample to Radiation Counting Laboratory.
April 26	Core tube moves from vacuum lab to Physical Chemical Lab.
April 26	MQF arrives; contingency sample goes to Physical Chemical Lab; rock description begun in vacuum lab.
April 27	Biosample rocks move from vacuum lab to Bioprep Lab; core tube prepared for biosample.
April 28	Spacecraft arrives.
April 29	Biosample compounded, thin-section chips sterilized out to Thin-Section Lab, remaining samples from Bioprep Lab canned.
May 1	Thin-section preparation complete, biosample prep complete, transfer to Physical-Chemical Lab complete, Bioprep Lab cleanup complete.
May 3	Biological protocols, Physical-Chemical Lab rock description begin.
May 8	Crew released from CRA
May 26	Rock description complete, Preliminary Examination Team data from Radiation Counting Lab and Gas Analysis Lab complete.
May 28	PET data write-up and sample catalog preparation begin.
May 30	Data summary for Lunar Sample Analysis Planning Team (LSAPT) complete.
June 1	LSAPT arrives.
June 2	LSAPT briefed on PET data, sample packaging begins.
June 6	Sample distribution plan complete, first batch monopole samples canned.
June 8	Monopole experiment begins.
June 10	Initial release of Apollo 13 samples; spacecraft release.
June 14	Spacecraft equipment released

SCHEDULE FOR TRANSPORT OF SAMPLES, SPACECRAFT AND CREW

amples

he first Apollo 13 sample return container (SRC) will be flown by helicopter from the deck of the USS Iwo ma to Christmas Island, from where it will be flown by C-130 aircraft to Hawaii. The SRC, half the mission nboard film and any medical samples ready at the time of helicopter departure from the Iwo Jima will be ansferred to an ARIA (Apollo Range Instrumented Aircraft) at Hawaii for the flight to Ellington AFB, six iles north of the Manned Spacecraft Center, with an estimated time of arrival at 11:30 a.m. EST April 22.

he second SRC and remainder of onboard film and medical samples will follow a similar sequence of flights ne following day and will arrive at Ellington AFB at an estimated time of 1 a.m. EST April 23. The SRCs will e moved by auto from Ellington AFB to the Lunar Receiving Laboratory.

pacecraft

he spacecraft should be aboard the Iwo Jima about two hours after crew recovery. The ship will arrive in lawaii at 2 p.m. EST April 25 and the spacecraft will be offloaded and transferred after deactivation to an ircraft for the flight to Ellington AFB, arriving April 28. The spacecraft will be trucked to the Lunar Receiving aboratory where it will enter quarantine.

rew

he flight crew is expected to enter the Mobile Quarantine Facility (MQF) on the Iwo Jima about 90 minutes fter splashdown. Upon arrival at Hawaii, the MQF will be offloaded and placed aboard a C-141 aircraft for ne flight to Ellington AFB, arriving at 1 a.m. EST April 25. A transporter truck will move the MQF from llington AFB to the Lunar Receiving Laboratory, about a two-hour trip.

APOLLO PROGRAM MANAGEMENT

he Apollo Program is the responsibility of the Office of Manned Space Flight (OMSF), National Aeronautics nd Space Administration, Washington, D.C. Dale D. Myers is Associate Administrator for Manned Space light.

JASA Manned, Spacecraft Center (MSC), Houston, is responsible for development of the Apollo Spacecraft, ight crew training, and flight control. Dr. Robert R. Gilruth is Center Director.

JASA Marshall Space Flight Center (MSFC), Huntsville, Ala., is responsible for development of the Saturn unch vehicles. Dr. Eberhard Rees is Center Director.

JASA John P. Kennedy Space Center (KSC), Fla., is responsible for Apollo/Saturn launch operations. Dr. Kurt I. Debus is Center Director.

he NASA Office of Tracking and Data Acquisition (OTDA) directs the program of tracking and data flow n Apollo. Gerald M. Truszynski is Associate Administrator for Tracking and Data Acquisition.

JASA Goddard Space Flight Center (GSFC), Greenbelt, Md., manages the Manned Space Flight Network and ommunications Network. Dr. John F. Clark is Center Director.

he Department of Defense is supporting NASA in Apollo 13 during launch, tracking and recovery perations. The Air Force Eastern Test Range is responsible for range activities during launch and downrange acking. Recovery operations include the use of recovery ships and Navy and Air Force aircraft.

Apollo/Saturn Officials

NASA Headquarters

Dr. Rocco A. Petrone	Apollo Program Director, OMSF
Chester M. Lee (Capt., USN, Ret.)	Apollo Mission Director, OMSF
Col. Thomas B. McMullen (USAF)	Apollo Assistant Mission Director, OMSF
John D. Stevenson (Maj. Gen., USAF, Ret.)	Director of Mission Operations, OMSF
Maj. Gen. James W. Humphreys, Jr.	Director of Space Medicine, OMSF (USAF, MC)
John K. Holcomb, (Capt., USN, Ret.)	Director of Apollo Operations, OMSF
Lee R. Scherer, (Capt., USN, Ret.)	Director of Apollo Lunar Exploration, OMSF
James C, Bavely	Chief of Network Operations Branch, OTDA

Marshall Space Flight Center

Lee B. James	Director, Program Management
Dr. F. A. Speer	Manager, Mission Operations Office
Roy E. Godfrey	Manager, Saturn Program Office
Matthew W. Urlaub	Manager, S-IC Stage, Saturn Program Office
William P. LaHatte	Manager, S-II Stage, Saturn Program Office
Charles H. Meyers	Manager (Acting), S-IVB Stage, Saturn Program Office
Frederich Duerr	Manager, Instrument Unit, Saturn Program Office
William D. Brown	Manager, Engine Program Office Kennedy Space Center
Walter J. Kapryan	Director of Launch Operations
Raymond L. Clark	Director of Technical Support
Edward R. Mathews	Apollo Program Manager
Dr. Hans F. Gruene	Director, Launch Vehicle Operations
John J. Williams	Director, Spacecraft Operations
Paul C. Donnelly	Launch Operations Manager
Isom A. Rigell	Deputy Director for Engineering

Manned Spacecraft Center

Col. James A. McDivitt, (USAF)	Manager, Apollo Spacecraft Program
Donald K. Slayton	Director, Flight Crew Operations
Sigurd A. Sjoberg	Director, Flight Operations
Milton L. Windler	Flight Director
Gerald Griffin	Flight Director
Glynn S. Lunney	Flight Director
Eugene F. Kranz	Flight Director
Dr. Charles A. Berry	Director, Medical Research and Operations

Goddard Space Flight Center

Ozro M. Covington	Director of Manned Flight Support
William P. Varson	Chief, Manned Flight Planning & Analysis Division
H. William Wood	Chief, Manned Flight Operations Division
Tecwyn Roberts	Chief, Manned Flight Engineering Division
L. R. Stelter	Chief, NASA Communications Division.

Department of Defense

Maj. Gen. David M. Jones, (USAF)	DOD Manager of Manned Space Flight Support Operations, Commander of USAF Eastern Test Range
Rear Adm. Wm. S. Guest, (USN)	Deputy DOD Manager of Manned Space Flight Support Operations, Commander Task Force 140, Atlantic Recovery Area
Rear Adm. Donald C. Davis, (USN)	Commander Task Force 130, Pacific Recovery Area
Col. Kenneth J. Mask, (USAF)	Director of DOD Manned Space Flight Support office
Maj. Gen. Allison C. Brooks,	Commander Aerospace Rescue and (USAF) Recovery Service

Major Apollo/Saturn V Contractors

Contractor	Item
Bellcomm, Washington, D. C.	Apollo Systems Engineering
The Boeing Co., Washington, D. C.	Technical Integration and Evaluation
General Electric, Daytona Beach, Fla.	Apollo Checkout and Apollo System, Quality and Reliability

North American Rockwell Corp. Space Div., Downey, Calif.	Command and Service Modules
Grumman Aircraft Engineering Corp., Bethpage, N.Y.	Lunar Module
Massachusetts Institute of Technology, Cambridge, Mass.	Guidance & Navigation (Technical Management)
General Motors Corp., AC Electronics Div., Milwaukee, Wis.	Guidance & Navigation (Manufacturing)
TRW Inc. System Group Redondo Beach, Calif.	Trajectory Analysis, LM Descent Engine LM Abort Guidance System
Avco Corp., Space System Div., Lowell, Mass.	Heat Shield Ablative Material
North American Rockwell Corp. Rocketdyne Div., Canoga Park, Calif.	J-2 Engines, F-1 Engines
The Boeing Co. New Orleans.	First Stage (SIC) of Saturn V Launch Vehicles, Saturn V Systems Engineering and Integration, Ground Support Equipment
North American Rockwell Corp. Space Div., Seal Beach Calif.	Development and Production of Saturn V Second Stage (S-II)
McDonnell Douglas Astronautics Co., Huntington Beach, Calif.	Development and Production of Saturn V Third Stage (S-IVB)
International Business Machines Federal System Div., Huntsville, Ala.	Instrument Unit
Bendix Corp. Navigation and Control Div. Teterboro, N.J.	Guidance components for Instrument Unit (Including ST-124M Stabilized Platform)
Federal Electric Corp.	Communications and Instrumentation Support, KSC
Bendix Field Engineering Corp.	Launch Operations/Complex Support KSC
Catalytic-Dow	Facilities Engineering and Modifications, KSC
Hamilton Standard Division	Portable Life Support System;
United Aircraft Corp., Windsor Locke, Conn.	LM ECS
ILC Industries, Dover, Del.	Space Suits
Radio Corp. of America Van NUYS, Calif.	110A Computer/ Saturn Checkout
Sanders Associates Nashua, N.H.	Operational Display System Saturn
Brown Engineering, Huntsville, Ala.	Discrete controls
Reynolds, Smith and Hill, Jacksonville, Fla.	Engineering Design of Mobile Launchers
Ingalls Iran Works, Birmingham Ala.	Mobile Launchers (ML) (Structural Work)
Smith/Ernst (Joint Venture) Tampa, Fla., Washington D. C.	Electrical Mechanical Portion of MLs
Power Shovel, Inc. Marion, Ohio	Transporter
Hayes International, Birmingham, Ala.	Mobile Launcher Service Arm
Bendix Aerospace System, Ann Arbor, Mich.	Apollo Lunar Surface Experiments Package (ALSEP)
AeroJet-Gen. Corp., El Monte, Calif.	Service Propulsion System Engine

Prelaunch Mission Operation Report

REPORT NO. M-932-70-13

APOLLO 13 (AS-508)

OFFICE OF MANNED SPACE FLIGHT
PREPARED BY: APOLLO PROGRAM OFFICE-MAO

No. M-932-70-13

'O: A/Administrator 31 March 1970

ROM: MA/Apollo Program Director

UBJECT: Apollo 13 Mission (AS-508)

)n 11 April 1970, we plan to launch Apollo 13 from Pad A of Launch Complex 39 at the Kennedy Space Center. This will be the third manned lunar landing mission and is targeted to a pre-selected point in the Fra Aauro Formation.

'rimary objectives of this mission include selenological inspection, survey, and sampling of the ejecta blanket hought to have been deposited during the formation of the Imbrium basin; deployment and activation of an Apollo Lunar Surface Experiments Package; continuing the development of man's capability to work in the unar environment; and obtaining photographs of candidate lunar exploration sites. Photographic records will e obtained and the extravehicular activities will be televised.

'he 10-day mission will be completed with landing in the Pacific Ocean. Recovery and transport of the crew, pacecraft, and lunar samples to the Lunar Receiving Laboratory at the Manned Spacecraft Center will be conducted under quarantine procedures that provide for biological isolation.

locco A. Petrone

)ale D. Myers
Associate Administrator for
Aanned Space Flight

FOREWORD

AISSION OPERATION REPORTS are published expressly for the use of NASA Senior Management, as equired by the Administrator in NASA Instruction 6-2-10, dated 15 August 1963. The purpose of these eports is to provide NASA Senior Management with timely, complete, and definitive information on flight nission plans, and to establish official mission objectives which provide the basis for assessment of mission ccomplishment.

nitial reports are prepared and issued for each flight project just prior to launch. Following launch, updating eports for each mission are issued to keep General Management currently informed of definitive mission esults as provided in NASA Instruction 6-2-10

'rimary distribution of these reports is intended for personnel having program/project management esponsibilities which sometimes results in a highly technical orientation. The Office of Public Affairs publishes comprehensive series of pre-launch and postlaunch reports on NASA flight missions which are available for lissemination to the Press.

APOLLO MISSION OPERATION REPORTS are published in two volumes: the MISSION OPERATION REPORT (MOR) and the MISSION OPERATION REPORT, APOLLO SUPPLEMENT. This format was designed o provide a mission-oriented document in the MOR, with supporting equipment and facility description in he MOR, APOLLO SUPPLEMENT. The MOR, APOLLO SUPPLEMENT is a program-oriented reference locument with a broad technical description of the space vehicle and associated equipment, the launch omplex, and minion control and support facilities.

'ublished and Distributed by PROGRAM and SPECIAL REPORTS DIVISION (XP) EXECUTIVE ECRETARIAT — NASA HEADQUARTERS

NASA OMSF PRIMARY MISSION OBJECTIVES FOR APOLLO 13

PRIMARY OBJECTIVES

Perform selenological inspection survey, and sampling of materials in a pre-selected region of the Fra Mauro Formation.

Deploy and activate an Apollo Lunar Surface Experiments Package (ALSEP).

Develop man's capability to work in the lunar environment.

Obtain photographs of candidate exploration sites.

Rocco A. Petrone
Apollo Program Director
Date: 24 March 1970

Dale D. Myers
Associate Administrator for
Manned Space Flight
Date: March 28, 1970

DETAILED OBJECTIVES AND EXPERIMENTS SPACECRAFT DETAILED OBJECTIVES AND EXPERIMENTS

Contingency Sample Collection.
Apollo Lunar Surface Experiments Package (ALSEP III) Deployment.
Selected Sample Collection.
Lunar Field Geology (S-059).
Photographs of Candidate Exploration Sites.
Evaluation of Landing Accuracy Techniques.
Television Coverage.
EVA Communication System Performance.
Lunar Soil Mechanics.
Selenodetic Reference Point Update.
Lunar Surface Close-Up Photography (S-184).
Thermal Coating Degradation.
CSM Orbital Science Photography.
Transearth Lunar Photography.
Solar Wind Composition (S-080).
Extravehicular Mobility Unit Water Consumption Measurement.
Gegenschein from Lunar Orbit (S-178).
Dim Light Photography.
CSM S-Band Transponder (S-164).
Downlink Bistatic Radar (VHF Only) (S-170).

SUMMARY OF APOLLO/SATURN MISSIONS

Mission	Launch Date	Launch Vehicle	Payload	Description
AS-201	2/26/66	SA-201	CSM-009	Launch vehicle and CSM development. Test of CSM subsystems and of the space vehicle. Demonstration of reentry adequacy of the CM at earth orbital conditions.
AS-203	7/5/66	SA-203	LH_2 in S-IVB	Launch vehicle development. Demonstration of control of LH_2 by continuous venting in orbit.
AS-202	8/25/66	SA-202	CSM-011	Launch vehicle and CSM development. Test of CSM subsystems and of the structural integrity and compatibility of the space vehicle. Demonstration of propulsion and entry control by G&N system. Demonstration of entry at 28,500 fps.
APOLLO 4	11/9/67	SA-501	CSM-017 LTA-10R	Launch vehicle and spacecraft development. Demonstration of Saturn V Launch Vehicle performance and of CM entry at lunar return velocity.
APOLLO 5	1/22/68	SA-204	LM-1 SLA-7	LM development. Verified operation of LM subsystems: ascent and descent propulsion systems (including restart) and structures. Evaluation of LM staging. Evaluation of S-IVB/IU orbital performance.
APOLLO 6	4/4/68	SA-502	CM-020 SM-014 LTA-2R SLA-9	Launch vehicle and spacecraft development. Demonstration of Saturn V Launch Vehicle performance.
APOLLO 7	10/11/68	SA-205	CM-101 SM-101 SLA-5	Manned CSM operations. Duration 10 days 20 hours.
APOLLO 8	12/21/68	SA-503	CM-103 SM-103 LTA-B SLA-11	Lunar orbital mission. Ten lunar orbits. Mission duration 6 days 3 hours. Manned CSM operations.
APOLLO 9	3/3/69	SA-504	CM-104 SM-104 LM-3 SLA-12	Earth orbital mission. Manned CSM/LM operations. Duration 10 days 1 hour.
APOLLO 10	5/18/69	SA-505	CM-106 SM-106 LM-4 SLA-13	Lunar orbital mission. Manned CSM/LM operations. Evaluation of LM performance in cislunar and lunar environment, following lunar landing profile. Mission duration 8 days.
APOLLO 11	7/16/69	SA-506	CM-107 SM-107 LM-5 SLA-14	First manned lunar landing mission. Lunar surface stay time 21.6 hours. Mission duration 8 days 3 hours.
APOLLO 12	11/14/69	SA-507	CM-108 SM-108 LM-6 SLA-15	Second manned lunar landing mission. Demonstration of point landing capability. Deployment of ALSEP I. Surveyor III investigation. Lunar Surface stay time 31.5 hours. Two dual EVA's (15.5 manhours). 89 hours in lunar orbit (45 orbits). Mission duration 10 days 4.6 hours.

LAUNCH VEHICLE DETAILED OBJECTIVES

Impact the expended S-IVB/IU on the lunar surface to excite ALSEP 1. Determine actual S-IVB/IU point of impact.

LAUNCH COUNTDOWN AND TURNAROUND CAPABILITY, AS-508

COUNTDOWN

Countdown for launch of the AS-508 Space Vehicle for the Apollo 13 Mission will begin with a pre-count starting at T-94 hours during which launch vehicle and spacecraft countdown activities will be conducted independently. Official coordinated spacecraft and launch vehicle countdown will begin at T-28 hours.

SCRUB/TURNAROUND

A scrub is a termination of the countdown. Turnaround is the time required to recycle and count down to launch (T-0) in a subsequent launch window assuming no serial repair activities are required. The scrub/turnaround plan will be placed in effect immediately following a scrub during the countdown. For a hold that results in a scrub prior to T-22 minutes, turnaround procedures are initiated from the point of hold. Should a hold occur from T-22 minutes (S-II start bottle chilldown) to T-16.2 seconds (S-IC forward umbilical disconnect), then a recycle to T-22 minutes, a hold, or a scrub is possible under conditions stated in the Launch Mission Rules. A hold between T-16.2 seconds and T-8.9 seconds (ignition) could result in either a recycle or a scrub depending upon the circumstances. An automatic or manual cutoff after T-8.9 seconds will result in a scrub.

30-DAY SCRUB/TURNAROUND

A 30-day turnaround capability exists in the event that a scrub occurs and there is no launch window available within the 24 or 48-hour turnaround capability. In the event of a 30-day scrub/turnaround a new countdown will be started at the beginning of pre-count. The Flight Readiness Test (FRT) and Countdown Demonstration Test (CDDT) will not be rerun.

48-HOUR SCRUB/TURNAROUND

The maximum scrub/turnaround time from any point in the launch countdown up to T-8.9 seconds is 48 hours. This maximum time assumes no serial repair activities are required and it provides for reservicing all space vehicle cryogenics.

24 HOUR SCRUB/TURNAROUND

A 24-hour turnaround capability exists as late in the countdown as T-8.9 seconds. This capability depends upon having sufficient spacecraft consumables margins above redline quantities stated in the Launch Mission Rules for the period remaining to the next launch window. Only one 24-hour scrub/turnaround can be accomplished.

FLIGHT MISSION DESCRIPTION

LANDING SITE

The landing site of the Apollo 13 Mission is a point 3°40'S latitude, 17°29'W longitude in the Fra Mauro Formation. The Fra Mauro Formation, an extensive geologic unit covering large portions of the lunar surface around Mare Imbrium, has been interpreted as the ejecta blanket deposited during the formation of the Imbrium basin. Sampling of the Fra Mauro Formation may provide information on ejecta blanket formation and modification, and yield samples of deep-seated crustal material giving information on the composition of the lunar interior and the processes active in its formation. Age dating the returned samples should establish the age of pre-mare deep-seated material and the age of the formation of the Imbrium basin and provide important points on the geologic time scale leading to an understanding of the early history of the moon.

LAUNCH WINDOWS

The launch windows for Fra Mauro are shown in Table 1 .

TABLE I

APOLLO 13 LAUNCH DATE	LAUNCH WINDOWS OPEN	CLOSE	SUN ELEVATION ANGLE*
April 11, 1970	14:13	17:37	9.9°
**May 9, 1970 (T-24)	13:25	16:43	7.8°
**May 10, 1970 (T-0)	13:35	16:44	7.8°
**May 11, 1970 (T+24)	13:32	16:38	18.5°

NOTE: Only one scrub/turnaround is feasible for May. April times are EST; all others are EDT.
* These values are subject to possible refinement.** The addition of the T-24 hour and T+24 hour windows to the optimum T-0 window provides increased flexibility in that all three opportunities are available for choice.

LAUNCH OPPORTUNITIES

The three opportunities established for May — in case the launch is postponed from 11 April — provide, in effect, the flexibility of a choice of two launch attempts. The optimum May launch window occurs on 10 May. The 3-day window permits a choice of attempting a launch 24 hours earlier than the optimum window and, if necessary, a further choice of a 24-hour or 48-hour recycle. It also permits a choice of making the first launch attempt on the optimum day with a 24-hour recycle capability. The 9 May window (T-24 hrs.) requires an additional 24 hours in lunar orbit before initiating powered descent to arrive at the landing site at the same time and hence have the same sun angle for landing as on 10 May. Should the 9 May window launch attempt be scrubbed, a decision will be made at that time, based on the reason for the scrub, status of spacecraft cryogenics and weather predictions, whether to recycle for 10 May (T-0 hrs.) or 11 May (T+24 hrs.) If launched on 11 May, the flight plan will be similar for the 10 May mission but the sun elevation angle at lunar landing will be 18.5° I instead of 7.8°.

HYBRID TRAJECTORY

The Apollo 13 Mission will use a hybrid trajectory that retains most of the safety features of the free-return trajectory but without the performance limitations. From earth orbit the spacecraft will be initially injected into a highly eccentric elliptical orbit (pericynthion of approximately 210 nautical miles (NM), which has a free-return characteristic, i.e., the spacecraft can return to the earth entry corridor without any further maneuvers. The spacecraft will not depart from the free-return ellipse until after the Lunar Module (LM) has been extracted from the launch vehicle and can provide a propulsion system backup to the Service Propulsion System (SPS). Approximately 28 hours after translunar injection (TLI), a midcourse maneuver will be performed by the SPS to place the spacecraft an a Lunar approach trajectory (non-free-return) having a pericynthion of 60 NM.

The use of a hybrid trajectory will permit;

Daylight launch/Pacific injection —This allows the crew to acquire the horizon as a backup attitude reference during high altitude abort, provides launch abort recovery visibility, and improves launch photographic coverage.

Desired lunar landing site sun elevation —The hybrid profile facilitates adjustment of translunar transit time which can be used to control sun angles on the landing site during lunar orbit and at landing.

Increased spacecraft performance —The energy of the spacecraft on a hybrid lunar approach trajectory is relatively low compared to what it would be on a full free-return trajectory thus reducing the differential velocity (DeltaV) required to achieve lunar orbit insertion.

Improved communication flexibility — This permits adjustment of the time of powered descent initiation (PDI) to occur within view of a 210-foot ground antenna.

LIGHTNING PRECAUTIONS

During the Apollo 12 Mission, the space vehicle was subjected to two distinct electrical discharge events. However, no serious damage occurred and the mission proceeded to a successful conclusion. Intensive investigation led to the conclusion that no hardware changes were necessary to protect the space vehicle from similar events. For Apollo 13 the Mission Rules have been revised to reduce the probability that the space vehicle will be launched into cloud formations that contain conditions conducive to initiating similar electrical discharges although flight into all clouds is not precluded.

FLIGHT PROFILE

Launch Through Earth Parking Orbit

The AS-508 Space Vehicle for the Apollo 13 Mission is planned to be launched at 14:13 EST on 11 April 1970 from Launch Complex 39A at the Kennedy Space Center Florida, on a flight azimuth of 72°. The Saturn V Launch Vehicle (LV) will insert the S-IVB/Instrument Unit (IU)/LM/CSM into a 103-NM, circular orbit. The S-IVB/IU and spacecraft checkout will be accomplished during the orbital coast phase. Figure 1 and Tables 2 through 4 summarize the flight profile events and space vehicle weight.

Translunar Injection

Approximately 2.6 hours after liftoff, the launch vehicle S-IVB stage will be reignited during the second parking orbit to perform the translunar injection (TLI) maneuver, placing the space vehicle on a free-return trajectory having a pericynthion of approximately 210 NM.

Translunar Coast

The CSM will separate from the S-IVB/IU/LM approximately 4 hours Ground Elapsed Time (GET), transpose dock, and initiate ejection of the LM/CSM from the S-IVB/IU. During these maneuvers, the LM and S-IVB/IU will be photographed to provide engineering data.

TABLE 2

APOLLO 13 SEQUENCE OF MAJOR EVENTS

EVENT	GET HR:MIN:SEC.	BURN DURATION (SYSTEM)	REMARKS
LAUNCH	00:00		PAD 39A, 4/11/70, 14:13 EST
TLI	2:35	358 (S-IVB)	PACIFIC INJECTION
MIDCOURSE CORRECTION-1	11:41	AS REQ'D	NOMINALLY ZERO
MIDCOURSE CORRECTION-2	30:41	2 (SPS)	HYBRID TRANSFER
MIDCOURSE CORRECTIONS (AS REQ'D)			74.9-HR. TRANSLUNAR COAST
LUNAR ORBIT INSERTION (LOI)	77:25	357 (SPS)	60 x 170 NM ORBIT
DESCENT ORBIT INSERTION	81:45	23 (SPS)	8 x 60-NM ORBIT (CSM/LM)
CSM/LM UNDOCKING & SEPARATION	99:16	6 (SM RCS)	SOFT UNDOCKING
CSM ORBIT CIRCULARIZATION	100:35	4 (SPS)	53 x 62 NM ORBIT (REV. 12)
POWERED DESCENT INITIATION (PDI)	103:31	687 (DPS)	LANDING POINT UPDATE
LANDING FRA MAURO	103:42		SEA = 7.1°, 4/15, 21:55 EST
EVA-1	108:00		4 HR. PLANNED
EVA-2	127:45		4 HR. PLANNED
ASCENT (LM LIFTOFF)	137:09	528 (APS)	4/17, 7:22 EST
DOCKING	140:45		CSM-ACTIVE
LM IMPACT BURN	144:32	75 (LM RCS)	ASCENT STAGE IMPACT AT 145:00
TRANSEARTH INJECTION (TEI)	167:29	135 (SPS)	73.6-HR. TRANSEARTH COAST
ENTRY INTERFACE	240:50		VELOCITY 36, 129 FPS
LANDING	241:03		PACIFIC, 4/21, 15:16 EST

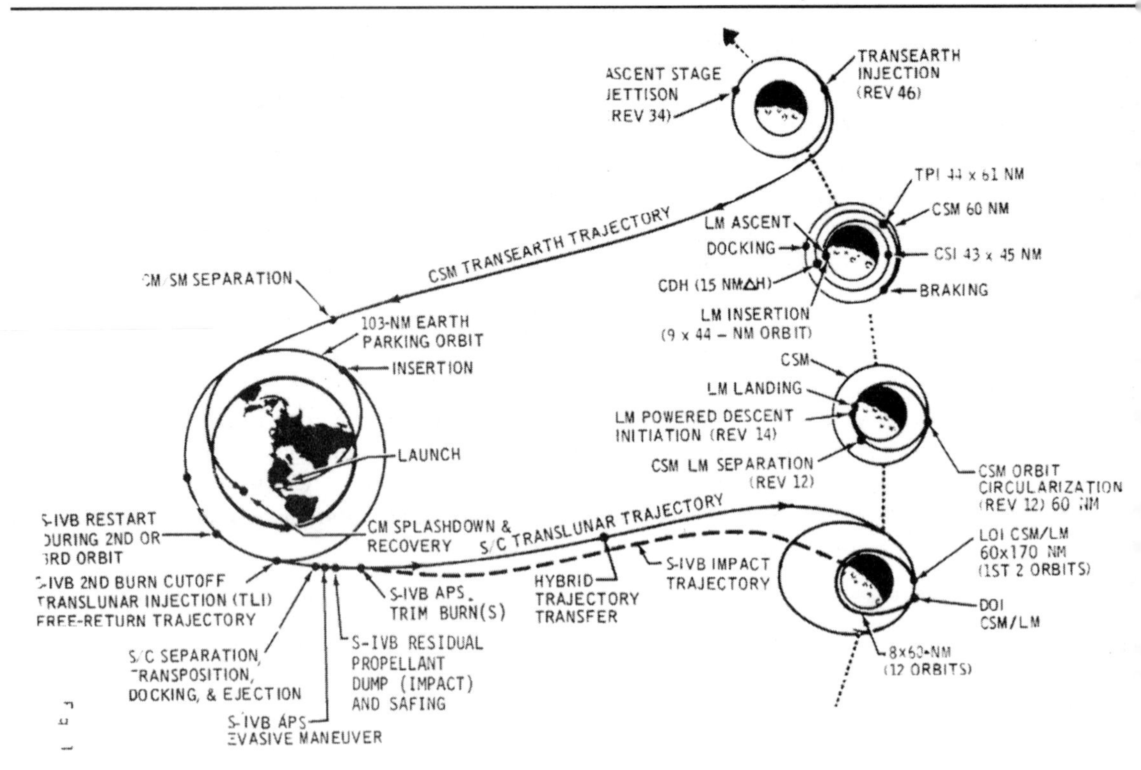

APOLLO 13 FLIGHT PROFILE

TABLE 3

APOLLO 13 TV SCHEDULE

DAY	DATE	EST	GET	DURATION	ACTIVITY/SUBJECT	VEH	STA
SATURDAY	APRIL 11	15:49	01:36	05 MIN	COLOR PHOTOS OF EARTH	CSM	KSC
SATURDAY	APRIL 11	17:28	03:15	1 HR 08 MIN	TRANSPOSITION & DOCKING	CSM	GDS
SUNDAY	APRIL 12	20:28	30:15	30 MIN	SPACECRAFT INTERIOR (MCC-2)	CSM	GDS
TUESDAY	APRIL 14	00:13	58:00	30 MIN	INTERIOR & IVT TO LM	CSM	GDS
WEDNESDAY	APRIL 15	14:03	95:50	15 MIN	FRA MAURO	CSM	MAD
THURSDAY	APRIL 16	02:23	108:10	3 HR 52 MIN	LUNAR SURFACE (EVA-1)	LM	GDS/HSK
THURSDAY	APRIL 16	22:03	127:50	6 HR 35 MIN	LUNAR SURFACE (EVA-2)	LM	GDS
FRIDAY	APRIL 17	10:36	140:23	12 MIN	DOCKING	CSM	MAD
SATURDAY	APRIL 18	12:23	166:10	40 MIN	LUNAR SURFACE	CSM	MAD*
SATURDAY	APRIL 18	14:13	168:00	25 MIN	LUNAR SURFACE (POST TEI)	CSM	MAD*
MONDAY	APRIL 20	19:58	221:45	15 MIN	EARTH & SPACECRAFT INTERIOR	CSM	GDS

* RECORDED ONLY

At 4.2 hours GET the S-IVB/IU will begin a series of programmed and ground-commanded operations which will alter the LV trajectory so that the S-IVB/IU will impact the lunar surface at the desired point providing a known energy source for the Apollo 12 ALSEP seismology equipment. The first is a programmed Auxiliary Propulsion System (APS) ullage motor retrograde burn evasive maneuver to provide initial launch vehicle/spacecraft separation to prevent recontact of the two vehicles. Second, by a combination of programmed liquid oxygen (LOX) dumping the S-IVB/IU will be placed on a lunar impact trajectory. A second APS ullage motor burn will be ground commanded at approximately 6.0 hours GET. The burn duration and attitude will be determined in real-time based on trajectory data. This burn is intended to place the S-IVB/IU on the trajectory for lunar impact at the desired point. A third APS ullage motor burn, also ground commanded, will be performed if necessary to refine the S-

TABLE 4

APOLLO 13 WEIGHT SUMMARY

(Weight in Pounds)

STAGE/MODULE	INERT WEIGHT	TOTAL EXPENDABLES	TOTAL WEIGHT	FINAL SEPARATION WEIGHT
S-IC	288,000	4,746,870	5,034,870	363,403
S-IC/S-II Interstage	11,464	---	11,464	---
S-II Stage	78,050	996,960	1,075,010	92,523
S-II/S-IVB Interstage	8,100	---	8,100	---
S-IVB Stage	25,050	236,671	261,721	35,526
Instrument Unit	4,482	---	4,482	---
Launch Vehicle at Ignition			6,395,647	
Spacecraft-LM Adapter	4,044	---	4,044	---
Lunar Module	9,915	23,568	33,483	*33,941
Service Module	10,532	40,567	51,099	**14,076
Command Module	12,572	---	12,572	**11,269 (Landing)
Launch Escape System	9,012	---	9,012	---
Spacecraft at Ignition			110,210	
Space Vehicle at Ignition			6,505,857	
S-IC Thrust Buildup			(-)84,598	
Space Vehicle at Liftoff			6,421,259	
Space Vehicle at Orbit Insertion			299,998	

* CSM/LM Separation

** CM/SM Separation

IVB/IU trajectory. This burn will occur approximately 9.0 hours GET. The desired impact will be within 350 kilometers of the target point, 3°S. latitude, 30°W. longitude. The impact will occur while the CSM/LM is on the backside of the moon. Later, the crew will photograph the S-IVB/IU target impact area. It is desired that post-flight determination of actual impact be within 5 kilometers in distance and 1 second in time. The spacecraft will be placed on a hybrid trajectory by performing a Service Propulsion System (SPS) maneuver at the time scheduled for the second Midcourse Correction (MCC), approximately 30.6 hours GET. The CSM/LM combination will be targeted far a pericynthion altitude of 60 NM and, as a result of the SPS maneuver, will be placed on a non-free return trajectory. The spacecraft will remain within the LM Descent Propulsion System (DPS) as well as the SPS return capability. Earth weather will be photographed for a 3 to 4 hour period beginning at approximately 7 hours GET. Lunar photography may be performed, at the crew's option, at 56 and 75 hours GET. MCC's will be made as required, using the Manned Space Flight Network (MSFN) for navigation.

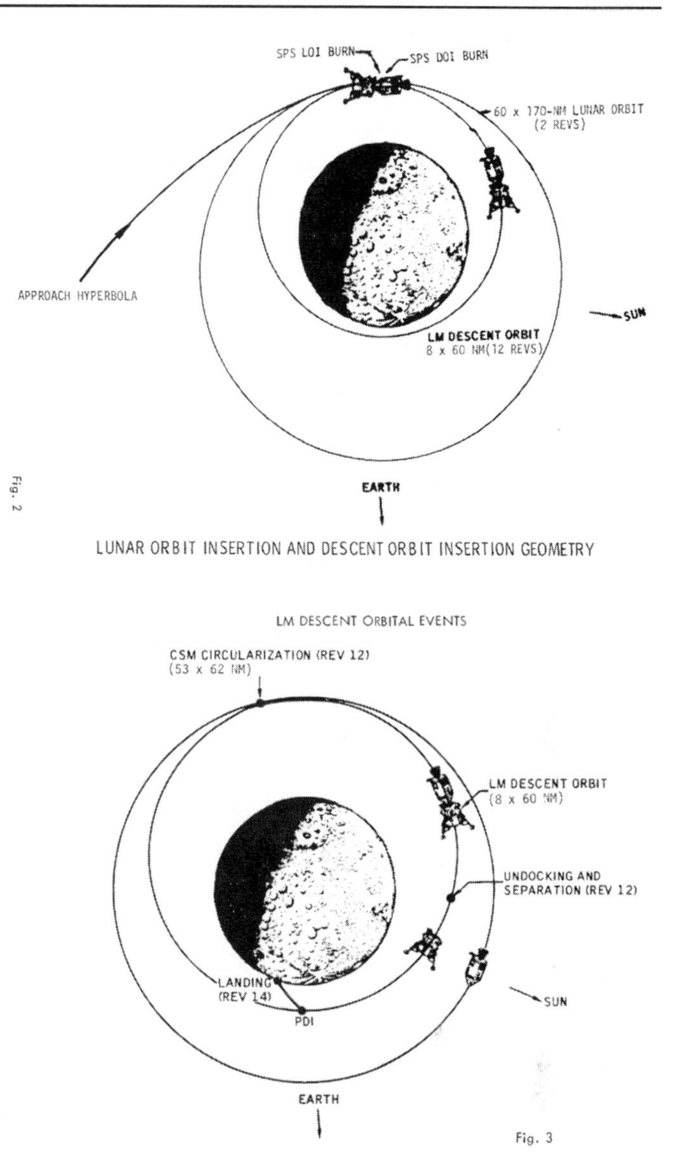

Fig. 2

LUNAR ORBIT INSERTION AND DESCENT ORBIT INSERTION GEOMETRY

Fig. 3

Lunar Orbit Insertion

The SPS will insert the spacecraft into an initial lunar orbit (approximately 60 x 170 NM) at 77.6 hours GET (Figure 2). The spacecraft will remain in a 60 x 170-NM orbit for approximately two revolutions.

Descent Orbit Insertion

After two revolutions in lunar orbit, the SPS will be used to insert the spacecraft in a 60 x 8-NM descent orbit. During the 4th revolution, the Command Module Pilot (CMP) will photograph the candidate exploration site Censorinus from the CSM at low altitude while the Commander (CDR) and Lunar Module Pilot (LMP) enter the LM for checkout and housekeeping. The crew will use approximately six revolutions for eat and rest periods, and then will prepare the LM for separation and powered descent. A soft docking will be made during the 12th revolution. Spacecraft separation will be executed by the Service Module Reaction Control System (SM RCS) with the CSM radially below the LM (figure 3.)

CSM Lunar Solo Rendezvous

During the 12th revolution (after undocking) the CSM will perform a circularization maneuver to a near-circular 60 NM orbit. The CSM will photograph selected sites that will include the candidate exploration site Censorinus. Lunar orbital science photography and dim light photography of zodiacal light, solar corona, and Gegenschein will be performed.

Lunar Module Descent

During the 14th revolution the DPS will be used for the powered descent maneuver, which will start approximately at pericynthion. The vertical descent portion of an automatic landing during the landing phase will start at an altitude of about 100 feet and will be terminated at touchdown on the lunar surface (Figure 3). The crew may elect to take over manually at an altitude of about 500 feet or below. Return to automatic control is a newly added capability of the LM guidance computer. During descent, the lunar surface will be photographed from the LMP's window to record LM movement and surface disturbances and to aid in determining the landed LM location.

Lunar Surface Operations

A summary of the lunar surface activities is shown in Figure 4.

Postlanding

Immediately upon landing, the LM crew will execute the lunar contact checklist and reach a stay/no stay decision. After reaching a decision to stay, the Inertial Measurement Unit (IMU) will be aligned, the Abort Guidance System (AGS) gyro will be calibrated and aligned, and the lunar surface will be photographed through the LM window. Following a crew eat period all loose items not required for extravehicular activity (EVA) will be stowed.

EVA-1

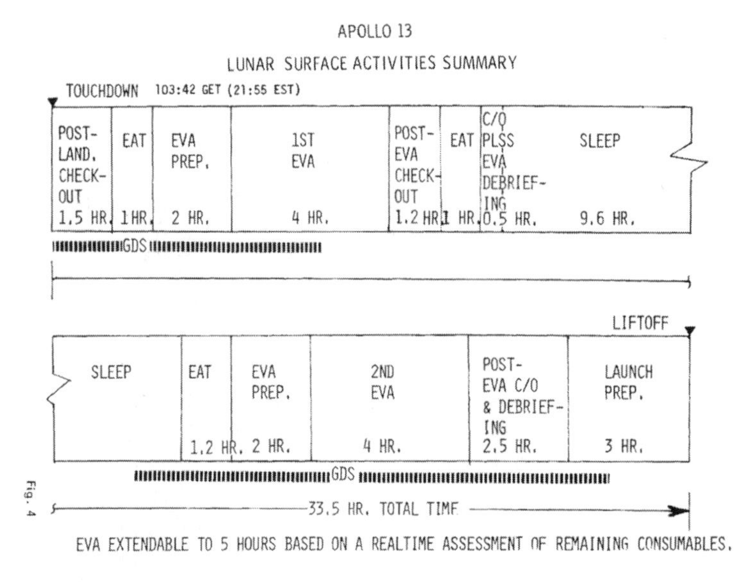

Fig. 4

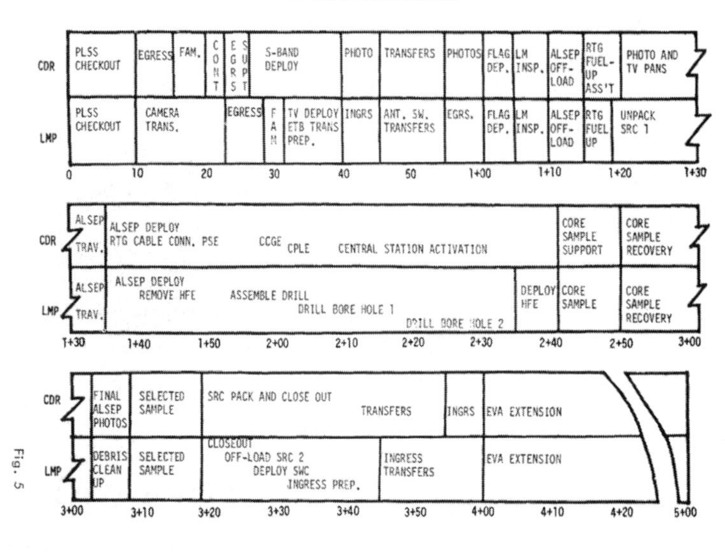

Fig. 5

The activity timeline for EVA-1 is shown in Figure 5. Both crew members will don helmets, gloves, Portable Life Support Systems (PLSS) and Oxygen Purge Systems (OPS) and the cabin will be depressurized. The CDR will move through the hatch, deploy the Lunar Equipment Conveyor (LEC), and move to the ladder where he will deploy the Modularized Equipment Stowage Assembly (MESA), Figure 6, which initiates television coverage from the MESA. He will then descend the ladder to the lunar surface. The LMP will monitor and photograph the CDR using a still camera (70mm Hasselblad Electric Camera) and the lunar geologic exploration sequence camera (16mm Data Acquisition Camera).

Environmental Familiarization — Contingency Sample Collection — After stepping to the surface and checking his mobility stability, and the Extravehicular Mobility Unit (EMU), the CDR will collect a contingency sample. This will make it possible to assess the nature of the Lunar surface material at the Apollo 13 landing site in the event the EVA were terminated at this point. The sample will be collected by quickly scooping up a loose sample of the lunar material (approximately 2 pounds), sealing it in a Contingency Sample Container, and placing the sample in the Equipment Transfer Bag

ETB), along with the lithium hydroxide (LiOH) canisters and PLSS batteries for later transfer into the LM ısing the LEC. The LMP will transfer the 70mm cameras to the surface with the LEC. The LMP will then lescend to the surface.

-band Antenna Deployment — The S-band antenna will be removed from the LM and carried to the site vhere the CDR will erect it, as shown in Figure 7, connect the antenna cable to the LM, and perform the equired alignment.

.unar TV Camera (Color) Deployment — Vhile the CDR deploys the S-band antenna, the .MP will unstow the TV camera and deploy it on he tripod approximately 50 feet from the LM. 'he LMP will then ingress the LM to activate and erify TV transmission with the Mission Control Center. The contingency sample and other quipment will be transferred into the LM and he 16mm lunar geologic exploration camera ransferred to the surface. The LMP will egress gain, leaving the hatch slightly ajar, and descend o the surface. The CDR will photograph the ontingency sample area and the LMP egress. ollowing this, the American flag will be leployed.

M Inspection — After repositioning the TV to iew the scientific equipment bay area, the LMP vill inspect and photograph the LM footpads and uadrants I, II, III, and IV with an EMU-mounted 0mm camera. Concurrently the CDR will also ıspect and photograph the LM. The Apollo unar Surface Close-Up Camera (ALSCC) will e removed from the MESA and placed down un.

ALSEP Deployment — After off loading ALSEP rom the LM, the Radioisotope Thermoelectric Generator (RTG) will be fueled (Figure 8), the ALSEP packages will be attached to a one-man arry bar for traverse in a barbell mode, and the 'V will be positioned to view the ALSEP site. The and tools will be loaded on the hand tool arrier. While the CDR obtains TV and hotographic panoramic views from the site, the MP will unload Sample Return Container umber I (SRC I) and remove the ALSCC. The CDR and LMP will then carry the ALSEP ackages, hand tool carrier, and Apollo Lunar urface Drill (ALSD) to the deployment site pproximately 500 feet from the LM. The crew vill survey the site and determine the desired ɔcation for the experiments. The following ıdividual experiment packages will then be eparated, assembled, and deployed to respective tes in the arrangement shown in Figure 9.

DEPLOYED MODULARIZED EQUIPMENT STOWAGE ASSEMBLY

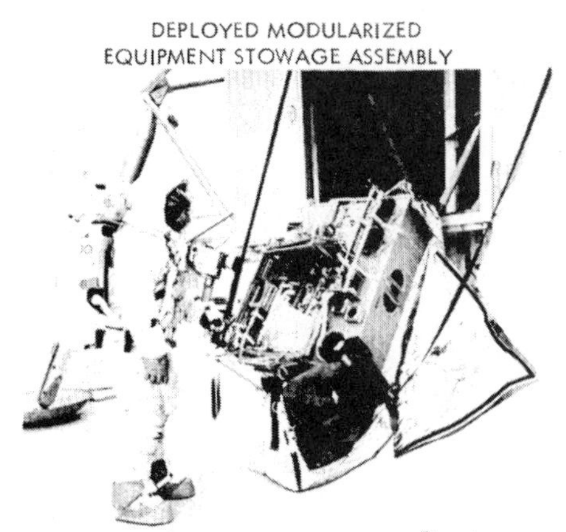

Fig. 6

DEPLOYED S-BAND ANTENNA

Fig. 7

APOLLO 13 ALSEP RADIOISOTOPE THERMOELECTRIC GENERATOR (UNFUELED)

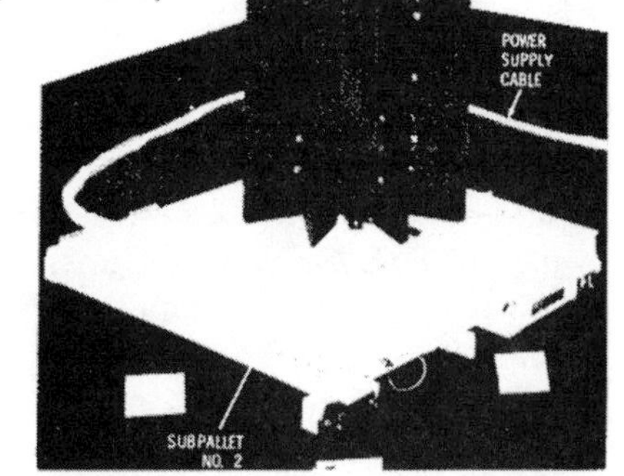

Fig. 8

(Apologies for the picture quality Ed.)

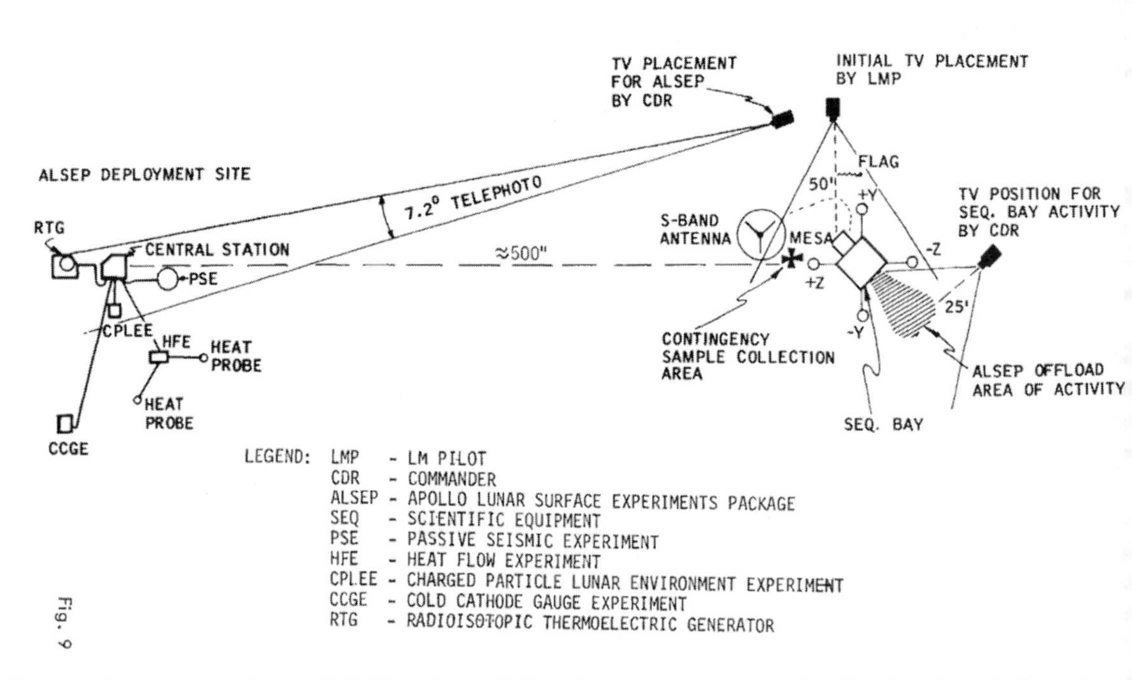

Fig. 9

Passive Seismic Experiment (S-031) — The CDR will deploy and set up the Passive Seismic Experiment package with its thermal cover.

Charged Particle Lunar Environment Experiment (S-038) — The CDR will deploy and orient the Charged Particle Lunar Environment Experiment Package while the LMP is assembling the ALSD and drilling the first hole for the heat flow experiment.

Cold Cathode Ion Gauge (S-058) — The CDR will deploy and orient the Cold Cathode Ion Gauge.

Heat Flow Experiment (S-037) — The LMP will assemble the battery-powered ALSD and will drill two three meter deep holes using hollow-center bore stems. Each bore stem will be left in place as an encasement into which the heat flow probes are inserted.

ALSEP Central Station — The CDR will level and align the central station which includes deployment of the sunshield. At this time the LMP will be drilling the second hole for the Heat Flow Experiment. The CDR will then assemble and align the ALSEP antenna. The CDR will activate the central station and photograph the ALSEP layout while the LMP is implanting the Heat Flow Experiment probes.

Core Sampling — The CDR will assist the LMP in modifying the ALSD, collecting core samples and photographing the operation.

Selected Sample Collection — The crew will begin the return traverse and initiate collection of the selected samples. AT the return to the LM, samples will be weighed and, with the core stems, stowed in SRC I and the SRC will be sealed. SRC 2 will be off loaded and SRC I will be transferred into the LM.

Solar Wind Composition Experiment (S-080) — The four-square-foot panel of aluminum foil will be deployed by the LMP. EMU cleaning and ingress into the LM will be accomplished by the LMP and the CDR. EVA-I will terminate when the LM cabin is repressurized.

<u>Post-EVA I Operations</u>

After configuring the LM system for Post-EVA-I operations, the PLSS's will be recharged. This includes filling the oxygen system to a minimum pressure of 875 pounds per square inch, filling the water reservoir, and replacing the battery and LiOH canister. A one-hour eating period is scheduled between the beginning and end of the PLSS recharge operations. The PLSS's and OPS's will be stowed, followed by a 9½ hour rest period and another eat period.

EVA-2

The LM will be configured for EVA activities and the CDR will egress. The LMP will again monitor and photograph the CDR and then transfer camera equipment in the ETB to the CDR with the LEC. The LMP will descend to the lunar surface leaving the LM hatch slightly ajar. A summary of EVA-2 activities is shown in figure 10.

Sample Collection and Camera Calibration — The crew will collect a thermal sample and a sieve sample from near the LM and a contaminated sample from under the LM. They will calibrate their cameras by photographing a special contrast chart on the hand tool carrier and then will position the TV camera for the field geology traverse.

Lunar Field Geology (S-059) Traverse — Both crewmen will conduct the field geology traverse, which is planned in detail prior to launch. Additional support and real-time planning will be provided from the ground based on features of the landing site obtained from crew descriptions and TV. Traversing outbound from the LM, the crew will obtain Close-Up Stereo Camera photos of selected areas. They will take panoramic photographs of the lunar surface and will use a special polarizing filter to photograph selected features. They will obtain subsurface samples, core samples, and surface samples. Special gas analysis, environmental, and magnetic lunar surface samples will be collected. Approximately ½ mile from the LM the CDR will dig a two-foot-deep trench for lunar soil mechanics evaluation. The LMP will collect a core sample and a special environmental sample from the trench and obtain photographic data of the boot prints in the trench material. Gas analysis and magnetic samples will also be obtained in the vicinity of the trench.

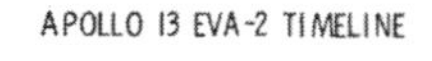

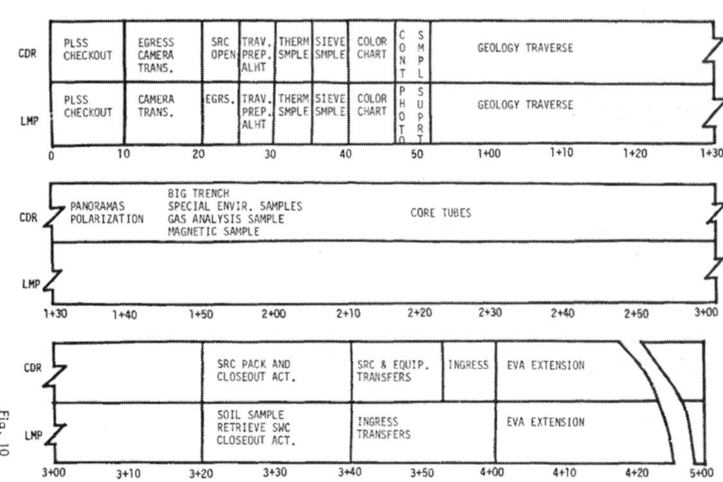

Fig. 10

The crew will begin traversing inbound to the LM by a different route to obtain additional documented samples as performed on the Outbound traverse. A typical documented sample procedure will include locating the gnomon up sun of the sample site, photography of the site and the sample description of the sample and stowage of the lunar surface material in a sample bag. The typical core sampling will consist of placing the gnomon up sun and photography of the sample site cross sun, driving the core tube into the surface, recovery and capping the sample within the tube.

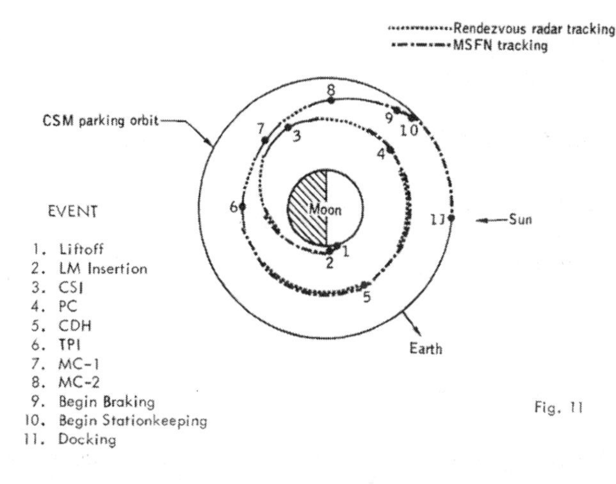

Fig. 11

Upon return to the LM, the CDR will off load the samples into SRC 2. The LMP will reposition the TV, collec a soil mechanics sample, and then take down and roll up the Solar Wind Composition experiment and plac it and the Close-Up Stereo Camera magazine in the SRC. He will then close and seal the SRC. The LMP wi clean his EMU, ingress into the LM, and hook up the LEC. The CDR and LMP will utilize the LEC to transfe samples and equipment into the LM. The CDR will clean his EMU, ascend into the LM, jettison the LEC an ingress. The LM will be repressurized, terminating EVA-2. Equipment and samples will be stowed an preparations made for equipment jettison. The LM will be depressurized, equipment jettisoned, and the L repressurized.

Lunar Module Ascent

The LM ascent (Figure 11) will begin after a lunar stay of approximately 33.5 hours. The Ascent Propulsio System (APS) powered ascent is divided into two phases. The first phase is a vertical rise, which is require to achieve terrain clearance, and the second phase is orbit insertion. After orbit insertion the LM will execut the coelliptic rendezvous sequence which nominally consists of four major maneuvers concentric sequenc initiation (CSI), constant delta height (CDH), terminal phase initiation (TPI), and terminal phase finalizatio (TPF). A nominally zero plane change (PC) maneuver will be scheduled between CSI and CDH, and tw nominally zero midcourse correction maneuvers will be scheduled between TPI and TPF; the TPF maneuve is actually divided into several braking maneuvers. All maneuvers after orbit insertion will be performed wit the LM RCS. Once docked to the CSM, the two crewmen will transfer to the CSM with equipment, luna samples, and exposed film. Decontamination operations will be performed, jettisonable items will be place in the Interim Stowage Assembly and transferred to the LM, and the LM will be configured for deorbit an lunar impact.

LM Ascent Stage Deorbit

The ascent stage will be de-orbited (Figure 12), during the 35th revolution, for lunar surface impact between the ALSEP I of Apollo 12 and the newly deployed ALSEP III, to provide a known energy source to produce signals for recording by the seismic experiments. The CSM will be separated radially from the ascent stage with a SM RCS retrograde burn approximately 2 hours after docking to the CSM. Following the LM jettison maneuver, the CSM will perform a pitch down maneuver. The LM deorbit maneuver will be a retrograde RCS burn initiated by ground control and the LM will be targeted to impact the lunar surface approximately 36.5 NM west northwest of the Apollo 13 landing site. The ascent stage jettison, ignition, and impacted lunar surface area will be photographed from the CSM.

LM ASCENT STAGE DEORBIT

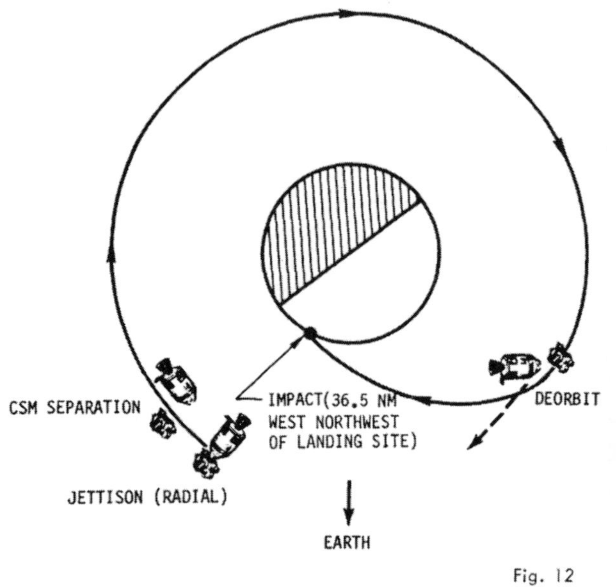

Fig. 12

CSM Lunar Orbit Operations

The CSM will execute an orbital plane change, soon after MSFN acquisition of signal in the 35th revolutio for approximately 9 hours of lunar reconnaissance photography. High resolution vertical and obliqu topographic photography, stereo strip photography, science photography and landmark tracking will b performed. The Censorinus, Descartes, and Davy Rille sites are of special photographic interest as candidat future landing sites.

Transearth Injection and Coast

At the end of the 46th revolution, and approximately 10 minutes prior to MSFN acquisition of signal, the SP will be used to inject the CSM onto the transearth trajectory. The return flight duration will be approximatel 73 hours (based on an 11 April launch) and the return inclination (to the earth's equator) will not exceed 4 degrees. Midcourse corrections will be made as required, using the MSFN for navigation.

ntry and Landing

rior to atmospheric entry, the CSM will maneuver to a heads-up attitude, the Command Module (CM) will ettison the SM and orient to the entry attitude. The nominal range from entry interface (EI) at 400,000 feet ltitude to landing will be approximately 1250 NM. Earth landing will nominally be in the Pacific Ocean at °34'S latitude and 157°30'W longitude (based on an 11 April launch) approximately 241 hours after liftoff.

rew Recovery and Quarantine

ollowing landing, the Apollo 13 crew will don the flight suits and face masks passed in to them through the pacecraft hatch by a recovery swimmer wearing standard scuba gear. Integral Biological Isolation Garments BIG's) will be available for use in case of an unexplained crew illness. The swimmer will swab the hatch and djacent areas with a liquid decontamination agent. The crew will then be carried by helicopter to the ecovery ship where they will enter a Mobile Quarantine Facility (MQF) and all subsequent crew quarantine rocedures will be the same as for the Apollo 11 and 12 Missions.

M and Data Retrieval Operations

fter flight crew pickup by the helicopter, the CM will be retrieved and placed on a dolly aboard the recovery hip. The CM will be mated to the MQF, and the lunar samples, film, flight logs, etc., will be retrieved and assed out through a decontamination lock for shipment to the Lunar Receiving Laboratory (LRL). The pacecraft will be off loaded from the ship at Pearl Harbor and transported to an area where deactivation f the CM pyrotechnics and propellant system will be accomplished. This operation will be confined to the xterior of the spacecraft. The spacecraft will then be flown to the LRL and placed in a special room for torage. Contingency plans call for sterilization and early release of the spacecraft if the situation so requires.

CONTINGENCY OPERATIONS

GENERAL

f an anomaly occurs after liftoff that would prevent the space vehicle from following its nominal flight plan, n abort or an alternate mission will be initiated. Aborts will provide for an acceptable flight crew and Command Module (CM) recovery while alternate missions will attempt to maximize the accomplishment of nission objectives as well as provide for an acceptable flight crew and CM recovery.

ABORTS

he following sections present the abort procedures and descriptions in order of the mission phase in which hey could occur.

aunch

here are six launch abort modes. The first three abort modes would result in termination of the launch equence and a CM landing in the launch abort area. The remaining three abort modes are essentially lternate launch procedures and result in insertion of the Command/Service Module (CSM) into earth orbit. ll of the launch abort modes are the same as those for the Apollo 11 Mission.

arth Parking Orbit

return to earth abort from earth parking orbit (EPO) will be performed by separating the CSM from the emainder of the space vehicle and performing a retrograde Service Propulsion System (SPS) burn to effect ntry. Should the SPS be inoperable, the Service Module Reaction Control System (SM RCS) will be used to erform the deorbit burn. After CM/SM separation and entry, the crew will fly a guided entry to a pre-elected target point, if available.

ranslunar Injection

ranslunar injection (TLI) will be continued to nominal cutoff, whenever possible, in order for the crew to erform malfunction analysis and determine the necessity of an abort.

ranslunar Coast

f ground control and the spacecraft crew determine that an abort situation exists, differential velocity DeltaV) targeting will be voiced to the crew or an onboard abort program will be used as required. In most cases, the Lunar Module (LM) will be jettisoned prior to the abort maneuver if a direct return is requirec An SPS burn will be initiated to achieve a direct return to a landing area. However, a real-time decisio capability will be exploited as necessary for a direct return or circumlunar trajectory by use of the severa CSM/LM propulsion systems in a docked configuration.

Lunar Orbit Insertion

In the event of an early shutdown of the SPS during lunar orbit insertion (LOI), contingency action wi depend on the condition which caused the shutdown. If the shutdown was inadvertent, and if specified SP limits have not been exceeded, an immediate restart will be attempted. Upon completion of the LOI burn real-time decision will be made on possible alternate missions. If, during the LOI burn, the SPS limits ar exceeded, a manual shutdown will be made. The LM Descent Propulsion System (DPS) will serve as a backu propulsion system.

Descent Orbit Insertion

In the event of a descent orbit insertion (DOI) overburn where, despite trim corrections, the pericynthio remains lower than desired, a so-called "bail-out" SPS burn will be performed to raise the pericynthion. Base on Mission Rules criteria, this situation could lead to a continuation of the nominal landing mission, selectio of an alternate mission, or early mission termination.

Transearth Injection

An SPS shutdown during transearth injection (TEI) my occur as the result of an inadvertent automati shutdown. Manual shutdowns are not recommended. If an automatic shutdown occurs, an immediate restar will be initiated.

ALTERNATE MISSION SUMMARY

The two general categories of alternate missions that can be performed during the Apollo 13 Mission are (1 earth orbital and (2) lunar. Both of these categories have several variations which depend upon the nature c the anomaly leading to the alternate mission and the resulting systems status of the LM and CSM. A brie description of these alternate missions is contained in the following paragraphs.

Earth Orbit

The CSM will dock with the LM, and the photographic equipment will be retrieved from the LM. Followin this, the LM will be de-orbited into the Pacific Ocean area to eliminate debris problems. The CSM wi perform SPS plane change maneuvers to achieve an orbital inclination of 40° with daylight coverage of all U passes. Earth orbital photography will then be conducted.

Lunar Orbit

CSM and LM

The nominal mission bootstrap photographic objectives will be accomplished. These objectives includ photographs of Censorinus, Descartes, and Davy Rille. The LM will normally be jettisoned prior t accomplishing photographic objectives to avoid CM window blockage.

CSM Alone

The hybrid transfer will be deleted in this case. If the hybrid transfer has been performed, the CSM will b placed back on a free return trajectory. A two-burn LOI sequence, as on Apollo 8, 11, and 12 will be use to place the vehicle in a 60-NM circular orbit. The LOI burn will also establish an orbit to pass ove Censorinus and Mosting C for photography and landmark tracking.

MISSION SUPPORT — GENERAL

Mission Support is provided by the Launch Control Center, the Mission Control Center, the Manned Spac Flight Network, and the recovery forces. A comprehensive description of the mission support elements is i the MOR Supplement.

CONTROL CENTERS

ne Launch Control Center (LCC), located at Kennedy Space Center, Florida, is the focal point for overall rection, control and monitoring of pre launch checkout, countdown and launch of Apollo/Saturn V Space hicles. The Mission Control Center (MCC), located at the Manned Spacecraft Center in Houston, Texas, ovides centralized mission control from the time the space vehicle clears the launch tower through tronaut and spacecraft recovery. The MCC functions within the framework of a Communications, ommand, and Telemetry System (CCATS); Real-Time Computer Complex (RTCC); Voice Communications stem; Display Control System; and a Mission Operations Control Room (MOCR) supported by Staff pport Rooms (SSR's). These system allow the flight control personnel to remain in contact with the space hicle, receive telemetry and operational data which can be processed by the CCATS and RTCC for rification of a safe mission, or compute alternatives. The MOCR and SSR's are staffed with specialists in all pects of the mission who provide the Flight Director and Mission Director with real-time evaluation of ission progress.

MANNED SPACE FLIGHT NETWORK

he Manned Space Flight Network (MSFN) is a worldwide communications and tracking network that is ntrolled by the MCC during Apollo missions. The network is composed of fixed stations supplemented by obile stations. The functions of these stations are to provide tracking, telemetry, updata, and voice mmunications between the acecraft and the MCC. Connection tween these many MSFN stations d the MCC is provided by the ASA Communications Network. gure 13 depicts communications ring lunar surface operations. gure 14 shows the MSFN nfiguration for Apollo 13.

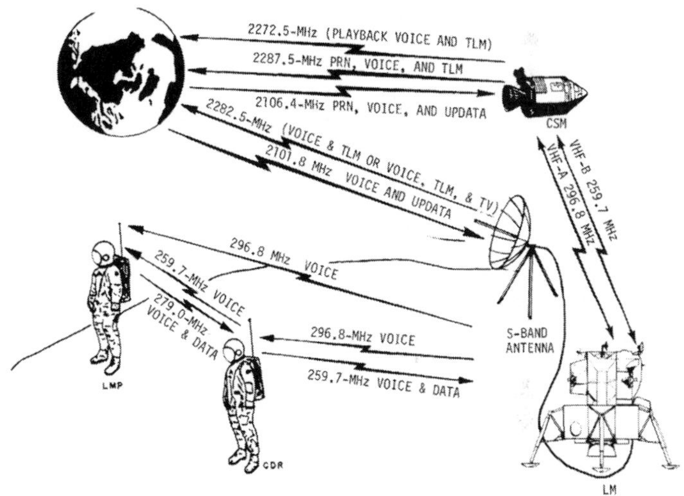

LUNAR SURFACE COMMUNICATIONS NETWORK

RECOVERY SUPPORT

eneral

he Apollo 13 flight crew and ommand Module (CM) will be covered as soon as possible after nding, while observing the nstraints required to maintain ological isolation of the flight crew, M, and materials removed from the M. After locating the CM, first nsideration will be given to etermining the condition of the tronauts and to providing first-level edical aid if required. The second nsideration will be recovery of the tronauts and CM. Retrieval of the M main parachutes, apex cover, and rogue parachutes, in that order, is ghly desirable if feasible and ractical. Special clothing, procedures, d the Mobile Quarantine Facility MQF) will be used to provide ological isolation of the astronauts d CM. The lunar soil and rock

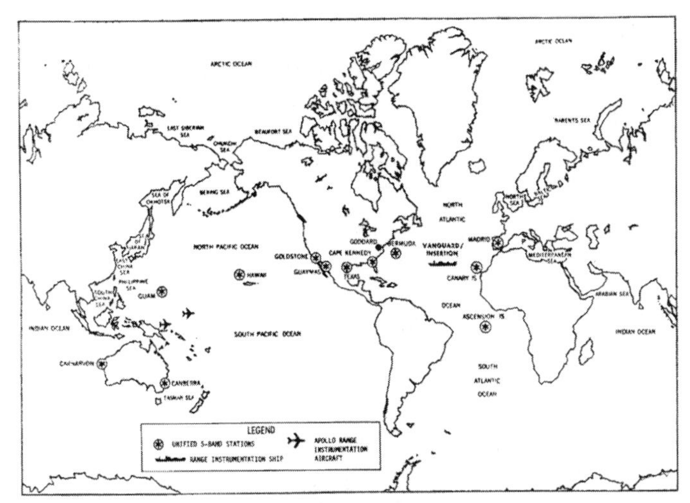

MANNED SPACE FLIGHT NETWORK (APOLLO 13)

samples will also be isolated for return to the Manned Spacecraft Center.

Primary Landing Area

The primary landing area, shown in Figure 15, is that area in which the CM will land following circumlunar or lunar orbital trajectories that are targeted to the mid-Pacific Ocean. The target point will normally be 1250 nautical miles (NM) downrange of the entry point (400,000 feet altitude). If the entry range is increased to avoid bad weather, the area moves along with the target point and contains all the high probability landing points as long as the entry range does not exceed 3500 NM.

Figures 16 and 17 show the primary landing area and worldwide recovery forces deployment. Recovery equipment and procedures changes for Apollo 13 are as follows:

- Recovery beacon CM antennas switched.
- Sea dye inside CM deployed on request.
- Navy type "Mae West" astronaut egress lifejacket.
- New design recovery life raft.

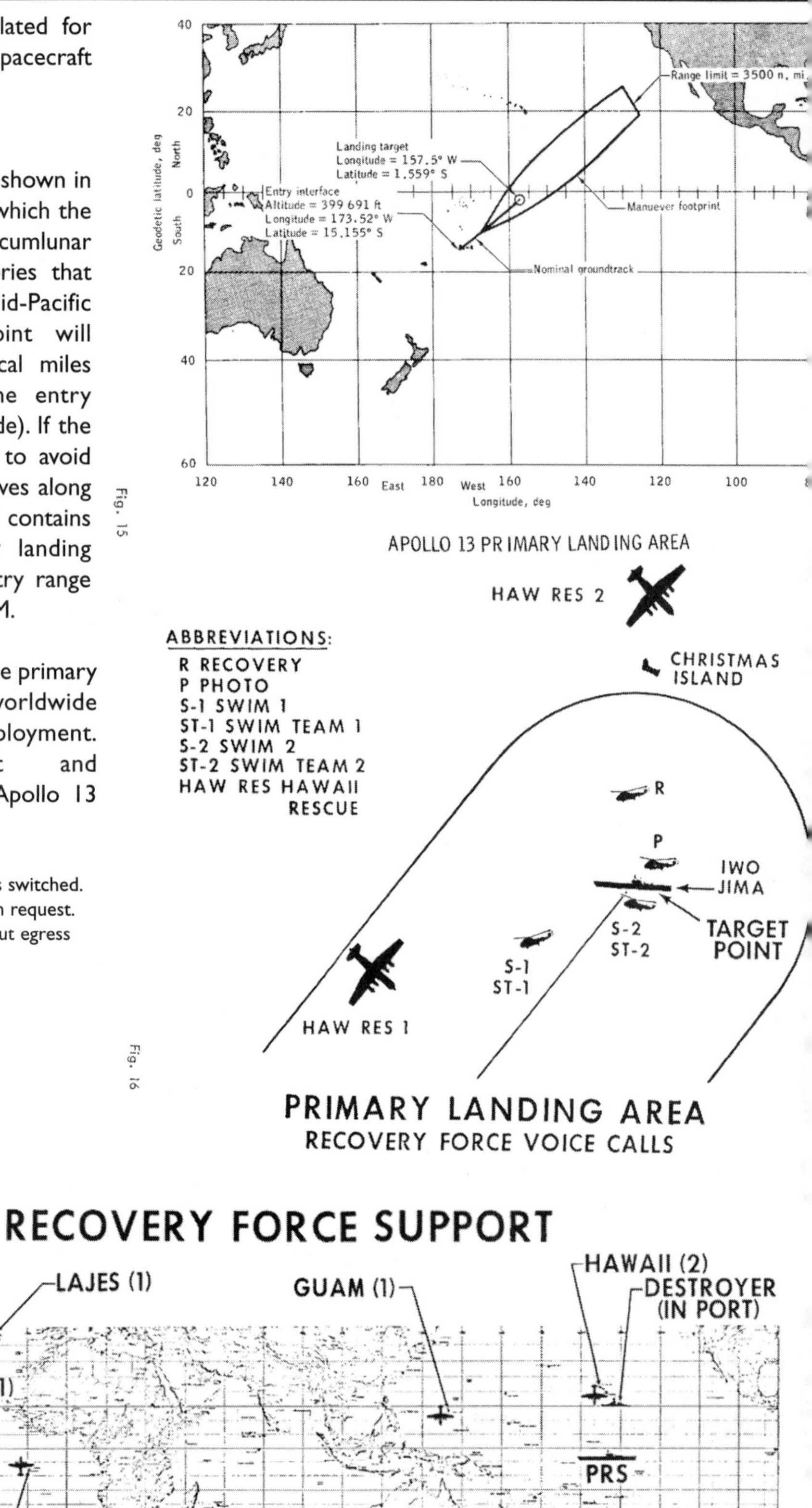

Fig. 15

APOLLO 13 PRIMARY LANDING AREA

Fig. 16

PRIMARY LANDING AREA
RECOVERY FORCE VOICE CALLS

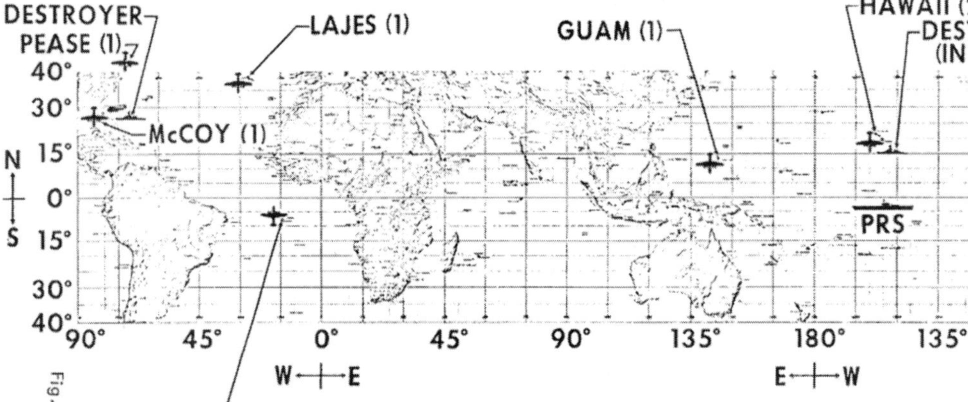

Fig. 17

CONFIGURATION DIFFERENCES

ACE VEHICLE	REMARKS
ommand/Service Module (CSM-109)	
Changed Service Module JettisonController (SMJC) timers.	Allows additional SM movement away from CM prior to RCS termination during separation provides decreased probability of CM/SM recontact during earth entry.
Added Lunar Topographic Camera System.	Provides for high resolution lunar topographic photography with image motion compensation.
Added cabin fan filter.	Removes lunar dust.
unar Module (LM-7) (Ascent Stage	
Added "auto" and "attitude hold" modes in P66 and iminated P65 and P67 software program.	Aids crew during lunar landing in obscured visibility,
Incorporated a non-clogging flow limiter in primary nvironmental Control System lithium hydroxide canister.	Improves crew comfort by eliminating water in suit loop.
unar Module (LM-7) (Descent Stage)	
Installed heat exchanger bypass line on descent stage fuel line.	Facilitates planned depressurization of fuel tanks after landing.
pacecraft-LM Adapter (SLA-16)	
(No significant differences.)	
AUNCH VEHICLE	
strument Unit (S-IU-508)	
Added fourth battery to IU.	Extends Command Communications System tracking to assist S-IVB/IU lunar impact trajectory corrections.
Relocated and added telemetry measurements for vibration vestigation of ST-124 inertial platform.	Provides data for analysis if a flight anomaly occurs on ST- 124.
Added four wires to IU Emergency etection System distributor.	Provides automatic vehicle ground command capability at spacecraft separation in event of a contingency separation.
IVB Stage (S-IVB-508)	
(No significant differences.)	
II Stage (S-11-8)	
Installed all spray foam insulation.	Reduces weight of the stage.
IC Stage (S-IC-	
(No significant differences.)	

FLIGHT CREW

LIGHT CREW ASSIGNMENTS

rime Crew (Figure 18)

ommander (CDR)	—	James A. Lovell, Jr. (Captain, USN)
ommand Module Pilot (CMP)	—	Thomas K. Mattingly, II (Lieutenant Commander, USN)
unar Module Pilot (LMP)	—	Fred W. Haise, Jr. (Civilian)

ackup Crew

ommander (CDR)	—	John W. Young (Commander USN)
ommand Module Pilot (CMP)	—	John L. Swigert, Jr. (Civilian)
unar Module Pilot (LMP)	—	Charles M. Duke, Jr. (Major, USAF)

he backup crew follows closely the training schedule for the prime crew and functions in three significant

Fig. 18

PRIME CREW OF SEVENTH MANNED APOLLO MISSION
JAMES A. LOVELL, JR. THOMAS K. MATTINGLY, II FRED W. HAISE, JR.

categories. One, they receive nearly complete mission training which becomes a valuable foundation for late assignments as a prime crew. Two, should the prime crew became unavailable, the backup crew is prepare to fly as prime crew up until the last few weeks prior to launch. Three, they are fully informed assistants wh help the prime crew organize the mission and check out the hardware.

During the final weeks before launch, the flight hardware and software, ground hardware and software, an flight crew and ground crews work as an integrated team to perform ground simulations and other tests o the upcoming mission. It is necessary that the flight crew that will conduct the mission take part in thes activities, which are not repeated for the benefit of the backup crew. To do so would add an additional costl and time consuming period to the pre launch schedule, which for a lunar mission would require reschedulin for a later lunar launch window.

PRIME CREW DATA

Commander

NAME: James A. Lovell, Jr. (Captain, USN)

SPACE FLIGHT EXPERIENCE: Captain Lovell was selected as an astronaut by NASA in September 1962. He has since served as backu pilot for the Gemini 4 flight and backup command pilot for the Gemini 9 flight.

On 4 December 1965, he and command pilot Frank Borman were launched on the Gemini 7 Mission. The flight lasted 330 hours, 3 minutes and included the first rendezvous of two manned maneuverable spacecraft as Gemini 7 was joined in orbit by Gemini 6.

As command pilot, Lovell flew the 4-day, 59-revolution, Gemini 12 Mission in November 1966. This flight, which included a thir revolution rendezvous with a previously launched Agena, marked the successful completion of the Gemini Program.

Lovell served as Command Module Pilot on the 6-day Apollo 8 (21-27 December 1968) first manned flight to the moon. Apollo performed 10 revolutions in lunar orbit and returned to an earth landing after a total flight time of 147 hours.

Captain Lovell has since served as the backup spacecraft Commander for Apollo 11, the first manned lunar landing. He has logged a tot of 572 hours, 10 minutes of space flight in three missions.

mmand Module Pilot

ME: Thomas K. Mattingly II (Lieutenant Commander, USN)

ACE FLIGHT EXPERIENCE; Lieutenant Commander Mattingly was selected as an astronaut by NASA in April 1966. He has served as member of the astronaut support crews for the Apollo 8 and 11 Missions.

nar Module Pilot

ME: Fred W. Haise Jr. (Civilian)

ACE FLIGHT EXPERIENCE: Mr. Haise was selected as an astronaut by NASA in April 1966. He has served as a member of the ronaut backup crew for the Apollo 8 and 11 Missions.

BACKUP CREW DATA

mmander

ME: John. W. Young (Commander, USN)

ACE FLIGHT EXPERIENCE: Commander Young was selected as an astronaut by NASA in September 1962. He served as pilot on the st manned Gemini flight — a 3 orbit mission launched on 23 March 1965. Following that assignment he was backup pilot for Gemini

18 July 1966, Young was the command pilot on the Gemini 10 Mission which made two successful rendezvous' and dockings with ena target vehicles. Gemini 10 was a 3-day, 44 revolution earth orbital flight.

was subsequently assigned as the backup Command Module Pilot for Apollo 7.

mmander Young flow as the Command Module Pilot on the Apollo 10 lunar orbital mission which performed all but the final minutes an actual lunar landing.

mmand Module Pilot

ME: John L. Swigert, Jr. (Civilian)

ACE FLIGHT EXPERIENCE: Mr. Swigert was selected as an astronaut by NASA in April 1966.

nar Module Pilot

ME: Charles M. Duke, Jr. (Major, USAF)

ACE FLIGHT EXPERIENCE: Major Duke was selected as an astronaut by NASA in April 1966.

MISSION MANAGEMENT RESPONSIBILITY

tle	Name	Organization
rector, Apollo Program	Dr. Rocco A. Petrone	NASA/OMSF
ssion Director	Capt. Chester M. Lee (Ret)	NASA/OMSF
sistant Mission Director	Col. Thomas H. McMullen	NASA/OMSF
rector, Mission Operations	Maj. Gen. John O. Stevenson (Ret)	NASA/OMSF
turn Program Manager	Mr. Roy E. Godfrey	NASA/MSFC
ssion Operations Manager	Dr. Fridtjof A. Speer	NASA/MSFC
pollo Spacecraft Program Manager	Col. James A. McDivitt	NASA/MSC
rector of Flight Operations	Mr. Sigurd A. Sjoberg	NASA/MSC
ght Directors	Mr. Milton Windler	NASA/KSC
	Mr. Gerald D. Griffin	
	Mr. Eugene F. Kranz	
	Mr. Glynn S. Lunney	
acecraft Commander (Prime)	Capt. James A. Lovell	NASA/MSC
acecraft Commander (Backup)	Cdr. John W. Young	NASA/MSC
pollo Program Manager KSC	Mr. Edward R. Mathews	NASA/KSC
rector of Launch Operations	Mr. Walter J. Kapryan	NASA/KSC
unch Operations Manager	Mr. Paul C. Donnelly	NASA/KSC

Post Launch Mission Operation Report

No. M-932-70-

TO: A/Administrator 28 April 1970

FROM: MA/Apollo Program Director
SUBJECT: Apollo 13 Mission (AS-508) Post Launch Mission Operation Report No. 1

The Apollo 13 Mission was successfully launched from Kennedy Space Center, Florida on Saturday, 11 Ap 1970. Apollo 13 was progressing smoothly to a planned lunar landing until about 56 hours into the flight wh a failure occurred in the Service Module cryogenic oxygen system. This resulted in a loss of capability generate electrical power, to provide oxygen, and to produce water in the Command/Service Module. T decision was made to not perform the lunar landing mission and to return to earth using the Lunar Modu for life support, power, propulsion, and guidance. Safe recovery of the crew and Command Module took pla in the Pacific Ocean recovery area on Friday, 17 April 1970. An intensive investigation has been initiated determine the cause of the anomaly.

The Mission Director's Summary Report for Apollo 13 is attached and submitted as Post Launch Missi Operation Report No. 1. Also attached are the NASA OMSF Primary Mission Objectives for Apollo 13. Sin these Primary Objectives could not be achieved without a lunar landing, I am recommending that the Apo 13 Mission be considered unsuccessful. Detailed analysis of all data will continue and appropriate refin results of the mission will be reported in the Manned Space Flight Centers technical reports.

Rocco A. Petrone

APPROVAL:

Dale D. Myers
Associate Administrator for
Manned Space Flight

NASA OMSF PRIMARY MISSION OBJECTIVES FOR APOLLO 13

PRIMARY OBJECTIVES

Perform selenological inspection, survey, and sampling of materials in a pre-selected region of the Fra Mau Formation.

Deploy and activate an Apollo Lunar Surface Experiments Package (ALSEP).

Develop man's capability to work in the lunar environment.

Obtain photographs of candidate exploration sites.

Rocco Petrone
Apollo Program Director

Date: 24 March 1970

Dale D. Myers
Associate Administrator
Manned Space Flight
Date: March 28 1970

RESULTS OF APOLLO 13 MISSION

pollo 13, launched 11 April 1970, was aborted after 56 hours of flight and terminated on 17 April 1970. The anned lunar landing was not accomplished and this mission is adjudged unsuccessful in accordance with the bjectives stated above.

occo A. Petrone	*Dale Myers*
pollo Program Director	Associate Administrator
	Manned Space Flight
ate: 24 April 1970	Date: Apr 28, 1970

) April 1970
O: Distribution
ROM: MA/ Apollo Mission Director
JBJECT: Mission Director's Summary Report, Apollo 13

INTRODUCTION

he Apollo 13 Mission was planned as a lunar landing mission but was aborted en route to the moon during the third ay of flight due to loss of Service Module cryogenic oxygen and consequent loss of capability to generate electrical ower, to provide oxygen, and to produce water in the Command/Service Module. Shortly after the anomaly, the ommand/Service Module was powered down and the remaining flight, except for entry, was made with the Lunar lodule providing necessary power, environmental control, guidance, and propulsion. Flight crew members were: ommander (CDR), Capt. James Lovell, Jr.; Command Module Pilot (CMP), Mr. John Swigert, Jr.; Lunar Module Pilot (LMP), 1r. Fred W. Haise, Jr. Swigert, officially the backup CMP for the Apollo 13 Mission, was substituted for LCDR Thomas K. lattingly II, the prime crew CMP, when it was feared that Mattingly had possibly contracted Rubella, and if so, could be dversely affected in performing his demanding duties. A vigorous simulation program was successfully completed prior o launch to ensure that Lovell, Swigert and Haise could function with unquestioned teamwork through even the most rduous and time-critical simulated emergency conditions. Significant detailed mission data are contained in Tables 1 hrough 4.

PRELAUNCH

lo problems occurred during space vehicle prelaunch operations to impact the countdown. However, the S-IC Stage No. liquid oxygen (LOX) vent valve did not close when commanded at T minus 1 hour 58 minutes. After cycling the valve everal times and flowing ambient nitrogen gas through the valve, it was successfully closed at T minus 1 hour 21 minutes. Veather conditions at launch were: overcast at 20,000 feet, visibility 10 miles, wind 10 knots.

LAUNCH AND EARTH PARKING ORBIT

pollo 13 was successfully launched on schedule from Launch Complex 39A, Kennedy Space Center, Florida, at 2:13 p.m. ST, 11 April 1970. The launch vehicle stages inserted the SIVB/Instrument Unit (IU)/spacecraft combination into an earth arking orbit with an apogee of 100.2 nautical miles (NM) and a perigee of 98.0 NM (100-NM circular planned). During econd stage boost the center engine of the S-II Stage cut off about 132 seconds early causing the remaining four engines o burn approximately 34 seconds longer than predicted. Space vehicle velocity after S-II boost was 223 feet per second fps) lower than planned. As a result, the S-IVB orbital insertion burn was approximately 9 seconds longer than predicted vith cutoff velocity within about 1.2 fps of planned. Total launch vehicle burn time was about 44 seconds longer than redicted. A greater than 3-sigma probability of meeting translunar injection (TLI) cutoff conditions existed with emaining S-IVB propellants.

fter orbital insertion, all launch vehicle and spacecraft systems were verified and preparations were made for TLI.)nboard television was initiated at 01:35 GET (hour: minutes ground elapsed time) For about 5.5 minutes. The second -IVB burn was initiated on schedule for TLI. All major systems operated satisfactorily and all end conditions were ominal for a free-return circumlunar trajectory.

TRANSLUNAR COAST

he Command/Service Module (CSM) separated from the Lunar Module (LM)/IU/S-IVB at about 03:07 GET. Onboard television was then initiated for about 72 minutes and clearly showed CSM "hard docking," ejection of the CSM/LM fro the S-IVB/IU at about 04:01 GET, and the S-IVB Auxiliary Propulsion System (APS) evasive maneuver as well as spacecra interior and exterior scenes. Service Module Reaction Control System (SM RCS) propellant usage for the separatio transposition, docking, and ejection was nominal. All launch vehicle safing activities were performed as scheduled.

The S-IVB APS evasive maneuver by an 8-second APS ullage burn was initiated at 04:18 GET and was successful completed. The LOX dump was initiated at 04:39 GET and was also successfully accomplished. The first S-IVB APS bu for lunar target point impact was initiated at 06:00 GET. The burn duration was 217 seconds producing a different velocity of approximately 28 fps. Tracking information available at 08:00 GET indicated that the S-IVB/IU would impact 6°53'S, 30°53'W versus the targeted 3°S, 30°W. Therefore, the second S-IVB APS (trim) burn was not required. Th gaseous nitrogen pressure dropped in the IU ST-124-M3 inertial platform at 18:25 GET and the S-IVB/IU no longer ha attitude control but began tumbling slowly. At approximately 19:17 GET, a step input in tracking data indicated a veloci increase of approximately 4 to 5 fps. No conclusions have been reached on the reason for this increase. The veloci change altered the lunar impact point closer to the target. The S-IVB/IU impacted the lunar surface at 77:56:40 GE (08:09:40 p.m. EST, 14 April) at 2.4°S, 27.9°W and the seismometer deployed during the Apollo 12 Mission successful detected the impact (see "MISSION SCIENCE"). The targeted impact point was 125 NM from the seismometer. Th actual impact point was 74 NM from the seismometer, well within the desired 189-NM (350-kilometer) radius.

The accuracy of the TLI maneuver was such that spacecraft midcourse correction No. 1 (MCC-1), scheduled for 11:4 GET, was not required. MCC-2 was performed as planned at 30:41 GET and resulted in placing the spacecraft on th desired, non-free-return circumlunar trajectory with a predicted closest approach to the moon of 62 NM. All Servic Propulsion System (SPS) burn parameters were normal . The accuracy of MCC-2 was such that MCC-3, scheduled fo 55:26 GET, was not performed. Good quality television coverage of the preparations and performance of MCC-2 wa received for 49 minutes beginning at 30:13 GET.

At approximately 55:55 GET (10:08 p.m. EST) the crew reported an under voltage alarm on the CSM Main Bus Pressure was rapidly lost in Service Module oxygen tank No. 2 and fuel cells 1 and 3 current dropped to zero due to los of their oxygen supply. A decision was made to abort the mission. The increased load on fuel cell 2 and decaying pressur in the remaining oxygen tank led to the decision to activate the LM, power down the CSM, and use the LM systems fo life support.

At 61:30 GET, a 38-fps midcourse maneuver (MCC-4) was performed by the LM Descent Propulsion System (DPS) t place the spacecraft in a free-return trajectory on which the Command Module (CM) would nominally land in the India Ocean south of Mauritius at approximately 152:00 GET.

TRANSEARTH COAST

At pericynthion plus 2 hours (79:28 GET), a LM DPS maneuver was performed to shorten the return trip time and mov the earth landing point. The 263.4-second burn produced a differential velocity of 860.5 fps and resulted in an initia predicted earth landing point in the mid-Pacific Ocean at 142:53 GET. Both LM guidance systems were powered up an the primary system was used for this maneuver. Following the maneuver, passive thermal control was established and th LM was powered down to conserve consumables; only the LM Environmental Control System (ECS) and communication and telemetry systems were kept powered up.

The LM DPS was used to perform MCC-5 at 105:19 GET. The 15-second burn (at 10% throttle) produced a velocit change of about 7.8 fps and successfully raised the entry flight path angle to -6.52°.

The CSM was partially powered up for a check of the thermal conditions of the CM with first reported receipt of S-ban signal at 101:53 GET. Thermal conditions on all CSM systems observed appeared to be in order for entry.

Due to the unusual spacecraft configuration, new procedures leading to entry were developed and verified i ground-based simulations. The resulting timeline called for a final midcourse correction (MCC-7) at entry interface (E -5 hours, jettison of the SM at EI -4.5 hours, then jettison of the LM at EI -1 hour prior to a normal atmospheric entr by the CM.

MCC-7 was successfully accomplished at 137:40 GET. The 22.4-second LM RCS maneuver resulted in a predicted entr flight path angle of -6.49°. The SM was jettisoned at 138:02 GET. The crew viewed and photographed the SM and reporte that an entire panel was missing near the S-band high-gain antenna and a great deal of debris was hanging out. The CM was powered up and then the LM was jettisoned at 141:30 GET. The EI at 400,000 feet was reached at 142:41 GET.

ENTRY AND RECOVERY

/eather in the prime recovery area was as follows: Broken stratus clouds at 2000 feet; visibility 10 miles; 6-knot ENE inds; and wave height 1 to 2 feet. Drogue and main parachutes deployed normally. Visual contact with the spacecraft as reported at 142:50 GET. Landing occurred at 142:54:41 GET (01:07:41 p.m. EST, 17 April). The landing point was in e mid-Pacific Ocean, approximately 21°40'S, 165°22'W. The CM landed in the stable 1 position about 3.5 NM from the ime recovery ship, USS IWO JIMA. The crew was picked up by a recovery helicopter and was safe aboard the ship at 53 p.m. EST, less than an hour after landing.

MISSION SCIENCE

he S-IVB Stage, weighing about 30,700 pounds, impacted the moon 74 NM from the Apollo 12 seismometer at an angle f about 80° to the horizontal with a velocity of 8465 fps and an energy equivalency of 11.5 tons of TNT. These data ompare with the Apollo 12 LM, which hit the moon at a distance of 42 NM from the seismometer at an angle of 3° to e horizontal, and an equivalent energy of approximately 1 ton of TNT.

he overall character of the seismic signal is similar to that of the LM impact signal, but the higher impact energy gave a eismic signal 20-30 times larger than the LM impact and 4 times longer in duration (approximately 4 hours vs. 1 hour). he signal was so large that the gain of the seismometer had to be reduced by ground command to keep the recording n scale. A clear signal was recorded on the three long period components so that it is possible to distinguish each event ith absolute certainty. Thirty seconds elapsed between time of impact and arrival of the seismic wave at the eismometer; peak amplitude occurred 7 minutes later.

he signal arrival time had been predicted on the basis of velocity measurements made on the Apollo 11 and 12 lunar ample materials in the laboratory. The average velocity of the seismic wave through the lunar material is 4.6 km/sec which compares favorably with the 3.2-km/sec velocity recorded by the LM impact. The depth of penetration of the S-IVB npact signal is believed to be 20-40 km (Vs 20 km for LM impact). This result implies that the outer shell of the moon, o depths of at least 20-40 km, may be formed of the same crystalline rock material as found at the surface. No evidence f a lower boundary to this material has been found in the seismic signal, although it is clear that it is too dense to form ne entire moon.

One puzzling feature of the signal is the unexpectedly rapid build-up from the beginning to its maximum. This part of the ignal, at least, cannot be satisfactorily explained by scattering of seismic waves in a rubble material as was thought ossible from the earlier LM impact data. Scattering of signals may explain the later part of the signal. Several alternate ypotheses are under study, but no firm conclusions have been reached. One possibility is that the expanding cloud of naterial from the impact produces seismic signals continuously as it sweeps across the lunar surface.

he fact that such precise targeting accuracy was achieved for the S-IVB and that the resulting seismic signals were so arge have greatly encouraged scientists to believe that planned future impacts can be extended to ranges of at least 500 m and that the data return will provide the means for determining the structure of the moon to depths approaching 00 km.

he Suprathermal Ion Detector Experiment (SIDE), also part of the ALSEP 1 experiments package deployed during the pollo 12 Mission, recorded a jump in the number of ion counts after the S-IVB impact. Since the instrument was in lunar hadow at the time of impact the ion count was essentially zero. A few ions were recorded 22 seconds after impact; a econd frame of data showed a jump to 250 ions, the third jumped to 2500 ions, the fourth dropped back to a few ions, hen the count fell back essentially to zero. These ions were in the 70 electron volt energy range. All of the counts were bserved over a period of 70 seconds. In addition to the ion counts, the mass analyzer of the instrument also recorded ons, almost all of which were in the 50-80 mass unit range (hydrogen = 1 mass unit).

Two possible mechanisms have been given for producing ions: (1) temperatures in the ranges 6000-10,000°C generated y the S-IVB impact could produce ionization; (2) particles that reach heights of 60 km could also be ionized by sunlight.

SYSTEMS PERFORMANCE

aturn V S-IC ignition, hold down arm release, and liftoff were accomplished within expected limits and indications are hat S-IC systems performed at or near nominal. LOX tank pressure was as expected throughout the burn.

All S-II Stage systems were nominal throughout S-II burn until the center J-2 engine shut down approximately 132 econds earlier than scheduled. Low frequency oscillations (14 to 16 hertz) experienced on the S-II Stage resulted in a 32-second premature center engine cutoff. Preliminary analysis indicates that a "Thrust OK" pressure switch cutoff occurred due to large pressure oscillations in the LOX system. No apparent engine or structural damage was incurre Oscillations in the stage and outboard engines decayed to a normal level following center engine cutoff. Preliminary da does not indicate any off-nominal performance of the four outboard engines.

All S-IVB systems operated within expected limits during both the first and second burns. The first burn was 9.2 secon longer than predicted, making up for the velocity deficit at S-II cutoff, The second (TLI) burn was approximately 5 secon longer than predicted from observed orbital conditions. A small vibration was reported by the crew approximately seconds prior to second burn cutoff.

All IU guidance and control functions were satisfactory and all systems performed as expected.

Performance of the CSM fuel cell and cryogenic systems was nominal until 55:53:36 GET when an unusual pressure ri was noted in oxygen tank No. 2. The pressure continued to rise to the relief valve crack pressure of 1004.1 psia (poun per square inch absolute). One second later 55:54:45 GET, the pressure reached a maximum of 1008.3 psia at which tin the tank vent valve apparently opened. The last valid tank pressure reading prior to loss of data was 995.7 psia at 55:54: GET. At 55:54:54 GET an under-voltage caution light occurred on Main Bus B, which was powered by fuel cell Concurrent with the abrupt loss of oxygen tank No. 2 pressure, oxygen tank No. 1 pressure showed a rapid decrease t about 373 psia in 87 seconds. Fuel cells 1 and 3 were removed from the line about 18 minutes after the anomaly. Fu cell 2 remained in operation for about 2 hours before the oxygen pressure in tank No. 1 had decreased to 61 psia ar the fuel cell was removed from the line. As a result of these occurrences, the CM was powered down and the LM w configured to supply the necessary power and other consumables.

Power down of the CSM began at 58:40 GET. The surge tank and repressurization package were isolated wi approximately 860 psi residual pressure (approximately 6.5 pounds of oxygen total). The primary water glycol syste was left with radiators bypassed. Indicated water tank residuals were 18.0 pounds in the waste tank and 37.5 pounds the potable tank. All SM RCS quads were powered down with heaters deactivated. All SPS parameters were nomin before and after the anomaly and no configuration changes took place after the anomaly.

All LM systems performed satisfactorily in providing the necessary power and environmental control to the spacecra The requirement for lithium hydroxide to remove carbon dioxide from the spacecraft atmosphere was met by combination of CM and LM cartridges since the LM cartridges alone would not satisfy the total requirement.

The crewmen, with instructions from Mission Control, built an adapter from the CM cartridges to accept the LM hose

The LM supercritical helium (SHe) tank pressure exhibited an increased rise rate after the second DPS firing. Prior t the burn, the rise rate had been 11 psi per hour. After the burn, the rate increased to 33 psi per hour. After the thir DPS maneuver, the SHe tank burst disc ruptured at 108:54 GET at a pressure of about 1940 psi, within the expecte range. The Passive thermal control mode in use at the time was affected by a small attitude rate change from the ventir SHe changing from a right yaw rate of 0.3°/sec to a left yaw rate of 3.0°/sec, but did not cause any problem.

The CSM was partially powered up at about 101:53 GET with the following results:

Telemetry — following a brief period of intermittent S-band reception, solid telemetry was received from 101:49 GE to system power-down at 102:03 GET. Telemetry system performance was nominal throughout the time period it wa powered up.

Instrumentation — A summary review indicated no discrepancies. The central timing equipment updated correctly i resetting to 0 (zero) and indicating accumulated time from the turn-on associated with the status check of the CSM a 101:53 GET.

Electrical Power — All system bus voltage and inverter performance was nominal. Only Main Bus B, Battery Bus B, an AC Bus 1 were powered up. Prior to instrumentation power-up, the three entry batteries had been on "true" open circu (i.e., no parasitic loads) since approximately 58:40 GET. All performance to that point had been nominal. CSM Main Bu B was powered up using Battery B and performance under load was nominal. Approximately 2.5 ampere-hours wer consumed. Battery A, which was used to supplement CM power immediately following the fuel cell anomaly, wa recharged from the LM ascent batteries. Battery B was also recharged.

Displays and Controls — No discrepancies noted.

Thermal/Propulsion — CM RCS helium tank temperatures were approximately as predicted with one about 4°F highe than predicted. SM RCS engine package and RCS and SPS temperatures indicated satisfactory passive thermal control

A CSM RCS engine heat-up procedure was required prior to separating the LM/CM combination from the SM. Other data available indicated the CM RCS system was nominal. All SPS parameters remained nominal during the powered-down portion of the flight. The oxidizer and fuel tank pressures decreased 6 psi each after the CSM was powered down, which can be attributed to helium absorption.

CREW PERFORMANCE

The Apollo 13 flight crew performance was outstanding throughout the flight. Most noteworthy was their calm, precise reaction to the emergency situation and their subsequent diligence in configuring and maintaining the LM for safe return to earth. Despite lack of adequate sleep and the low temperature in the spacecraft, neither their performance nor their spirits ever faltered throughout the flight. Similarly, the flight operations team exhibited outstanding performance throughout the flight in planning and aiding the crew to a safe return.

All information and data in this report are preliminary and subject to revision by the normal Manned Space Flight Center technical reports.

C.M. Lee

TABLE I

APOLLO 13 LAUNCH VEHICLE SEQUENCE OF EVENTS

EVENT	*PLANNED (GET) HR:MIN:SEC	ACTUAL (GET) HR:MIN:SEC
Range Zero (02:13:00.0 p.m. EST, 11 April)	00:00:00.0	00:00:00.0
Liftoff Signal (Timebase 1)	00:00:00.7	00:00:00.6
Pitch and Roll Start	00:00:12.5	00:00:12.6
Roll Complete	00:00:30.4	00:00:32.1
S-IC Center Engine Cutoff (TB-2 minus .1 sec)	00:02:15.3	00:02:15.2
Begin Tilt Arrest	00:02:42.0	00:02:43.3
S-IC Outboard Engine Cutoff (TB-3)	00:02:44.0	00:02:43.6
S-IC/S-II Separation	00:02:44.7	00:02:44.3
S-II Ignition (Command)	00:02:45.4	00:02:45.0
S-II Second Plane Separation	00:03:14.7	00:03:14.3
Launch Escape Tower Jettison	00:03:20.4	00:03:20.0
S-II Center Engine Cutoff	00:07:43.0	00:05:30.6
S-II Outboard Engine Cutoff (TB-4 minus .1 sec)	00:09:18.1	00:09:52.6
S-II/S-IVB Separation	00:09:19.0	00:09:53.5
S-IVB Ignition	00:09:22.1	00:09:56.9
S-IVB Cutoff (TB-5 minus .2 sec)	00:11:45.8	00:12:29.8
Earth Parking Orbit insertion	00:11:55.8	00:12:39.8
Begin Restart Preparation (TB-6)	02:25:49.9	02:26:08.1
Second S-IVB ignition	02:35:27.9	02:35:46.4
Second S-IVB Cutoff (TB-7 minus .2 sec)	02:41:23.6	02:41:37.2
Translunar Injection	02:41:33.6	02:41:47.2
CSM/S-IVB Separation	03:06:27.8	03:06:36.9
Spacecraft Ejection from S-IVB	04:01:23.8	04:01:03.0
S-IVB APS Evasive Maneuver	04:19:25.0	04:18:00.5
LOX Dump	04:40:43.8	04:39:19.3
S-IVB APS Maneuver for Lunar Impact	06:00:00.0	05:59:59.0
S-IVB Lunar Impact	77:48:32.0	77:56:40.0

*Prelaunch planned times are based on MSFC launch vehicle operational trajectory

TABLE 2

APOLLO 13 MISSION SEQUENCE OF EVENTS

EVENT	GROUND ELAPSED TIME (HR:MIN:SEC)
Range Zero (02:13:00.0 p.m. EST, 11 April)	00:00:00
Earth Parking Orbit Insertion	00:12:40
Second S-IVB Ignition	02:35:46
Translunar Injection	02:41:47
CSM/S-IVB Separation	03:06:39
Spacecraft Ejection from S-IVB	04:01:03
S-IVB APS Evasive Maneuver	04:18:01
S-IVB APS Maneuver for Lunar Impact	05:59:59
Midcourse Correction - 2 (Hybrid Transfer)	30:40:50
Liquid Oxygen Tank Anomaly	55:54:53
Midcourse Correction - 4	61:29:43
S-IVB Lunar Impact	77:56:40
Pericynthion Plus 2 Hour Maneuver	79:27:39
Midcourse Correction - 5	105:18:32
Midcourse Correction - 7	137:39:49
Service Module Jettison	138:02:06
Lunar Module Jettison	141:30:02
Entry Interface	142:40:47
Landing	142:54:41

TABLE 3

APOLLO 13 TRANSLUNAR AND TRANSEARTH MANEUVER SUMMARY

MANEUVER	GROUND ELAPSED TIME (GET) AT IGNITION (hr:min:sec)			BURN TIME (seconds)			VELOCITY CHANGE (feet per second - fps)			GET OF CLOSEST APPROACH / HT. (NM) CLOSEST APPROACH		
	PRE-LAUNCH	REAL-TIME	ACTUAL	PRE-LAUNCH PLAN	REAL-TIME PLAN	ACTUAL	PRE-LAUNCH PLAN	REAL-TIME PLAN	ACTUAL	PRE-LAUNCH PLAN	REAL-TIME PLAN	* ACTUAL
TLI (S-IVB)	02:35:27.9	Not Available	2:35:46.4	355.7	346	350.7	10,437.1	Not Available	Not Available	77:40:21.9 210	Not Available Not Available	77:51:17 415.8
MCC-1	11:41:23.5	N.A.	N. P.	0.0	N.A.	N. P.	0.0	N.A.	N. P.	77:40:21.9 210	Not Available Not Available	N. P. N. P.
MCC-2 (SPS)	30:40:49.0	30:40:49.0	30:40:50	2.2	3.39	3.37	14.7	23.2	23.1	77:15:00 57.3	77:28:34 60.22	77:28:37 64.87
MCC-3	55:26:02	N.A.	N. P.	0.0	N.A.	N. P.	0.0	N.A.	N. P.	77:15:00 57.3	N. A. N. A.	N. P. N. P.
Nominal mission aborted at this point--remaining maneuvers planned in real time for return to earth										GET entry interface (EI) Velocity (fps) at EI Flight path angle at EI		
MCC-4 (DPS)	N. A.	61:29:42.8	61:29:42.8	N. A.	30.7	30.4	N. A.	38.0	37.8	N. A. N. A. N. A.	151:45:06 36,141.2 -6.53	151:45:27 36,141.1 -6.53
PC+2 (DPS)	N. A.	79:27:38.3	79:27:39	N. A.	263.7	263.4	N. A.	861.5	860.5	N. A. N. A. N. A.	142:38:52 36,209.6 -6.50	142:39:00 36,210.6 -6.53
MCC-5 (DPS)	N. A.	105:30:00	105:18:32	N. A.	15:38	15:38	N. A.	7.8	7.8	N. A. N. A. N. A.	142:40:35 36,211 -6.51	142:40:34 36,210.61 -6.52
MCC-7 (LM RCS)	N. A.	137:39:48.4	137:39:49.4	N. A.	23.2	22.4	N. A.	3.1	2.9	N. A. N. A. N. A.	142:40:40 36,210 -6.50	142:40:41 36,210 -6.49

N. A. - Not Applicable   N. P. - Not Performed   * Actual values are as determined shortly after maneuver.

TABLE 4

APOLLO 13 DISCREPANCY SUMMARY

AUNCH VEHICLE (SA-508)

. EARLY S-II CENTER ENGINE CUTOFF/S-II LOW FREQUENCY OSCILLATIONS.

COMMAND/SERVICE MODULE (CSM-109)

. SUIT PRESSURE TRANSDUCER READING APPROXIMATELY .5 PSI BELOW CABIN PRESSURE.

. POTABLE WATER QUANTITY READING — READING ERRATIC. DROPPED APPROXIMATELY 20% AT APPROXIMATELY 22:42:50 GET, THEN RETURNED TO 100%.

3. 40:00 GET — OPTICS COUPLING DISPLAY UNIT FLUCTUATING 0.16 DEGREES IN ZERO OPTICS MODE.

4. OXYGEN TANK NO. 2 QUANTITY WENT TO OFF-SCALE HIGH AT 46:45 GET, CABIN METER CONFIRMED AS OFF-SCALE HIGH BY THE CREW AT 47:42 GET. PROBLEM OCCURRED AFTER THE CRYOGENIC FANS WERE ACTIVATED.

5. OXYGEN TANK NO. 2 PRESSURE DROPPED TO ZERO PSI AT 55:54:52 GET. OXYGEN TANK NO. 1 PRESSURE BEGAN TO DECAY AT THE SAME TIME, DROPPING 373 PSI IN 1 MINUTE 14 SECONDS.

LUNAR MODULE (LM-7)

. BATTERY 2 SENSOR MALFUNCTION ON CAUTION AND WARNING AT 99:57 GET.

. PROPULSIVE SUPERCRITICAL HELIUM TANK VENT.

3. CHANGE IN SUPERCRITICAL HELIUM PRESSURE RISE AFTER DPS FIRING NO. 1 (FROM 6.9 PSI/HR TO 11.5 PSI/HR) AND AGAIN AFTER DPS FIRING NO. 2 (FROM 11.5 PSI/HR TO 33 PSI/HR).

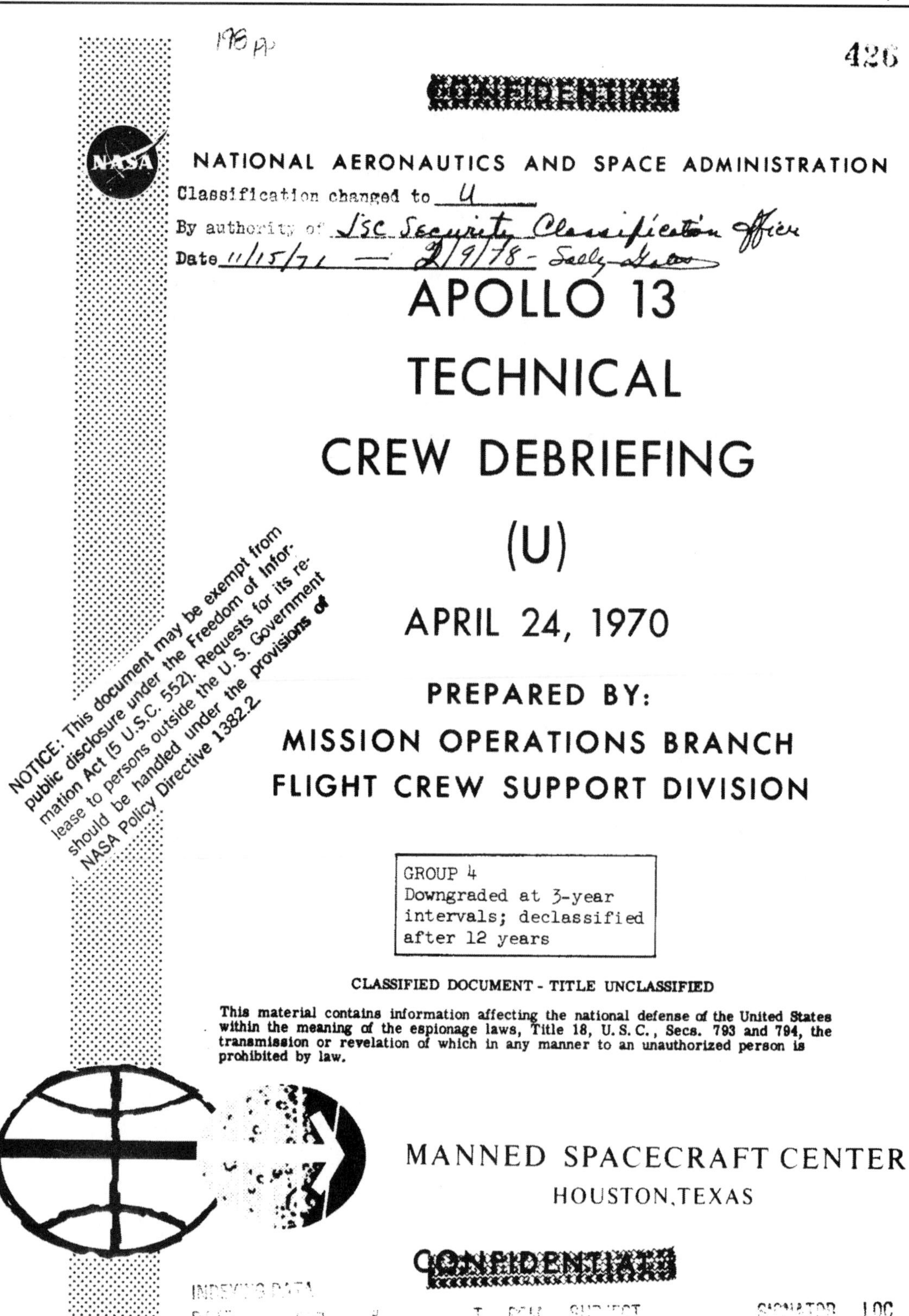

178 AP

426

CONFIDENTIAL

NATIONAL AERONAUTICS AND SPACE ADMINISTRATION

Classification changed to U

By authority of JSC Security Classification Office

Date 11/15/71 — 2/9/78 - Sally Gates

APOLLO 13 TECHNICAL CREW DEBRIEFING (U)

APRIL 24, 1970

PREPARED BY:

MISSION OPERATIONS BRANCH

FLIGHT CREW SUPPORT DIVISION

NOTICE: This document may be exempt from public disclosure under the Freedom of Information Act (5 U.S.C. 552). Requests for its release to persons outside the U.S. Government should be handled under the provisions of NASA Policy Directive 1382.2

GROUP 4
Downgraded at 3-year intervals; declassified after 12 years

CLASSIFIED DOCUMENT - TITLE UNCLASSIFIED

This material contains information affecting the national defense of the United States within the meaning of the espionage laws, Title 18, U. S. C., Secs. 793 and 794, the transmission or revelation of which in any manner to an unauthorized person is prohibited by law.

MANNED SPACECRAFT CENTER
HOUSTON, TEXAS

CONFIDENTIAL

SECURITY CLASSIFICATION The material contained herein has been transcribed into a working paper in order to facilitate review by interested MSC elements. This document, or portions thereof, may be declassified subject to the following guidelines: Portions of this document will be classified CONFIDENTIAL, Group 4, to the extent that they: (1) define quantitative performance characteristics of the Apollo Spacecraft, (2) detail critical performance characteristics of Apollo crew systems and equipment, (3) provide technical details of significant launch vehicle malfunctions in actual flight or reveal actual launch trajectory data, (4) reveal medical data on flight crew members which can be considered privileged data, or (5) reveal other data which can be individually determined to require classification under the authority of the Apollo Program Security Classification Guide, SCG-11, Rev. 1, 1/1/66.

1.0 SUITING AND INGRESS

LOVELL — I thought the time was adequate. I had no particular problems with the suiting or the ingress procedures. The only thing I did notice was that we had our protective visors on a lot longer than I expected.

SWIGERT — We did not get the protective visors off until after we had ingressed and were all strapped down. The fact that we were strapped down made removing the visors difficult. Everything after that went according to the checklist, and we had adequate time.

2.0 STATUS CHECKS AND COUNTDOWN

LOVELL — Ground communications were very good, and the countdown proceeded smoothly. The controls and displays were as shown to us and as we had experienced in the countdown demonstration. I experienced no particular unusual sounds in the launch vehicle sequence before the nominal engine ignition.

70-H-436

2.5 LAUNCH VEHICLE SEQUENCE

SWIGERT — The only thing that both Fred and I noticed was a fluctuation in fuel cell flows. When I switched fuel cells, the flows would be stable for 2 or 3 seconds and then would begin fluctuating. This occurred in all three fuel cells; because it occurred in all three, we attributed it to some sort of signal-conditioning problem. The fluctuations were 1 cycle/sec, wouldn't you say, Fred?

70-H-501

HAISE — I'm not sure they were even that regular. When you switched from one to the other, the reading would first be very stable; then, after a few seconds it would start drifting up a couple of Machs on the scale above, and then drift back down to the normal reading. Every now and then, it would repeat this sort of cycle. I'm not sure it had an exact frequency tied to it. As Jack said, it was the same in all three, so we actually assumed that it was in the signal conditioning. I had one other thing to add on the launch vehicle. It's very subtle, but I thought when they said they'd put the hydraulics to the S-IVB, I could feel a little tremble below us at that time; but, other than that, there were no booster actions that I could ever detect.

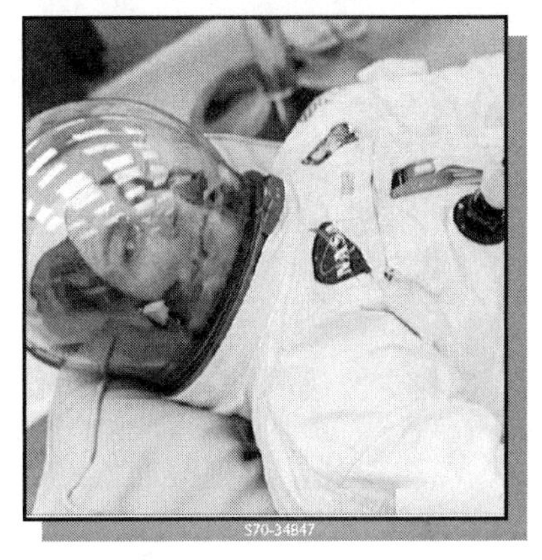

S70-34847

LOVELL — In comparing this part of the flight preparation with Apollo 8, I can say that it was a lot more comfortable an Apollo 13. On Apollo 8, I was very cold during this period, and I suspect they've changed the environmental control system. It was very comfortable this time.

HAISE — I have one other thing to add on the crew station controls. We spend so much time in the simulators that we forget the contrast between the simulator hardware and the real hardware, which isn't used to any degree. It was very apparent that all the switches move very hard in the spacecraft compared to the simulator. In fact, the three position switches went to the intermediate position and then I actually had to force them down into position. The same was true for the rotary knob.

3.0 POWERED FLIGHT

SWIGERT — I have just one comment. I think all of us felt the PU shifts.

LOVELL — I want to emphasize that communications were a lot better than I expecte They were a lot better than they were on Apollo 8. The simulation of the powered fligh matched very closely to the actual case. I was much more aware of what was going o on Apollo 13 than I was on Apollo 8, but maybe it was the different seat. The PU shift, as Jack mentioned, was quite evident; certainly the change in acceleration was apparent.

70-HC-293

SWIGERT — All of us immediately looked over at the engine light. It was quite apparent.

LOVELL — I think we discussed the early engine out on the second stage during the inflight debriefing quite adequately.

HAISE — On the first-stage separation, I saw a flash out to my left. It didn't appear to extend ahead of us. After the second stage staging, there was a lot of debris that went out in front of us that we subsequently flew right on through. It looked like frozen particles or something in that state, but I didn't notice any of this attaching itself to the windows.

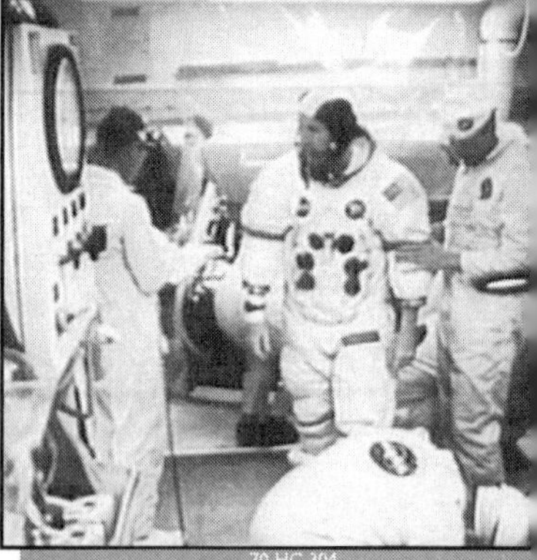
70-HC-304

SWIGERT — Our windows came through in good order. I was expecting frozen particles from the water under the BPC, but we didn't have any of that.

4.0 EARTH ORBIT AND SYSTEMS CHECKOUT

4.1 EVALUATION OF INSERTION PARAMETERS

SWIGERT — I hit a VERB on ECO and copied down the parameters, which were nominal. We were right on the trajectory until we lost the center engine. We regained most of the velocity, but our time was longer.

LOVELL — Our insertion time was about 1 minute longer at that point than nominal.

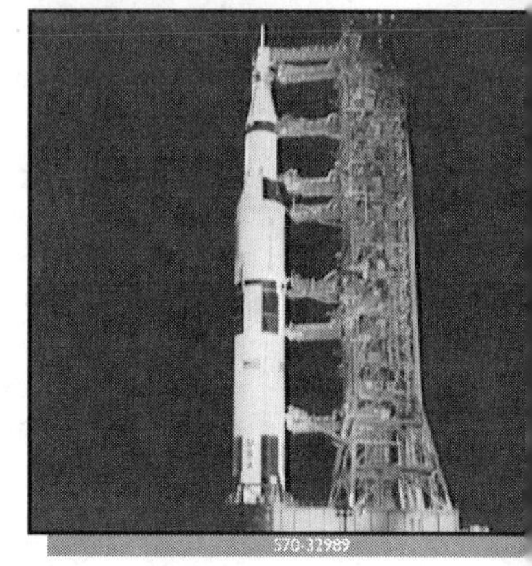
S70-32989

4.3 ORDEAL

LOVELL — I had no problems with the ORDEAL. I was able to unstow that by myself. This is something you can't do in a simulator. I actually improved our insertion schedule.

SWIGERT — When I got out of the couch, Jim told me to move slowly and take it easy. I had no problems at any time. I adapted myse and proceeded just as we had done in the simulator, at full speed through the thing an never had any problems - no dizziness, no uneasy feelings at all.

4.4 OPTICS COVER JETTISON (DEBRIS)

LOVELL — Jack, how about this optics coverage?

SWIGERT — We had a problem with that. I read off the procedure and then did it. I tol you I wasn't seeing stars; so, I entered P52 and didn't feel the optics cover jettison until the optics drove in P52 to the first star. Well, I did the optics jettison procedure twice, reading down the checklist item by item, and I didn't feel they jettisoned either time until I entered P52. I felt they jettisoned all at once in the P52.

LOVELL — Were you looking through the telescope at the time? You can usually see debris go off in that thing.

SWIGERT — You know, it's just completely black and then all of a sudden there are beautiful stars. And it's just like night and day.

S70-34853

LOVELL — It might have been a hang-up of some sort.

4.5 COAS AND HORIZON CHECK

LOVELL — I have no comments there. The check was nominal. The S-IVB held the local horizontal.

SWIGERT — I think that the checklist was adequate. We had adequate time to do everything. I think we were well ahead of the time line.

HAISE — Yes. We were sitting around waiting there for one period for approximately 30 minutes for the next event to take place.

S70-34854

4.7 COMMUNICATIONS

LOVELL — I noticed no communications problems. Did you?

SWIGERT — None.

HAISE — Under that heading, I guess, our proposed TV show was a complete bust. The whole Gulf Coast was cloudy and what we had hoped to show was the nice coastline and there wasn't any to be seen.

4.8 TLI PREPARATION

LOVELL — We used the nominal TLI procedure. We had adequate time. There were no hang-ups. The ground gave us a change in data to use based on the insertion of the booster, which was riding high all the time. The change worked out quite well and was covered briefly in the inflight briefing regarding powered flight. They gave us a new angle of 20 degrees for 57 minutes. And at T the ball was zero. So it worked out.

S70-34855

4.9 SUBJECTIVE REACTIONS TO WEIGHTLESSNESS

LOVELL — My feelings were as I've had previously. When we first get subjected to zero g, I feel I'm upside down, my head is full, and blood is rushing to my head; this lasts several hours. I think this sensation lasted approximately 6 hours. But basically, that's the only sensation I felt in zero g. After that, it went away.

SWIGERT — I think Fred and I felt the same fullness of the head.

HAISE — We both mentioned it about the same time. I don't know who mentioned it first. We both had it go away about the same time. Offhand, I don't remember how many hours had elapsed.

SWIGERT — It was around 8 hours; we both mentioned that the fullness of the head was gone.

HAISE — I had one other different reaction. On the morning of the second day, I woke up with a pretty severe headache. I drank some juice and ate some bacon cubes. That didn't sit right and I upchucked about 2 ounces of my juice. I sat still for about half a day pretty much; I never had any symptoms again after that.

LOVELL — I think a general comment concerning space flight is in order. The fact that when you first get inserted, what you do for the first day (especially if we go into Skylab or something like that) should be held down. We should not try to do too many different things per day. No matter who you are, it's going to take a while to get used to zero gravity. Towards the end of our flight, we didn't know we were in zero or one g. You get so used to it. But, in the beginning, zero gravity is different. You do feel different so, you've got to just take it easy until you get accustomed to it.

5.0 TLI THROUGH S-IVB CLOSEOUT

5.1 TLI BURN

LOVELL — From the left seat, the TLI burn was completely nominal. Attitudes held. The psi progressed according to the chart. We had about a 3-second overburn, if I recall.

HAISE — Three and three-fourths.

LOVELL — The overburn of 3-3/4 seconds was based on our clocks on board. We had no anomalies concerning the TLI burn.

5.2 S-IVB PERFORMANCE AND ECO

LOVELL — S-IVB performance and ECO were nominal.

5.3 S-IVB MANEUVER TO SEPARATION ATTITUDE

LOVELL — Jack, why don't you discuss the S-IVB maneuver to separation attitude? You were over there about that time.

SWIGERT — The S-IVB began its maneuver on time; it maneuvered similar to what we have observed in the simulator; and held T&D attitude well.

5.4 S-IVB MANEUVER TO T&D ATTITUDE

SWIGERT — We proceeded to use the normal procedure for T&D, and this worked out well. Pitchover was very favorable compared to what I've observed in the simulator, with the exception of translation control movements which I felt were somewhat different from the simulator in that, in the simulator, you can just tweak the translation controller a small amount and you get a small amount of translation. Here, it seemed to work in jerks. A small tweak didn't produce anything, and I actually had to hold it in. Then my Y and Z translation appeared to be made in a jerky fashion rather than a smooth translation like I had experienced in the simulator; but we had no problem docking. I would expect that the S-IVB pitch around was about 80 or 90 feet out. Does that seem like a good number to you, Fred?

HAISE — Yes. About 80.

SWIGERT — About 80 feet out, which was about what I was observing in the simulator on my pitch around. I felt that the closure rate was slow, maybe 0.2 fps on the contact; and we didn't try to hurry. We had adequate time, and I think the majority of the fuel I expended was trying to get stable. We had drifted around quite a bit after we got contact, and I was trying to get things stabilized.

5.9 DOCKING

SWIGERT — When we went into hard dock, the latches ripple fired; they didn't all go at one time. I think that's because we had a slight yaw rate about the time the latches fired. There was no problem with sunlight.

5.11 SUNLIGHT AND CSM DOCKING LIGHTS

SWIGERT — The S-IVB was immediately visible. The sunlight on the docking target did wash out the COAS. I had the COAS full bright, and it made sighting the target a little bit difficult. Right in the final phases of docking, we did get into the shadow where the shadow of the CM blocked out the Sun and the docking target was fully visible. I guess that occurred at about 5 feet on in.

LOVELL — The hatch removal was nominal. The usual odor was up in the hatch. The odor had been reported before, and I had forgotten about it; but, when I got up there, I could smell it. There is a burnt odor in the docking area after the hatch is removed. I don't know what it's caused from - probably the docking sequences or something like that. There were two latches that were not engaged completely.

SWIGERT — I had to recock them.

LOVELL — You recocked them and got them back in position. Other than that, there was nothing unusual about the tunnel area.

SWIGERT — I think they were latches 1 and 4.

LOVELL — We connected the LM power cables, which was no problem.

5.13 EMS BEHAVIOR DURING TD&E

SWIGERT — The EMS was just about what we had experienced in previous flights. Our bias test got continuously worse, and we did have a bias in there. I didn't particularly use the EMS except merely as a rough guide. I used my translation predominantly on time, and it decreased very rapidly during the pitch around; but that has been observed on previous flights, so it didn't bother me.

AS13-60-8581

5.12 EXTRACTION (SPRING EJECTION)

SWIGERT — The extraction was performed according to the checklist, and we had no problems at all. It went just exactly as we had experienced in the simulator.

LOVELL — I might mention that the procedure that Jack used was different from the one Ken used. However, the procedures worked out perfectly as far as our crew coordination was concerned. We had no problems that way and we were in good shape through transposition and docking. I saw really nothing unusual during the whole procedure. This is one procedure that I thought required a lot of close coordination, a lot of working, because there were many things happening here. Everything worked out perfectly. One thing that we did do - we had the TV up, and that took a lot of Fred's time. ... to hold that TV to get the pictures of the docking. You might want to comment on that, Fred, and also on the high-gain antenna.

5.15 PHOTOGRAPHY OF TD&E

HAISE — It's probably not as appropriate here as it was during the next TV session where we were trying to do the midcourse, but if you are going to play with it in the opposite focus and worry about the lighting and contrast and that sort of thing, it does take about three quarters of your time fiddling with it. I guess my only other job during this period was to make Jack feel that his estimates were right, when he would ask me about how far out it was, and to take some pictures. For this particular sequence of events, I didn't feel I was shortening myself too much in what I was supposed to be doing, which wasn't that much. About the only picture I missed was halfway through the turnaround. I was still worrying about getting the TV set up, and I missed the same picture that 12 had already shot, which was one SLA panel drifting off with the Earth for a background; but I didn't have the camera handy right then. Other than that, I felt that I got the number of pictures they wanted with the LM coming in at varying distances. Lighting was surprisingly good to me. I don't know if it was a different attitude, different Sun angle, or what, but at least from an eyeballing standpoint I thought the lighting on the LM, on the S-IVB, and in the IU was very good.

LOVELL — We've probably got some pretty good pictures of the S-IVB. One general comment concerning that: unless there's a definite engineering requirement, I would suggest that we review using the TV during docking and the midcourse burn because I think that we've overdone that.

5.18 S-BAND PERFORMANCE

HAISE — On the S-band performance, I had one goof-up there. I thought I had the angle set for docking attitude, but I had left a switch in MANUAL. I thought I had it in REACQ. We came around and locked up beautifully and had good gain; but, when Jack went to the next set of attitudes, we started losing signal strength. That's when I found out that I didn't have it where I thought I had it; so I put it down to REACQ, and it immediately AUTO TRACKed, got its gain back, and worked beautifully thereafter. The only S-band problem was an operator error.

SWIGERT — As far as sounds go, I think the RCS sounds were much like they've been reported previously - that you can hear the sound of the valves opening - and I didn't notice any difference from the simulator.

HAISE — The closest I could reproduce the sounds of the thruster was by sticking the pad of Velcro on my foot to the lower bulkhead and then snapping in and out the bulkhead. That kind of made a sound like the thrusters, which upset Jack now and then too.

SWIGERT — Because I would have the switches off, and I'd say we're not supposed to be firing. What's firing?

HAISE — That was my foot firing.

LOVELL — There was nothing unusual. I thought that the contact had more of a jolt to it than I thought it was going to have. That's why I asked you what our closing rate was.

SWIGERT — It was slow; it was very, very slow.

LOVELL — The man in the middle seat is really blind. He's worse off than the people back on the ground who can see the television because you can't see anything from the middle seat, and Fred and Jack could see everything from their rendezvous windows.

5.21 WORKLOAD AND TIME LINES

LOVELL — The workloads and time lines, I thought, were nominal. I don't think we have to have any changes there. I think that the crew can handle those with no problem.

SLAYTON — Any comments on photography other than the TV?

HAISE — I shot whatever the flight plan called for. I think it was either five or 10 pictures of the LM during both docking and extraction, and then we shot some of the

S-IVB after we did our maneuvering. We had the camera in the center hatch.

SWIGERT — We also had the sequence camera going, as the flight plan called for. We followed the flight plan completely.

HAISE — That's why I made the comment about the lighting a while ago. It looked pretty good for the settings we had, so I expect the pictures to be all right.

6.0 TRANSLUNAR COAST

6.1 IMU REALIGNMENT AND OPTICS CALIBRATION

SWIGERT — The first P52 was done with PICAPAR. We put the star right in there; we had no problem. The optics calibrations for the first P23s were nominal. I think I only did four of them and three of them were the same value, so I used that value

LOVELL — About 300ths or something like that.

SWIGERT — Minus 300ths - 89997. I guess while I'm on this thing I could talk about that first set of cislunar navigation sightings. All the stars were completely visible. You and I had a coordination exercise there that worked out well.

LOVELL — Yes.

SWIGERT — We got those done within the time allotted.

LOVELL — That was one thing that I didn't think we were going to do, really. I gave cislunar navigation a secondary priority. I thought that, if we didn't get finished in the time line, I was just going to drop it because it really wasn't required on the way out. It was merely training. I thought we would try to get the DELTA-H for Jack's calibration, but we got through all the stars. In fact, we repeated one.

SWIGERT — I'll tell you also that I had done an awful lot of P23s and I became very proficient. I knew I had a good hack on fuel. During the simulator sessions, it had taken me 15 pounds to do that first set of P23s, and it took exactly 15 pounds in the flight. They called up the fuel used, and it was exactly 15 pounds; so it compared very well. They relayed back that the DELTA-H was very constant - within 2 kilometers, I think, which was 17 kilometers plus or minus 2, I think. They were very happy with it.

6.3 PASSIVE THERMAL CONTROL

LOVELL — We had a small problem with the first attempt at passive thermal control. I'm not too sure what our reason was for that. We didn't null out the rates, though.

SWIGERT — We nulled out the rates okay, but remember Ken's checklist had a red mark in there that says, "Enable opposite pairs." In the checklist, where it is headed, "Enable all jets," it had that crossed out and had "Enable opposite or couples" — "Opposite of opposing couples." We followed that and we didn't turn an all the quads. We just turned on the couples on those particular quads. As a result, we were off on that, and Houston called back up and said, "Have you enabled all the jets?" We discarded that part of the checklist and went back enabling all our jets. The second time, we used Houston to tell us when our rates were null, so we knew our rates were stable when we started. The second one worked out very well. I think we went some 20 hours without firing the jets at all.

6.5 MIDCOURSE CORRECTIONS

LOVELL — The MCC that we did was nominal in every respect. I saw nothing wrong with the procedures. We used the card that we had rather than the checklist.

6.6 PHOTOGRAPHY AND TELEVISION

LOVELL — My only comment concerns the next line which, if I had it to do over again, I would request not to have it televised because it cuts into our normal crew flow of activities. I didn't think that Fred was going to spend that much time on the television camera trying to get things done. This was the first time that engine was ever burned, and I thought it was kind of important. I would have probably just eliminated it. So I would eliminate that the next time unless they want it for engineering purposes and then we'd just put the camera up somewhere.

6.7 HIGH GAIN ANTENNA PERFORMANCE

HAISE — Actually, on the translunar coast, we didn't use it except during the periods of TV. For the most part, Houston just had us select OMNI B; and, as we went around through the pole switching, they would just cycle back to D or not D. The ground really handled all the switching on the OMNIs. We didn't have any COMM problems at all. There are a couple of things I ought to say. One of the things the simulator guys wanted me to notice in particular was the effect of turning on the gimbal motors on the O2 flow. In the simulator, you get an enormous jump in the flows in the fuel cells when you turn on the gimbal motors; and, in the vehicle, you don't. The fuel cell flows barely moved. You do get a very rapid jump on the ammeter. If you're looking at the appropriate fuel cell for the bus of the gimbal motors, you're turning on - about 8 to 10 amps.

LOVELL — Did we get a light?

HAISE — We never got an UNDERVOLT light, which is normally true in the simulator. The other distinction I noticed was, when the burn started, that the ball valves opened very, very slowly. In the simulator, they snap open. In the real vehicle, it's almost like you can see the worm gear turning, and they're slowly grinding open. I would guess it's probably a 0.25 second or so, but it was quite a bit slower than in the simulator.

6.8 DAYLIGHT IMU REALIGN AND STAR CHECK

LOVELL — The star check for the burn was nominal.

6.9 CM/LM DELTA-P

SWIGERT — The DELTA-P between the LM and the CM — remember we started out, and by the time we went to open the hatch, we had 1.1. Remember it was part of our procedure; we had to vent the tunnel down to 1.7 or greater. We had a pretty good tight tunnel connection.

LOVELL — There was one question I asked Houston. The answer was to get a better purge in the LM before we went into it. I guess that was missed in the training someplace along the line. I didn't see it in the flight plan when I went through it. Okay, that was no problem.

6.10 LM AND TUNNEL PRESSURE

LOVELL — LM and tunnel pressures were nominal.

6.11 REMOVAL OF PROBE AND DROGUE

SWIGERT — On that, I followed the decals printed on the tunnel wall. I think this was our first time through it, and I think it took us slightly under 15 minutes to do it.

HAISE About 12 minutes.

SWIGERT — I thought that was pretty good for the first time. We never reinstalled them, but I'm sure the second time would have been significantly less because we were purposely going very slowly, trying to do it right the first time.

LOVELL — When we took that drogue and probe out, we slowly realized we were going to be living with it for the next 5 days.

SWIGERT — Right. We had three bodies on the couch. We had one hatch, one probe, and one drogue strapped down to the couch for all the rest of the flight in the CM.

6.12 ODORS

LOVELL — When I removed the hatch, all of a sudden I smelled this burnt smell. I guess it must have been caused by the docking with the connecting of things and the rubbing and friction.

SWIGERT — But you know, I carefully looked there when I took out the probe and the drogue. I looked for scratches, and there were none. We hit it pretty much dead center.

LOVELL — You mean in the drogue?

SWIGERT — Yes, in the drogue. I looked at the probe, the head of the probe also. There weren't any scratches at all.

LOVELL — I do recall, though, putting my hand up against the probe when I first removed the hatch, and it was still pretty warm because it had been sitting out there in the sunlight.

7.0 LUNAR MODULE FAMILIARIZATION

LOVELL — You might want to start this, Fred. You went in there first. We had one thing to do in this thing that wasn't on the flight plan and that was the SHe tank.

HAISE — Our communications were yelling back and forth through the tunnel and we lived with that the next 4 days. It was really entirely adequate; particularly this time, because we didn't have all the pumps going in the LM. I didn't find any real problem in going into the LM. The shift in orientation did seem sort of strange. Although I had done it in the water tank, I found myself standing on the ceiling in the LM; when I got down in there, I had to do a 180 turn around. The LM itself was very clean. I found two washers floating around and I found the plastic cap from the sequence camera. It had drifted off and was lodged behind one of the ED switches, over on Jim's side of panel 8. That was the only thing out of place in the whole vehicle. We went through the regular checklist of housekeeping items. Then we threw in the extra addendum page that Houston had read up to us. They wanted a reading on a SHe tank, which, for the record, turned out to be exactly what Houston predicted. We didn't have much of a SHe tank problem.

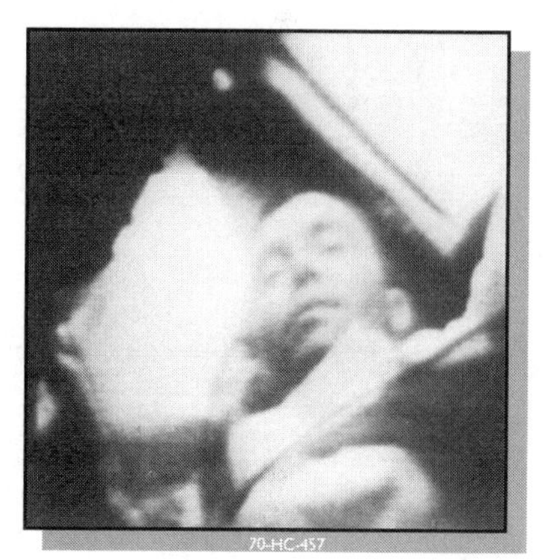
70-HC-457

LOVELL — At that time it was between 7:10 and 7:20.

HAISE — We had no COMM checks during this LM visit. I think the only transfer of equipment we made was the G&N Dictionary. I took the LM Time-Line and the LM Activation books back with me because we were going to discuss our power-up and descent operations with Jack and coordinate those with respect to the CSM solo book. I did all the housekeeping items with the exception of the 16mm camera items. They had been deleted back to PDI day because it would have interfered with getting the hatch down and tucked away.

LOVELL — Did we carry the film in?

HAISE — Not the 16mm film.

SPEAKER — How about the 70-mm film?

HAISE — We carried all the 70-mm film, but no 16mm film. Other than that, I added a little tape to the right side of the crash bar and that was about it on the housekeeping side. We spent the rest of that visit in the LM putting on the TV show.

LOVELL — I'd like to make one important point. We received a GO to enter the LM 3 hours early because we were ahead of the time line.

LOVELL — I think that was fortunate in several ways. Aside from the incident that occurred right after this, we could have gotten the nominal things finished and not have had the TV interfere with us. One man could operate the camera and do all that work. That is a lot more effective than if he had had to do the TV at the same time he was looking at the SHe tank pressure. People wouldn't understand what was going on. We had the TV concurrent with going into the LM. That is not the way to do it. After normal LM housekeeping, we should have set aside a time for nothing but TV. We had the time.

HAISE — Neither do justice to the other - they detract from each other. One should do one or the other. One should plan a TV show and put it on; then run the spacecraft when needed.

SLAYTON — Explain the SHe tank use.

LOVELL — We never did it. I did an IVT to the CM for about 8 minutes. I wasn't up there very long. This is where the PRESS vent went. I guess it was about 8:40 when we finished the TV show and the next time I looked at a watch, it was 3:00 in the morning. The time went pretty fast after the emergency. I might mention the TV show was just over and the scene was set for the incident.

LOVELL — We were geared to bangs because Fred had actuated the REPRESS valve a couple of times. These caused a bang in the spacecraft. The first time he forgot to tell us about it, so Jack and I were spring loaded to loud bangs. When the actual bang came, I didn't know exactly what the situation was. I thought maybe Fred had actuated the valve again.

HAISE — I was sitting down in the LM.

SLAYTON — Is this the LM REPRESS valve you're talking about?

LOVELL — Yes, it's the LM REPRESS valve.

8.0 SPACECRAFT EMERGENCY

LOVELL — To the best of my knowledge, Jack, you were in the left-hand seat.

SWIGERT — I was in the left-hand seat.

LOVELL — I was in the LEB, and Fred was somewhere up in the LM. We all heard the explosion together. I said to Fred, "Do you know what that noise was?" Fred said he didn't. Then, Jack said, "Remember the 80-amp glitch we had in training?"

SWIGERT — You explained the 90 amps short on MAIN B.

LOVELL — Then you said, "The MAIN B LIGHT is on."

SWIGERT — That was my concern.

LOVELL — That's right. Then I went over to look at the instruments. I don't think you even closed the hatch on the LM side, did you?

HAISE — I left the LM hatch open and came down to look at the systems.

LOVELL — When I heard the explosion, I thought I saw a light someplace along the side

It might have been just a reflection off the hatch door when you were closing it. That's what made me believe you had your hand on the hatch. At that time, Fred came back to the right seat to look at the systems. I moved over to the center. Jack was in the left-hand seat.

SWIGERT — Then you called Houston about our problem.

HAISE — Yes, that was our first transmission.

LOVELL — Then I called again and said we had a serious problem. The MAIN B BUS UNDERVOLT light was on, and we had a FUEL CELL light on. Jack, tell them what you saw.

S70-34628

SWIGERT — I heard the explosion. It was about 1 or 2 seconds until we had a MASTER ALARM and a MAIN B UNDERVOLT light. I immediately left the left-hand couch and floated over to the right-hand side and looked at MAIN BUS B. We had normal voltage, normal current, and normal fuel cell flows. At this time, I came to the conclusion that whatever had occurred was a transient on main bus B because the performance of main bus B had returned to normal. At that time, I figured something had happened to the LM. My concern was the open hatch. I wanted to get the hatch installed and then take stock of what was happening. I went to get the hatch. I transmitted to Houston that, "We have a problem here." At that time, I went back to get the hatch.

S70-34904

LOVELL — The LM hatch was still open. We were going to put the CM hatch back on.

SWIGERT — While Jim and I were trying to do this, I misaligned it in the tunnel and didn't get it in the first time. While Jim and I were doing this, Fred slithered down and started to look at the systems.

HAISE — I'm not sure how many seconds I was behind Jack. When I looked at main B, the volt meter was pegged full-scale low. About that time an AC 2 light came on. Shortly thereafter, an AC overload light came on. I turned off inverter 2, but that didn't change anything. The meter only reads down to 23 volts. It could have been 22 or less, but as far as I knew it was zero. I looked at fuel cell 3, and its flows were showing full-scale low. This meant that this fuel cell wasn't carrying any load. That meant the whole bus was gone. I admitted that LOI was going to be NO GO about now. I didn't even think to look at the other two fuel cells at this time. I started switching AC loads to get all those things that were on AC 2 over to AC 1. The first couple of items I cycled I had a MAIN A UNDERVOLT. Then I looked at main A and it was down around 25 volts. I cycled through the other two fuel cells. Fuel cell 1 was showing no flow. It was not producing anything so we had only one fuel cell on the line at that time. About that time, Houston wanted all the regulated pressures of O2, and N2, and H2. Jack read them down to Houston.

HAISE — It turned out that N2 was pretty sick on fuel cell 3; and O2 was the one that was off nominal on fuel cell 1. Those were the two readings that were not looking very good.

LOVELL — Before the incident, we did have a transducer failure in O2 tank 2 quantity. Then we started looking through our systems again. We saw the pressure on the O2 tank 2 was zero. I never saw any transients at all just zero. Number 1 tank was down to 500 Psi.

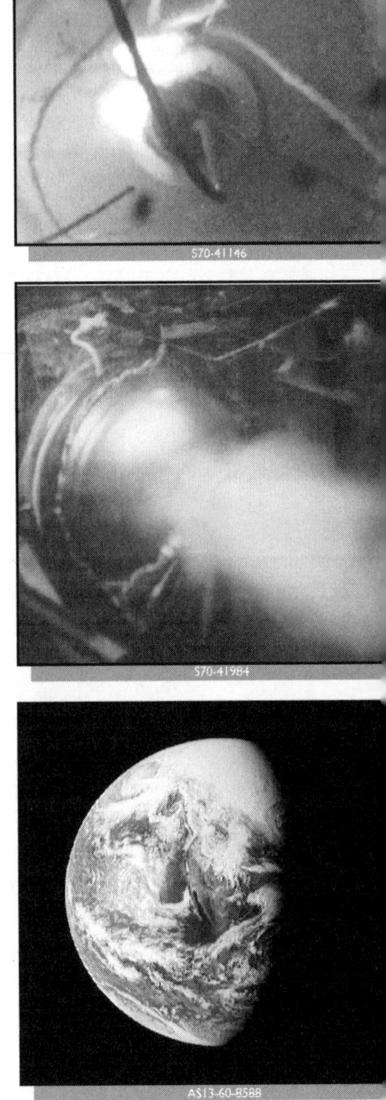
S70-41146
S70-41984
AS13-60-8588

SLAYTON — The O2 tank 2 quantity failed prior to this.

LOVELL — Yes, it failed off-scale high.

SWIGERT — We had been having some stratification problems when we cycled the fans. During our scheduled periods of fan cycling we would get a CRYO PRESS light which is an indication of stratification in the tanks. During one of these fan cycles, the 02 tank 2 quantity indicator pegged full-scale high. We did another fan exercise to try to see if we could jar it back the other way. It never did. It stayed full-scale high for the remainder of the flight.

HAISE — The next thing that showed up was the surge tank continuing to go down. When it kept going down below the pressure needed I

HAISE — for the remaining fuel cell, I knew the remaining fuel cell was going to go the same way as the others. I left the CM about that time.

SWIGERT — At this time, I called Houston and suggested that perhaps we should get somebody in the LM and start coarse aligning the platform. Then Houston asked us to shut down fuel cell 3.

LOVELL — Yes.

SWIGERT — I read the procedure to you and you did it item by item.

LOVELL — We had questions on the REACS valves. Once we threw the switch on th REACS valves, we couldn't get the fuel cells back again. It wasn't obvious to us at th time, but we should have known we didn't have any fuel cells then because we didn have any oxygen. Throwing the REACS valves was just merely a formality.

SWIGERT — We came back and shut down fuel cell 1. We asked Houston to confir that decision. They did, and we proceeded with the procedure to shut down fuel cell

Then we started activating the LM to get our platform coarse aligned.

LOVELL — It was none too soon.

SWIGERT — It wasn't much later that Houston came back with the advisory that we had about 15 minutes of life left on fuel cell 2 as a result of the decreasing pressure in tank 1. Jim and Fred proceeded into the LM to power it up. They did it expeditiously and we got the platform aligned. I did two VERB 06, NOUN 20, ENTERS and read the angles down. This gave Houston some fine torquing angles to give to you and you got the platform fine aligned, and in good order. We had good coordination here.

LOVELL — One of the big turning points in the flight was the fact that we got the LM platform up. We received the coarse align from Jack and the torquing angles from MSFN. The one VERB 06, NOUN 20 that we got isn't what we normally do. During a normal activation, one gets a better angle out of it. We did get the platform aligned, enough to do the burn. I think where we made a mistake was going into the normal activation checklist. We should have gone into a quicker activation checklist. There is a lesson to be learned here. To get that LM powered up, one has to get the platform up because it is the heart of the whole thing.

SWIGERT — At that time, I had BAT tie AC on to help with the load. About 15 minutes later, the fuel cell flows on fuel cell 2 went to zero. The LM was powered up at that time.

AS13-60-8707

SLAYTON — How did COMM work? When did you get the LM COMM and was there a problem there?

HAISE — No. The only problem was that we were in the hot-mike mode for a long time without knowing it. We didn't go into the activation checklist on our own. We went under the direction of Houston.

LOVELL — Yes, that's right.

HAISE — They gave us the sections of the activation checklist to use.

LOVELL — That was good. It cut down on the time to get the platform up.

HAISE — It really wasn't faster, but there was less chaff in it. We had to use the 2-hour PGNS turn-on in the Contingency book. That is the only one that gets one a good platform. A 30-minute activation doesn't get a platform.

HAISE — They would have had less to weed out if they had jumped into the Contingency book. Thereafter, that is all we used.

LOVELL — One of the turning points was that we did get that LM platform aligned enough before we lost the CM platform.

SWIGERT — We failed to mention the venting outside the SM.

LOVELL — Yes, a tremendous amount of venting could be seen out the left-hand window.

SLAYTON — You called that out almost instantaneously with no alarms.

LOVELL — Yes, it was just pouring out. We could see it because the Sun angle was jus right. Another thing along with the physical sensations was the debris. The oxyge venting disappeared almost immediately, but the debris around the spacecraft wa tremendous. An early discussion we had with Houston was to use the stars to get a alignment. It was very difficult to see anything out the window with all the debris. Ther were all kinds of debris out there.

8.1 COMMAND AND SERVICE MODULE

LOVELL — We went over there and saw the venting; I knew that we were losin something at the time. I really wasn't too sure what it was. I suspected that it was oxyge because I saw the pressures were down.

SWIGERT — The pressure was going down. We all came to that conclusion.

LOVELL — Yes, right.

SWIGERT — I don't think any of us quite realized the extensiveness of it until we shu down the second fuel cell and the pressure was still decreasing.

LOVELL — About that time, we realized that there wasn't any sense in putting in th CM hatch, and we put it back down again. About that time, Fred was going into the L anyway, and Houston came up and finally said we'd better activate the LM systems. W activated the communications, the power, and the guidance system.

QUERY— And didn't you have some trouble with brakes in here, Jack? Controls?

SWIGERT — Yes, but I don't feel that that's strictly because of the same problem th you had with pitch that the RHC is no good with the stack on. You have to use the TH to get adequate pitch control. Remember that you had the same problem when yo were using the THC.

LOVELL — I don't think we ever knew, though, whether our SM RCS system wa completely working or not.

SWIGERT — Yes, I did. I had good thruster control with it. I think I did get some rate I don't recall now exactly how much.

LOVELL — That might be nominal with the hand controller when you have the whol stack on.

HAISE — We had just put on the TV show. We were in a stabilized attitude for high-ga angles; the vehicle wasn't moving. All at once, Jack got negative pitch problems. He fire a thruster, and I remember your telling the ground about it as I was coming back throug the tunnel. I could hear jets firing. I thought you mentioned you had rates in two axes

SWIGERT — Yes.

LOVELL — Okay, but we never did figure that one out completely. That could very we have been due to the venting of the oxygen, because when the SHe tank blew, it change the motion. Most of the mass is back in the SM, so it wouldn't make that much of change.

SLAYTON — At some point there, quad C malfunctioned. Do you remember anythir about quad C specifically?

SWIGERT — I didn't. Quad C was one that gave me pitch. I thought that perhaps I didn't have any quad C. I don't really have any absolute proof to substantiate either its loss or its performance, because I didn't try to control pitch to any large degree with the THC. I didn't really try to stabilize it out. We did get the rates down somewhat. I know I had direct thruster control; I used the DIRECT switches, and that led me to believe that quad C was okay. I do recall calling up channel 31 and looking at the computer to see that the breakout switch was okay.

LOVELL — That's right, we were in that malfunction procedure at the time, weren't we?

S70-34627

SWIGERT — Yes, but, looking back now, I don't really have any substantiating evidence either to prove or to disprove the operation of the normal switches in quad C.

LOVELL — That's an area that we're still a little bit hazy on, mainly because we shut down power in the CM to conserve the batteries; and we really didn't have enough time to psyche it out.

SWIGERT — I think that's the reason that I had channel 31 called up. I did look at that, and that appeared normal. About that time, we got the emergency power down procedure for the CSM; and that was about all the troubleshooting we did on the quad. The emergency power down procedure was a very simple six-step procedure. The CM power was completely killed. We pulled all the circuit breakers on panel 250 except the sequence circuit breaker. The CM was like a tomb.

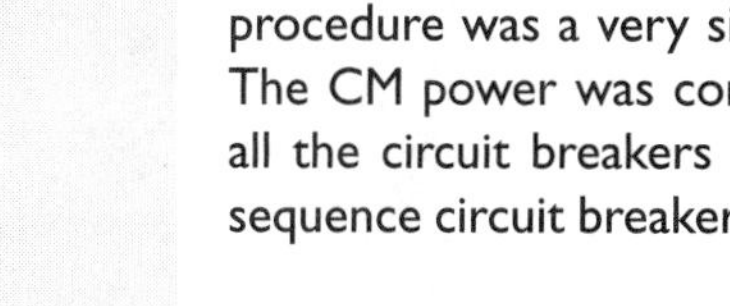

S70-34902

8.1.2 Noises and Flashes

LOVELL — There was a dull but definite bang - not much of a vibration, though. I didn't think there was much vibration - just a noise.

SWIGERT — Just a noise.

LOVELL — Probably came through the structure.

HAISE — I felt just a slight shudder.

LOVELL — Maybe I was floating at the time; I didn't feel it.

S70-34986

8.1.3 Debris

LOVELL — There was much debris all around outside the spacecraft; we couldn't even see stars.

HAISE — The debris particles weren't very large. They were small, like frozen particles, or what maybe look like floating metal. A couple of inches or less.

LOVELL — I saw one piece of wrapping float by.

HAISE — Yes, but there was nothing that was extraordinarily large.

LOVELL — We had no indication from the debris as to the extent of the damage to th back end.

SLAYTON — You commented at one point that you saw about a 4-inch chunk floatin by.

LOVELL — Yes.

SLAYTON — Is that about the biggest piece you saw?

8.1.4 Physical Sensations (Attitudes, Rates, Vibrations)

LOVELL — Yes, it was a piece of wrapping or something.

LOVELL — We think that the venting did impart a velocity to the spacecraft stack.

8.1.5 Communications

LOVELL — We had no problems with COMM during the emergency.

8.1.6 Tunnel Operations

LOVELL — We had difficulty putting the hatch on, but I think it was due to our rush because we went back and checked it again before entry and it worked fine. We decide to leave the whole tunnel system open because we determined there was nothin wrong with the LM, finally.

8.1.7 Cabin Atmosphere

LOVELL — There was no problem there.

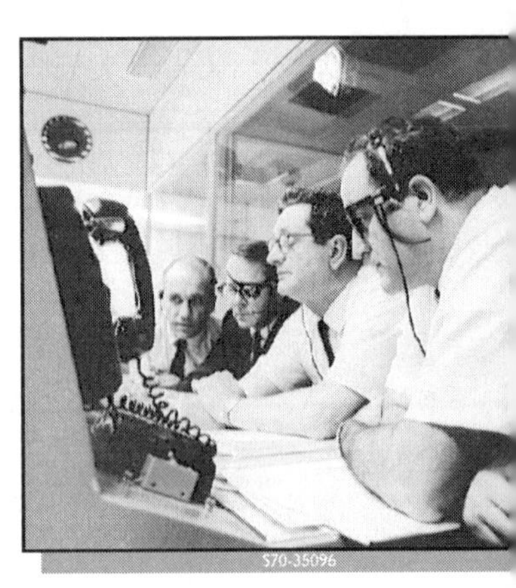

S70-35096

8.1.8 Ground Updates (Procedures)

LOVELL — Everybody, including ourselves, was trying to figure out what the story was. We didn't know exactly. The basic thing was to get the LM powered up and get the PGNS on the line.

8.1.9 Emergency CSM Power down

LOVELL — Emergency CSM power down went along according to the checklist.

8.2 LUNAR MODULE

8.2.1 Power Transfer Activation and Checkout

LOVELL — We went through the Activation checklist as we mentioned before. If we had gone through the contingency 2-hour checklist, we'd have had a little less to work with. In any kind of 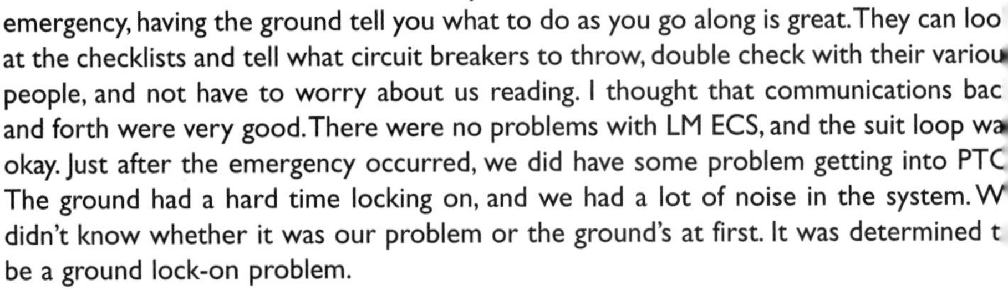
emergency, having the ground tell you what to do as you go along is great. They can loo at the checklists and tell what circuit breakers to throw, double check with their variou people, and not have to worry about us reading. I thought that communications bac and forth were very good. There were no problems with LM ECS, and the suit loop wa okay. Just after the emergency occurred, we did have some problem getting into PTC The ground had a hard time locking on, and we had a lot of noise in the system. W didn't know whether it was our problem or the ground's at first. It was determined t be a ground lock-on problem.

8.2.4 PGNS Activation

LOVELL — The PGNS activation went okay. We got only one set of gyrotorquing angle: We could not do a really fine activation like we normally would have done. Because had made mistakes in the arithmetic several times during SIMs, I wanted to be sure w got the right arithmetic in. So, before I put it in the computer, I asked the ground t confirm my math. When they said it was okay, we would put it in.

8.2.5 Update Pads, Alignments, and Calibration

LOVELL — In general, the update pads were very good. I think the technique of takin the existing checklists and having the ground modify them to fit the particula emergency was fairly good. It eliminated running down a lot of complete checklists. I wa a little worried that we would have people on the ground that would be interested onl in a certain part of a system and would not see the overall picture. I was interested i keeping everything as short and as simple as possible. I didn't want to get a lot of stuff up there that really wasn't required. That is why I made some comments on the way, to just make sure we did only the essentials.

SLAYTON — We spent so much time on that final activation because we wanted to make sure that we used what you had on board as much as possible and that we did not give you a whole bunch of unnecessary stuff.

LOVELL — That technique is good. What we didn't have, and we never thought that we would ever use one, was a CSM Activation checklist. We should look at some of our contingency books and include some of these items in it. I don't think that we will ever get away from having to modify some checklist. The systems guys are going to have to look at what you have and what you don't have, and how to work around it. I certainly never thought about powering up or charging CSM batteries with LM power. It never occurred to me that it could be done. Jack and Fred thought that they could do that. The ground had the technique, and it worked well. That was a big help. We have our Contingency checklist; our method of doing a DPS burn; and how to control using the TTCAs for attitude control, pitch, and roll, and ACA for yaw control. This technique did work and was adequate. In fact, that is the way we flew the vehicle all the time. Our only big problem was when we shut down the FDAI to save power and went to the computer flashing 16 20, which gave us yaw, pitch, and roll, actually outer, inner, middle gimbal angles. We wanted to keep the middle gimbal angle out of gimbal lock. The technique that is in the contingency checklist is not valid. You can't use the TTCA and fly the computer the way we fly the 8-ball.

S70-35368

HAISE — That's right.

LOVELL — As a matter of fact, we spent hours trying to do it. I still don't have the technique. You just have to try to figure out by experimenting which way to hit the thruster. It changes depending on where you are, what quadrant you are in, and what the angles are as to which way to throw that translation controller to stop the angle from going toward gimbal lock. We were trying to keep it at a gimbal angle as close to zero as possible. I wasn't too worried about the other two gimbal angles. It was a continual battle to find it. Maybe we ought to do some more research into using that technique. In the future in event of such a contingency, we ought to look at keeping the 8-ball powered up and powering down the DSKY, or something like that. I think our PTC mode was finally the AGS ATT HOLD, which held the vehicle once it was in position.

HAISE — The problem could be handled the same way that we did it. We taped over each ball top and side and wrote in what the representative TTCA gave in terms of pitch-up, pitchdown, roll right, and roll left. This is a nice handy reference. You didn't want to think about the geometry of things if you could just look at this piece of tape and tell you which way to do it.

8.2.7 ORDEAL

HAISE — We did not use the ORDEAL in the LM.

8.2.8 DAP Loads

LOVELL — DAP loads were sent up by the ground. We didn't use DAP at all; we used AGS almost all of the time.

8.2.9 MSFN Relay

LOVELL — We did not use the MSFN relay itself.

8.2.10 DPS Maneuver

LOVELL — As soon as we got the LM powered up and got our alignment, the ground which was quite correct, planned to get us back on our free-return trajectory. We did the first DPS maneuver in AGS ATT HOLD. We had to maneuver manually to the proper attitude, and then PGNS ATT HOLD held us at that attitude. We couldn't do an AUTO maneuver to it. We maneuvered manually to the attitude, nulling out the needles, and PGNS ATT HOLD held us there. We went through the DPS throttle check.

HAISE — Pericynthion plus 2? About 25 ft/sec.

LOVELL — Is it in the LM contingency checklist?

HAISE — The 38 DELTA-VR. Okay.

LOVELL — After the 31-second burn, we reinitialized the gimbal angles; we put new gimbal angles in based on the stack. The whole burn worked out okay.

SWIGERT — Was that the one where I called out the times — 5 seconds at the 10 percent?

LOVELL — Yes.

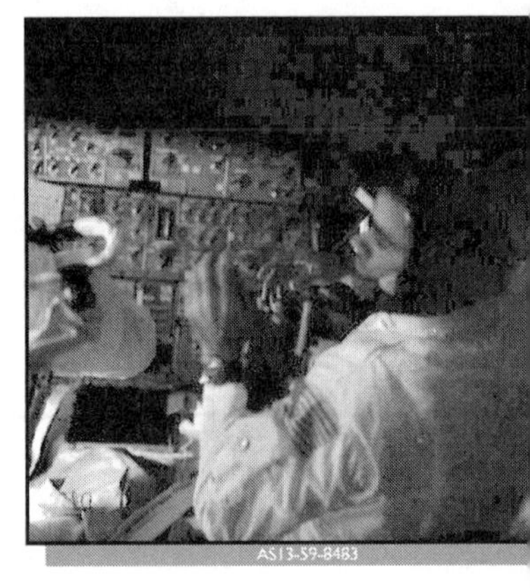
AS13-59-8483

HAISE — It was right on the money on the time. The residuals slipped to 0.2.

LOVELL — Yes. That was a very good burn, as far as the DPS goes.

SWIGERT — It was beautiful. The attitude excursions were nil.

LOVELL — Yes.

8.2.11 LM Power down

HAISE — We didn't do a lot of LM powering down. We kept the PGNS up.

LOVELL — We got rid of the FDAIs; we kept the PGNS activated until the pericynthion plus-2 burn.

HAISE — We never powered up the AGS.

LOVELL — The powered-down configuration is probably listed in the contingency checklist. We can decipher it from the other power-downs

8.2.14 Procedures (Onboard and Ground)

LOVELL — The procedures were completely changed as we went along. It was a case of never going back and doing exactly what we planned to do but looking for the ground to do what was required and passing those modified procedures up to us so we could do the job. The best indication that they were adequate is the fact that we're back here.

SWIGERT — The ground passed me up a basic switch configuration for the CM, which I set up. We just went on down the launch checklist and set every switch per the ground instructions. We started out with the basic CM switch list.

HAISE — The power down we did was the one that we ad-libbed. We went down the rows and gave up what we thought we could give up, lights and things that were very obvious. We didn't do too much on the power down. I think the crew and the ground were both hoping to keep the platform going to get the next PC-plus-2 burn done.

LOVELL — We powered down everything that we knew we wouldn't need, and we just pulled the circuit breakers on it. We almost had a problem, though, because we almost pulled the PGNS circuit breaker, which we did not do, fortunately.

8.2.15 Passive Thermal Control

LOVELL — PTC was done primarily on the computer, and it was difficult to fly the gimbal angles on the computer in this configuration. In the future, we should prepare for that type of flying.

HAISE — The PTC you're talking about here is where, you were turning approximately 90 degrees.

SWIGERT — We would turn 90 degrees, then sit an hour, then go on 90 degrees, stopping for an hour, et cetera, rather than in a normal, continuous PTC motion.

8.2.16 Spacecraft Stability

LOVELL — There was no trouble controlling the spacecraft motions. That was one thing I was worried about. If we'd ever got uncontrolled, we'd have been in deep trouble. You can control the motion a lot better if you have a body to orient on, like the Moon or the Earth.

8.2.17 Cabin Environment

HAISE — The CM was dark and unpowered and was just going to go down slowly in temperature. At this time, I thought the LM was fairly comfortable. I don't recall any cabin temperature readings, but I don't remember being really uncomfortable in the LM or in the CM during that first period of activity; they were both quite reasonable.

AS13-59-8484

SWIGERT — Down to that first PC plus 2, both spacecraft were comfortable. That first night we did like before and put the window shades UP.

LOVELL — That was a mistake.

SWIGERT — That cooled the CM down, and we decided from then on that we'd leave the window shades off.

LOVELL — We'd put the window shades up, and it would really cool it down faster than we wanted it to just in the CM. We used that as a bedroom and so we had the window shades down to keep it dark in there.

SWIGERT — It was still pretty reasonable though. One could sleep up there. Prior to this, when we were on the normal flight plan, we had kept accurate records as to exactly what we had eaten and we had transmitted to the ground our sleep, and the quality of the sleep. Urination was no problem; we had been very regular. After the mishap, it was a problem in that we couldn't get our water. We were told not to use the LM water, and about this time they passed a procedure for activating the CM to obtain water. I did this several times and filled a number of the juice bags to try to get ahead. I figured that we needed at least one 8-ounce bag of juice per meal, so at one time I filled 13, and at another time about a dozen. At one time, we filled 20 or 22 bags.

LOVELL — The procedure gave us a lot of water at one time, but if you didn't use it th pressure would bleed off.

SWIGERT — I kept filling juice bags until the pressure had bled off and I couldn't ge any more water out of it. However, the only problem there was that I had no idea how much oxygen I was using out of the surge tank every time I did this. I didn't think it wa extreme, but I just didn't know when I was going to limit our CM. At that time, we didn know how long the LM would last, and I wanted to have lot of CM O2 left.

HAISE — You might think that you have to stop overboard dump because of the loss o the electric power and the heaters on the urine dump, but Jack actually rigged up th AUX urine dump through the forward hatch and I think he tried it.

SWIGERT — Jim did it once.

LOVELL — You should tell Houston, too, that dumping overboard was a bad thing t do.

HAISE — The point I was making was, other than the problems of tracking - I don't thin you need a heater for that overboard dump, and I think you can use it forever and eve without having a heater. With the tracking, though, that made the picture entirel different. When we couldn't dump it anymore, I think we improvised some place to stor all this good fluid.

SWIGERT — One comment on using that auxiliary dump for urine is that it doe completely cloud up the hatch window. If you ever use that, you might as well forge about photography. We used it for one urine dump, and there were particles on th window from then on.

LOVELL — That is a good point. Using that auxiliary dump either for waste water o urine is strictly for backup. We kept the urine on board, and we had to figure out way of keeping it.

HAISE — We used both bags. We filled both of those bags we showed on TV, which wer the bags we were going to fill with water from the PLSS, and we used all the Gemi bags out of the CM. We used all the urine bags in the LM; I think there were six. W were down to where we were contemplating next using our old drink bags.

LOVELL — We had urine all over the place, stacked in places we never even though about. The nice thing about it, though, is that we found enough quick disconnects an rigged up lines to get urine into things that normally we were putting other stuff int or taking stuff out of. So, it worked out that we could store a lot more urine than w thought we could.

HAISE — In fact, while we were thinking about the water, and talking about feeding PLS water into the sublimator, I had a way figured out to get the urine through th sublimator.

LOVELL — Of course, at this time, we were also thinking about the PLSSs, about usin the water for the sublimator and then using the fans and the batteries and the oxyge in case the LM system failed.

SWIGERT — It's really surprising that a lot of the things the ground sent up, we ha discussed. Fred had immediately done some calculating and figuring on the life of th batteries which proved to be very accurate, compared to what the ground had. C course, at this time, we hadn't figured on the power down.

HAISE — Both of us missed it; both the ground and I were initially conservative. The LM did its emergency power down better than either I or the ground figured. In fact, I think we even got down to 10.3 amps there. On the water, I missed. I had figured about an average of 3½ or 4 pounds an hour of water for the whole time and ended up having 1 hour to spare. Of course, we came out a lot better.

LOVELL — It was very uncomfortable. Basically, the cold made it uncomfortable.

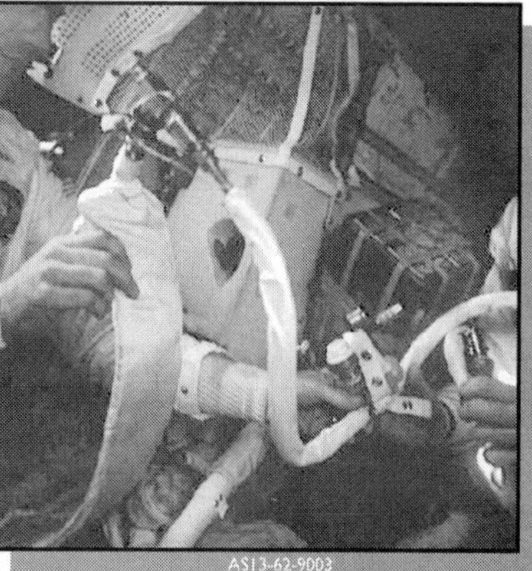
AS13-62-9003

SWIGERT — First of all, even though the temperatures were comfortable, the humidity started to climb. The LM, obviously, couldn't extract the water out of both spacecraft. We began getting condensation on the CM windows right away, even though we still had comfortable temperatures in both vehicles. And then the temperature in the CM and LM started to lower.

LOVELL — So it was a case of having a cold, high-humidity environment. The cabin pressure was no problem. The CO_2 buildup - that's a story in itself. Houston came through with a technique for using CM LiOH canisters in the LM, which worked probably as well as the basic system. We ended up with a complete primary LiOH canister that we didn't use. And that was 40 hours worth of running.

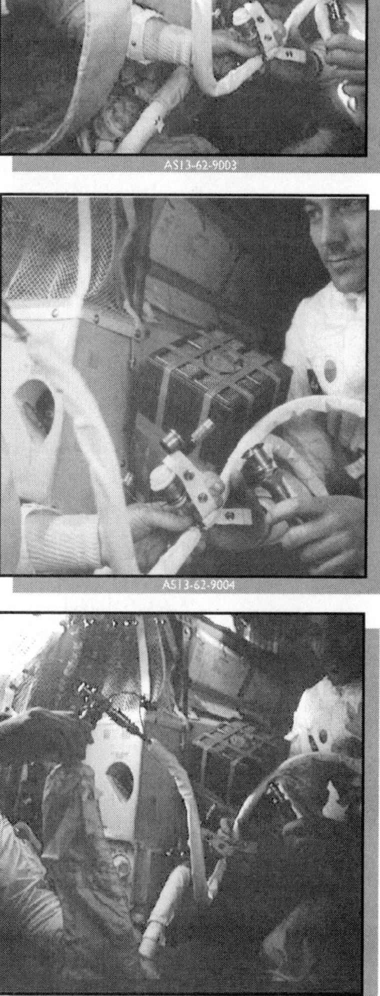
AS13-62-9004

AS13-62-9005

SWIGERT — We had more canisters in the CM which we could have just added onto this thing. I felt we had an unlimited supply of LiOH canisters.

LOVELL — So even though we probably didn't even have to go through that other mode, it kept us busy. It's like putting up the antenna in a life raft. Does it work? Maybe it'll keep you busy for a while.

SWIGERT — It's worthwhile mentioning that on the first canister, they did allow us to go up to 15 millimeters and, qualitatively, I didn't notice any change in the environment at all.

LOVELL — I was worried that when we started the sleep-rest-work cycle we would forget about these CO_2 buildups. We had adequate ventilation in the CM, too, by putting the hose through the tunnel.

SWIGERT — We put that vacuum hose on the CDR's hoses. That reached up into the tunnel and was one thing that contributed to the CM getting cold.

LOVELL — Yes. We had one hose in the LM and one hose as far up the tunnel as possible to ventilate the CM. That kept it going.

SWIGERT — Like a snorkel sticking up there. The sleep-rest cycle, the first couple c cycles, seemed regular; but, after that, I really kind of lost track of who was on watch.

LOVELL — They tried to set up something, but I couldn't go to sleep after the acciden I don't know how many hours after that it was when I just quit working. You know, I jus had to see if things were going right or not. It was a little later I realized we couldn't d it for long, and so then we tried to get some- thing going on the work-rest cycl However, I don't think we really did accomplish that objective.

SLAYTON — No. The flight-planning guys were trying to work out one, but we final decided it was better to block out periods and somebody could be sleeping; that is, l you men figure out who should be sleeping.

LOVELL — Yes. That's the only way we could do it. When a guy felt tired, he tried to ge some sleep and another guy would take over. But we just couldn't look in advance. W knew that we had to get some sleep; we couldn't last forever. So, we didn't get muc sleep at all, maybe an hour at a time, I think. Actually, Fred, you got some good sleep?

HAISE — Yes, I did.

SWIGERT — Yes. You slept that one time in the tunnel very well.

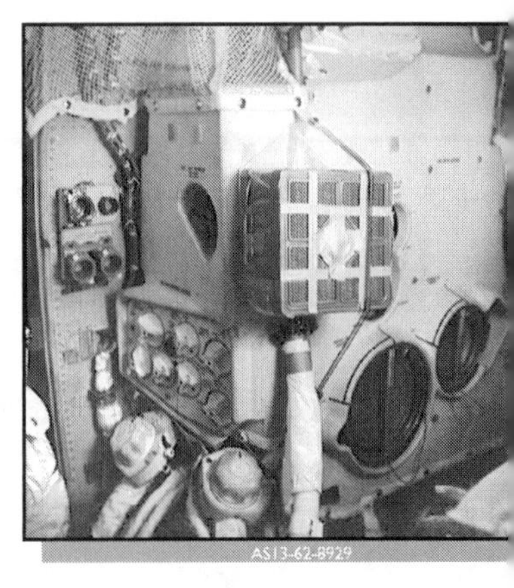
AS13-62-8929

LOVELL — And one time in the CM, I think you got about 4 or 5 hour's good sleep.

SWIGERT — My sleep was very sporadic.

HAISE — I would sleep in the tunnel right next to the ECS unit. This was the warmest place. And I got in the sleep restraint and slept upside down in the tunnel with my face back toward the hatch. I zipped up the sleep restraint and used a string on it to hook myself to the latch handle on the LM hatch so I wouldn't drift away. It must have looked very strange. And the food business was another thing. We may have cut ou selves, without thinking about it, a little short on the liquids. With the water problen we stopped reconstituting. So, the only food we ate after the incident was cubes, th wet packs, and the sandwich spread. We didn't reconstitute another bit of food afte that.

LOVELL — Well, I wanted to save the water.

HAISE — We also didn't have any hot water. Some of the reconstituted food was n too good without hot water.

LOVELL — I might be wrong, but I thought that using all the water we had for juice and then using that other food, was better then trying to reconstitute some of that foo

HAISE — I meant for regular water. You'd drank all the juice we drank and you'd st eaten reconstituted food also.

LOVELL — Oh, yes.

HAISE — We didn't have much water, so, we were short.

LOVELL — Anything else on the comfort and eating?

SWIGERT — I think the only thing as far as eating was that we filled about 35 juice bags when we ran out of CM water. We had gotten about that quantity out when we ran out of water.

SLAYTON — Were you consciously thirsty at any point?

LOVELL — We were, right after the accident. My mouth was dry.

SWIGERT — I don't think we were really thirsty. I think the last day I was thirsty.4

LOVELL — Yes. That was all.

SWIGERT — And then, at that time (about 12 hours after) we ran out of CM water and we had used all the juice bags.

HAISE — Then we knew we were kind of bad on water, I was thirsty. But when I started having urine-burning problem, I drank excessively. That was the old school medicine I remember, which says you ought to flush the system. So, I started drinking to do just that.

SWIGERT — Something that we didn't mention was a leak in the LM water gun at one time, which deposited a considerable amount of water in the LM.

LOVELL — That's right.

SWIGERT — We disconnected the LM water gun and then used the CM water gun.

HAISE — Would you guess about a quart of water, maybe?

SWIGERT — Yes. I think so. It had quite a bit of adhesion and it stuck all the way around the ascent-engine bell cover and then around the part where the LM water gun attaches. It took six towels to sop it up.

SLAYTON — You replaced that with the CM water gun?

SWIGERT — We replaced it with the CM water gun.

SLAYTON — Was it leaking constantly?

SWIGERT — I'll tell you how I noticed it. All of a sudden my feet were wet. My feet were so damn cold. It took me 2 days to get my feet dry. It had completely soaked through my booties and my CWGs and that was my first indication.

HAISE — Jack was in his usual LM crew position, straddling the ascent- engine can with his feet draped in the water pool.

SWIGERT — But, anyway I went back up in the CM. You disconnected the water gun.

HAISE — I shut off the descent O2 valve and no more water could get down the tube; then we just disconnected the gun. Some time later it was brought down.

9.0 LUNAR FLYBY THROUGH 2-HOUR MANEUVER

LOVELL — The next thing we should talk about is the PC-plus-2-hour burn. That w the PGNS DPS stack burn. We got two updates and an update to the original burn. W updated a whole change in the DPS profile. We maneuvered manually and went in PGNS ATT HOLD. We powered up early. It was a mistake, but I wanted to make sure was a little worried about getting into the proper attitude. So I asked if we could pow up and we went through the contingency DPS burn faster in the checklist than w thought we were going to be able to. We were sitting there for almost an hour power up. I kicked myself, I don't know how many times, for powering up early and using that power when we didn't have to.

AS13-60-8703

SWIGERT — We were so concerned about getting this burn off.

LOVELL — We wanted to get the burn off and wanted to make sure of the proper attitude to do it. We powered up early. The ground didn't say anything, so I was thinking seriously of shutting down and starting up again; but I thought, we're already here and everything is all squared away, why don't we just do it?

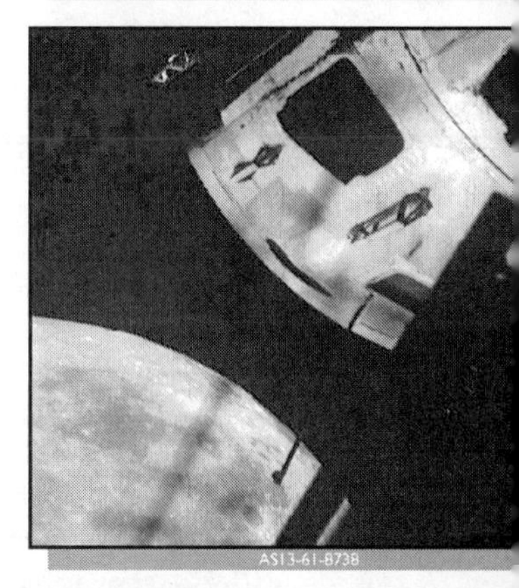

AS13-61-8738

SWIGERT — All our consumables and everything else were based on the fact we were going to be powered up through that point.

LOVELL — So, I guess those were okay. Jack tried to keep track of time for us when the engine started. We were 5 seconds at idle, 21 seconds at 40-percent throttle, and the remainder of the burn at full throttle. But the way it was configured, at 26 seconds, it goes to full throttle. So, it actually beat me going to full throttle at that time. When the DPS burned, it was exactly like flying the simulator. The attitudes were very stable - no oscillations - and the engine was very quiet, very smooth. At the time, I wished I was doing the landing with it. It was a beautiful burn.

HAISE — The only way I could tell that engine was lit was to watch the ENGINE THRUST gage. It was extremely quiet.

AS13-60-8700

LOVELL — I don't have any more information on the PC plus 2. Does anybody else have anything on that burn?

HAISE — No, except during the first one, apparently the gimbals had settled down pretty well and the attitude was extreme good.

LOVELL — We did not change gimbal angle this time on the engine. We went with wh was left over from the last burn. We used a VERB 49 to get the needles to fly by and t which we had to nudge manually. PGNS would hold the stack at this position. It wouldn get it there. It would hold the stack at that position. We also did something else. Didn we power down - Rather, didn't we shut off some thrusters so we wouldn't impinge on the CM?

HAISE — Oh, yes. We did NORMAL, VERB 65 before the burn. I had AGS up to this time and its DELTA-V readout was within the COMP cycle. It was right with the PGNS all the way.

SLAYTON — That was the other factor in this thing; if you had burned the DPS engine for that period of time without the CM. They were afraid to fire it up again because they had no data to indicate it was a safe thing to do, because of the soakback.

AS13-60-8648

SWIGERT — About the SM, you mean?

SLAYTON — No. Just the performance on the DPS itself. On the DPS engine with the shorter burn, nobody was concerned about cranking that UP.

SWIGERT — Oh, I see. For the long burn, they were concerned about it burning up.

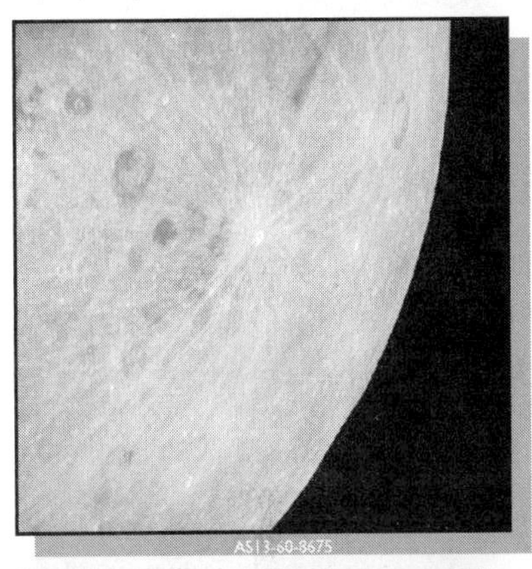
AS13-60-8675

SLAYTON — It was almost to fuel depletion, although we figured you had approximately a 7-percent margin on it. They would have been afraid to fire using it again.

LOVELL — We were looking at that for maybe the last time.

SLAYTON — The next area we should talk about is what occurred in lunar sphere of influence. That is, in terms of observation, photography, and anything like that.

LOVELL — Well, we got photography. Jack and Fred took the cameras on our pass around behind the Moon and took pictures and confirmed that at this stage, the Moon was really gray. They're free to discuss it themselves.

AS13-62-8918

HAISE — Jack and I were crying all the way around and shooting pictures like crazy. I guess we were up in the CM, first shooting out of - window I as we came upon the back-side terminator and subsequently ended up at the right window of the LM as we came around the corner. Then we were also shooting out of the overhead (docking) windows.

LOVELL — I'll be perfectly frank. I wasn't interested in photography at the time.

HAISE — In fact, we keep getting comments from our Commander, "Hurry up, we have a burn to do. Hurry up." And we said, "Relax, Jim, you've already been here before, but we haven't." We were just taking pictures like crazy. But we had LOS at the proper tin and we had sunrise at the proper time and we had AOS at the proper time, so we h a fairly good idea that the ground had good tracking on us.

LOVELL — I might mention the one thing I was worried about in controlling th spacecraft before we did the midcourse. It was back on the free return and that w when using the TTCAs. I didn't know what kind of trajectory changes it would give m After we did our very first midcourse, tracking indicated that we had a 60-mi pericynthion, and I wasn't too sure whether control of the stack with the TTCA w going to reduce that or increase it. I didn't know what it was going to do. I made th comment to Houston. I guess it didn't make that much difference.

HAISE — Well, after our free return, we now had 137.

LOVELL — Well, I wasn't worried then.

HAISE — We went around it and we had lots of fat.

LOVELL — Okay. After PC plus 2, we did a PTC maneuver using PGNS. The procedur was called up by Houston and I can't really recall what it was. I don't have the checkli with me. After the burn, they gave us a method of using the PGNS to do a PTC maneuver. So, then they were going to shut down the PGNS.

AS13-62-8923

LOVELL — I'll have to renege on making any exact comment on that; I don't remember what it was exactly.

SLAYTON — That's the one he had trouble getting into, though. That is when we were really sweating your fuel consumables.

LOVELL — Yes. I was anxious to get the power turned off, too, and to get it cut down again. Right now, I don't recall exactly the type of procedure I actually had to do.

LOVELL — Anyway, I think the procedure worked very well. Two things about this time we also powered up the PGNS and we went to MANUAL on the antenna. That wa basically the procedure we used throughout the entire transearth coast.

HAISE — We did go into the power down on page 5 in the Contingency book. That' the first real power down we had; that's after you got that PTC down.

LOVELL — Yes, we went through it to where we had gone down to minimum power.

HAISE — I thought it was pretty nice the way they went to a section and page in thi book - which was clearly appropriate for it - and made the deviations from tha according to the situation they wanted and had us update that. It really wasn't ver extensive. We just followed the script and powered it down; it was very simple.

10.0 TRANSEARTH COAST

LOVELL — Basically, the transearth coast consisted of the spacecraft in a powered down situation; it was in somewhat of a PTC attitude with a rate that would keep th thermal conditions consistently even. We performed one manual midcourse maneuve at 105 hours and then went to a power-up situation and entry.

10.1 SYSTEMS

LOVELL — During the transearth coast, all systems were powered down, except for the communications system, and the ECS in the LM that was necessary to keep us alive.

10.2 NAVIGATION

LOVELL — Navigation was performed by ground tracking, and by the midcourse correction maneuvers. The midcourse corrections used a procedure that had been generated earlier and, from the crew point of view, was very simple to perform. The procedure is to use the terminator of the Earth to align the spacecraft to either retrograde or posigrade position and then to perform either a retrograde or a posigrade burn to change the entry angle. It was very simple to perform this procedure in the configuration we were in. We accomplished midcourse correction 5 on the AGS, using the TTCA to control roll and pitch.

LOVELL — It was almost a three-man operation. Fred did the ullage; I started it on time; and Jack called the time for stop. We also set up a timer. The correction was performed with 10-percent throttle. I controlled roll with my TTCA, and Fred controlled the pitch. Jack yelled "Shutdown," and I stopped the engine. Jack brought up a good point to mention while we are discussing this particular burn - an attitude check using the position of the Sun. If we had been in the proper attitude with respect to the position of the Earth, a pitch would have been valid because of the position of the Sun at that time. So, the idea was to position the Sun at the top of the reticle in the AOT. This procedure worked well, and that's how we got our pitch alignment with roll and yaw - using the terminator.

HAISE — I checked both the COAS and the AOT, and both were right where they should have been. It was a beautiful job.

LOVELL — That technique - a manual burn using the AGS - does work. We ought to think about that kind of burn for the future.

10.4 PASSIVE THERMAL CONTROL

SWIGERT — According to Houston, the earliest possible time for the SHe tank to blow was 107 hours, shortly after the burn and after we had gone to PTC. The latest time they expected it to blow was 110 hours. So, we began to watch the tank after this burn, and we discussed it among ourselves whether we would hear it blow, how we would notice it. I don't think we came to any conclusion, though. We never did hear it blow. I think Fred was sleeping at the time it went.

HAISE — It didn't bother me.

LOVELL — Another thing that was particularly good about this manual burn was that by using AGS, we got a good ball alignment after we got into position. So, we actually did not use an outside reference for the burn. We got in position using the position of the Earth, but then we got the ball aligned. with the AGS, and we used the AGS ball for attitude control during the burn. We also used the attitude deviation needles; then, we went back to PULSE. They told us to roll 90 degrees to get us back in the proper attitude for PTC, which we did. They said to null rates within 0.05 deg/sec. I didn't see how this was possible, but we nulled them as much as we could. We got the attitude down, and then we put in 21 clicks, either 21 or 12 clicks, of yaw. There was a little roll and yaw, but no pitch. There was also some coupling when we got started. There was some debate with Houston about whether we should start or not. I think the decision was that we could go. This gave us a pretty good reference, because the Earth and Moon would appear in our windows. After a while, Houston came up with angles. This PTC attitude was very good until the SHe tank vented. At the time of venting, I think Jack Lousma asked if we saw anything. We did see it out the right window, the LMP's window. It reversed yaw completely and gave us a coupling in pitch and roll. That was the attitude in which we remained for the PTC.

HAISE — Also, the venting about doubled the rates.

LOVELL — Yes, it really spun us up.

HAISE — It not only stopped our rate in one direction, but it doubled our rates in th
other direction.

LOVELL — That was supposed to be a non-propulsive vent.

SWIGERT — We were really switching antennas quite rapidly for a while.

LOVELL — But there was one interesting thing. From the time we started it, we did
touch the thrusters at all; attitude control was

LOVELL — strictly on its own, except that the rates slowed down. By 5 hours befo
entry, it had slowed down to where the Earth would come by the window only on
every 12 minutes or so.

SWIGERT — I attributed that to the sublimator.

LOVELL — There was something else venting, too.

SWIGERT — Yes, we had some venting from the CM periodically.

LOVELL — We forgot to mention that. There was something all during this period, whi
we were checking through the AOT, that was venting out the SM. We attributed th
venting at that time to the hydrogen tank.

SWIGERT — What I thought was happening was that the hydrogen tanks would go u
pop the relief valves, vent for a period of time, and then, the venting would stop. Jim an
I were trying to see whether we could see stars as we went around. We found tha
during periods of no venting, there were attitudes from which we could pick out who
constellations. Jim picked out Scorpio and Nunki. I picked out Acrux, the Alpha and Bet
Centauri, and the Southern Cross. We could see whole constellations, but, when th
venting started, it was immediately apparent that the stars were gone.

LOVELL — So, we had SM venting of some sort, which we thought was hydrogen.

SWIGERT — It was just before or just after this midcourse correction that Housto
passed up some procedures for powering up the CM with the CM batteries, so that the
could get some telemetry and read some instruments. I did this, and I also read som
voltages. They read the telemetry, and then we shut off the TM. So there was a perio
of about 5 minutes that we had the CM powered up.

LOVELL — When did we start doing the battery charge?

SWIGERT — After the midcourse correction. Houston passed up a procedure to powe
up the CM using LM power. We powered up main bus B, and that procedure worked lik
a charm. Shortly after that, we began charging battery A, and Houston estimated tha
charging would take 15 hours. We checked out differential current, and, sure enough
there was an 8-amp difference. Then, we finally had a little confidence that Housto
knew what they were doing. Actually, we had confidence all along, but it was ver
comforting to know that they were that accurate on the amount of amp-hour
consumed.

LOVELL — At the same time on the transearth coast, we were passed a procedure t configure the CM canisters to scrub CO2 out of the LM ECS system. That procedure worked very well, and we had no problems. We powered up the LM just before entry. We got it powered up 2½ hours early. One reason for early power-up was heating. Again, we were very cold. We thought of using the window heaters, but I was very reluctant to use them. They use quite a lot of power. Also, they were really cold and wet, and we were worried about somebody applying heat to them.

SWIGERT — We tried not to disturb the environment. We had talked about pressurizing internally with the PLSS or the OPSs to make sure that we didn't use the CM REGs or cabin REGs on descent. We had determined that we would not do anything to disturb the environment on the inside.

LOVELL — Along with that, too, was the question, "Should we break out the suits and put them on?" Right now, it's still a little bit hazy in my mind whether we should have donned the suits. Without the suits, we were so much more maneuverable, especially in getting rid of urine and moving around, that I was reluctant to put on the suits.

HAISE — The problem with suits is that your body can't breath in them. With no hoses plugged in, there is no flow. Even as cold as it was, inside the suit one starts getting hot and sweaty. You've got to crawl out of them about every 2 hours. Then, you're exposed soaking wet to that chilly atmosphere.

10.9 STAR/EARTH HORIZONS

LOVELL — So, we powered up. It took about 30 minutes before the LM started getting warm. The windows cleared, and we never did use the heaters on the windows. We stopped the PTC attitude. The first attitude maneuver was to the Earth. I wanted to make sure I got the Earth in sight again because I knew I was going to do midcourse correction 7. We squared away whether we would do an alignment or not, and we did. We did Moon/Sun alignment with the LM. I'm not so proud of the alignment, but I really don't know what the situation was. It was a stack. We had been doing it with the TTCA. The way we finally did it was: Fred maneuvered, and I told him how to maneuver so that we could get the Sun and the Moon across from the hairlines. I tried to put the mark in as soon as it went to the center. We got about 1-degree star angle difference.

HAISE — 1.1 degrees.

LOVELL — Yes, something like that. It's a pretty big torquing angle; but, for what we were going to use it for, I thought it was completely adequate. Again, we used the filter for the Sun. We ought seriously to consider using the Sun and Moon for alignment, because when you're out there, you just don't see stars. You just can't rely on getting good star alignment if something's wrong with the CMC. You have to use something, and then the only thing you've got is the three bodies.

HAISE — You can't do them in a simulator.

SWIGERT — You cannot do Sun and Moon alignment; no planets are available in the simulator.

LOVELL — Because the Sun shield is so thick, looking at the reticle is very difficult. It's hard to see and hard to read.

HAISE — It's hard to pick up the reticles.

LOVELL — Yes, it's very hard to pick up the reticles with the Sun. Maybe we're going to have to be satisfied with rough alignments, with the AOT when you have a maneuver stack like that.

HAISE — It made us feel very good that we had picked up Jack's alignment before we'd powered down the CM.

LOVELL — That's how we did midcourse correction 5. I wasn't really worried about that also because it could burn it on the Earth.

LOVELL — But, I wanted to make sure that he got a good alignment from the CM especially because the LM was getting the rough alignment and then doing the transformation backwards, which we had never done before. Going back and giving Jack the angles to put into the CM allowed him to get a rough alignment in the CMC so that we could do a P52. That's what we wanted to do.

SWIGERT — Houston calculated those angles and passed them to-us. One key thing to this whole time line was doing that Sun/Moon alignment. That gave me a lot of confidence; even if I never saw any stars or we didn't get my alignment, we had a good enough alignment to get in.

LOVELL — This really wasn't the original procedure. Normally, Jack would have been on his own to get an alignment.

SLAYTON — When we discovered that we had about 100-percent margins at the time we told you to power up. That's when you started up. The best way to warm you up was to power up, and once you'd done that, you might as well go the other route also.

LOVELL — Well, I was a little bit worried about having to go to a Moon attitude and then a Sun attitude for Jack. It was a lot easier for me to go to those attitudes and then do this rough alignment, because everything was right in the LM cockpit. Then too, Jack gave me some angles to go to. I thought that approach was best.

HAISE — The technique was to align in the same manner we usually do on the terminator of the Earth, just a pure pitch. I was looking through the AOT, and I'd tell him when it was right in the plane of all the bodies. Then I'd tell him when the next one trickled in. We'd stop around that one and go to work aligning on that one. Then, it was just another pure pitch from there to pick up the next body. It was pretty straightforward attacking it that way.

LOVELL — We finally got our rough alignment for midcourse correction 7. Houston called up and asked if we would like to do a PGNS burn, and we said "Fine." That's where I really got confused. I guess Dr. Berry thought I was tired. Well, maybe I was tired. I got Fred to go over and check those switches, too, and I think Charlie finally told me what was wrong. We maneuvered manually to what I thought the attitude would be, based on the angles we were reading. Then, I went to PGNS AUTO, and it drove the spacecraft there, but the needles never nulled out.

HAISE — Two of them did not null.

LOVELL — Yes, two of them didn't. I was worried about whether I should null the needles to get the proper attitude, or whether I should hold what I had, because the computer knew what the attitude was. As it turned out, it really didn't make any difference; because, no matter how I burned, it would have been okay. I think you know I forgot to PROCEED 50 18 or something like that. Charlie said that I should have. Anyway, that's what got me confused. That's why we had a delay. I wanted Houston to find out what the situation was. I really preferred to do the old AGS burn again, because we had done it one time and I knew that it worked. But, this burn was okay. We burned ECS this time. I guess it was 3.1 ft/sec. That worked fine.

HAISE — That ended up being an AGS burn, Jim.

LOVELL — Yes. Houston finally told us to go to the AGS. There was a lot of confusion about this time. I guess we'll get to it a little bit later on in discussing the entry. I guess there was some confusion on the ground, too. But, anyway, that burn was performed with the AGS, and there was no problem with attitude control once we made the burn. There was some confusion on my part about the exact attitude I should be in. I was also worried about the fact that the Earth wasn't perpendicular the way it should have been. I found out it should have been 8 degrees off in attitude. Our midcourse correction 5 really should have been that way too, but to make it easy on us, they wanted to do it perpendicular.

10.16 EATING, REST, SLEEP, FATIGUE

LOVELL — Fred woke up with the chills before we did midcourse correction 7.

HAISE — Yes. I wasn't sure what gave me the chills. I was back in the CM at about that time, and I had to go to the bathroom. I stripped naked in the 42 degree temperature and ricocheted around touching bare metal, and it just chilled me to the bone every time I'd touch anything. You can't help but bounce all around in there. I was really cold for about the next 4 hours. From that time on, it sort of began to catch up with me. I began to feel tired. Before that, I really didn't feel much effect at all.

10.23 FINAL STOWAGE

SWIGERT — One thing, Jim. You and I had gone down, and we had practiced installing the CM hatch.

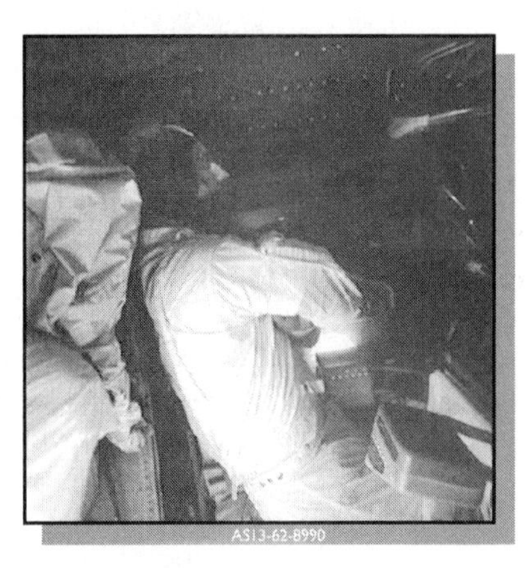
AS13-62-8990

LOVELL — Yes. That's another thing.

SWIGERT — It also completed the stowage list, which was read up to us from Houston. So, we had everything done. I had completely stowed and tied everything down in the CM, I had gotten the strut lanyards in place, and we had the CM all ready to go before 6 hours 30 minutes before entry.

LOVELL — That's why I called down and tried to simplify the procedures. We wanted to run through them and make sure we didn't have any conflicts between what Jack was doing in the CM and what Fred and I were doing in the LM.

SWIGERT — We had the procedures worked out. I copied down that long procedure that Ken read up to me, and Fred and Jim copied down the LM procedures. Then, we sat down and went through each procedure item by item to make sure that we interfaced correctly, and we found that everything worked pretty well. There were only one or two items that we had to question Houston about. They had us pulling one more circuit breaker in the CM than we had. But, generally, it was a well-followed procedure, it was well read out, and we had no problems at all integrating the procedure.

LOVELL — The last couple of hours after midcourse correction 7, Jack brought in the probe and drogue, and we stashed those in the LM. We also got a lot of the debris out of the CM, and we put the trash in bags in the LM. We latched down the ISA because we put a lot of stuff in there. We latched it on top of the PLSS on the floor in the LM. In the last few hours, we had everything we were going to jettison in the LM already there.

HAISE — We took a lot of pictures of this. It was pretty interesting looking, although the lighting is not very good inside the LM.

LOVELL — The midcourse correction was performed at EI minus 5 hours, and at 4½ hours, we went to the SM jettison procedure.

SWIGERT — At this point, I had to pressurize the CM RCS system.

11.0 ENTRY

LOVELL — We'll go to that point between the midcourse 7, which was the las midcourse, to SM jettison. We had to power up the RCS system and do the checkout

SWIGERT — Yes. I did this in ACCEL COMMAND. The thrusters sounded just like th simulator. I followed the checklist. I checked everything off, all these items; I checke them off with a pencil. We had good thrusters on both rings; all 12 thrusters fired.

LOVELL — We heard them from the LM.

SWIGERT — We could probably have seen some of them. I wondered because some c those thrusters pointed almost directly at us.

LOVELL — The AOT had them in sight, too.

11.2 CM/SM SEPARATION

LOVELL — The separation procedure, which was called up to us for separating from th SM, was very good. I don't know the details of the checklist that Jack went through When we got to the point to jettison the SM, I thrusted up. Then, Fred went to verif that Jack was going to throw the right switch.

SWIGERT — I wanted Fred there to make sure that I raised the CM/SM SEP switch an not the CM/LM SEP switches.

HAISE — I did go, but he had gray tape over the LM SEP switches. I figured that wa enough of a safeguard, and the way Jim thrusted, I needed to be there to control th pitch again with the TTCA.

SWIGERT — You should have seen Fred when we got back there. I was all ready to go I had the logic up and I was ready for pyro arm. Fred said he would get a GO from MSFN. Then I reminded him that we didn't have any telemetry and MSFN couldn't giv us a GO. When I asked if he was ready, he looked at me with a wistful sigh, as if, "Well go ahead." I put power up, and I could hear the relays clicking.

LOVELL — We debated putting the hatches on, but we thought we might as well go al the way.

SWIGERT — I was worried if we'd had some sort of relay, but both power system armed beautifully. I was sitting there all ready to go, and Jim thrusted and yelled, "Fire," and I hit the switches, and the SM went.

LOVELL — Did you hear me from all the way down in the LM?

SWIGERT — Yes. I safed the pyros immediately, put the guards down on the CM/SM SEP switches, and went over to window 5 because I was supposed to be the first one to se it. I kept watching while Jim was pitching around.

LOVELL — The SM jettison part of the maneuver pitched me down instead of pitchin me up, which was the wrong direction. I was trying to get back in control to pitch up again. And, of course, we were in that CM/LM configuration, which we have never SIMed

That was the first time I ever had an ACA that would operate. Finally, when I pitched up, I saw it go by, and I grabbed one Hasselblad and took pictures through the overhead window. I don't think I had all the minus-x thrusting that I wanted.

HAISE — We got about 1 ft/sec, which in my mind, I didn't argue with at the time. It would have been nice to have had a little separation right there, and they didn't allow for any with the procedure that they gave us. If Jim had been fast on that TTCA, we'd have pitched up there and the SM would have been 6 feet away. How far is it from the LM?

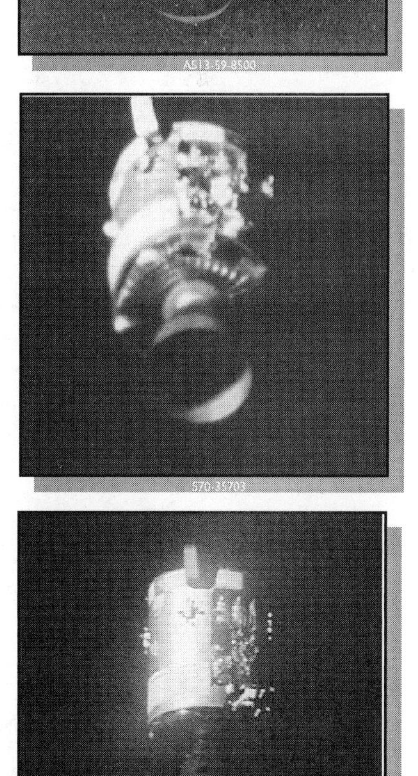

AS13-59-8500

S70-35703

AS13-58-8458

SLAYTON — It was figured that you'd have about 70 to 80 feet by the time you'd pitched through.

LOVELL — Well, it was about that when I looked up, and it was straight ahead. However, when I got to the forward windows, it was farther away than that.

SWIGERT — It was good. When Fred called, I came on down because I had the 250-mm lens on the Hasselblad.

SLAYTON — Did you get pictures with all three cameras?

SWIGERT — We got pictures with all three cameras.

LOVELL — The pictures I got were through the overhead. After that, the SM floated in front of the windows over on the right-hand side, so I didn't see it again. Fred and Jack got pictures then.

HAISE — They told me to use the lunar surface camera and gave me f-stops and speed, but didn't really specify a magazine. I got a lunar surface camera and slapped on lunar surface film. What they really wanted was CM ASA 64.

LOVELL — When I first saw it, I saw that the whole panel, the core panel, was missing off the SM. I could see the interior. I couldn't see any specific damage, but I didn't really know exactly what I was looking at, although there seemed to be a lot of debris hanging out. It looked like insulation-type material hanging out, and the panel went all the way back to the high gain antenna. We saw a streak on the engine bell, and that's about all I saw before I got the camera and started taking pictures of it.

HAISE — I guess the two things that were identified very promptly as specific objects sitting out there were two barrel-looking things. I could see one set of tanks that looked to be in place. The streak on the engine was a kind of a green-gray color. When I first looked at the bell, I actually said that it looked like it was cracked. Then it turned around in a yaw maneuver and I looked straight up the bell. It was in good shape; it was not cracked.

SWIGERT — I didn't get down there until much later in the time line and it was at quite a distance. I didn't distinguish any of the streaking. I could distinguish that a panel was missing because of the color difference in the other panels and that particular panel. The SM was in a very slow yaw maneuver, which gave us time to observe it all the way around. I did take about 28 pictures with the 250-mm lens. I used the settings Houston gave me, which was f:8 at 1/250th, and it appeared to me when I saw it that the SPS bell was intact. I did see some debris hanging out of the side and even hanging off the high gain antenna. When the SM turned around, either the debris was on the high gain antenna or was sufficiently far out to the side that it appeared to be hanging off the high gain antenna.

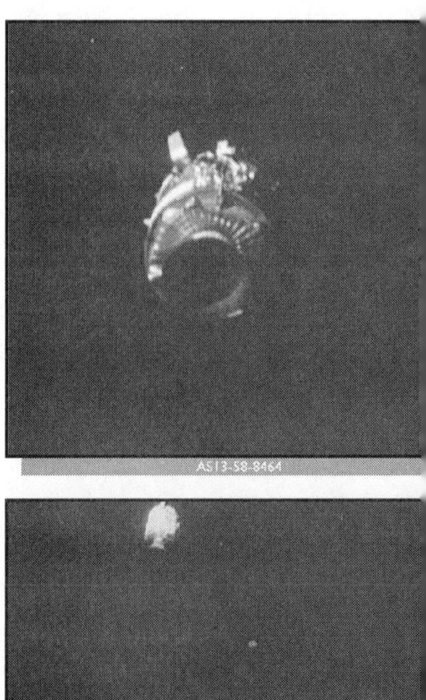

AS13-58-8464

LOVELL — That's what I thought. Something got to the high gain antenna because it did not look natural back there.

SLAYTON — Did you notice whether the barrel was a fuel cell or hydrogen tank? Did it look like it was displaced, or did it look like it was in the proper position?

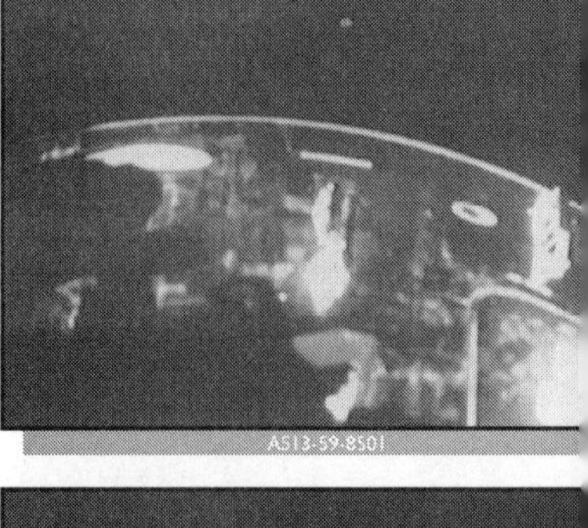

AS13-59-8501

HAISE — No, it just looked like it was - where I would expect it. I guess, from a few schematic pictures I've seen, it probably was a fuel cell. However, it looked physically mounted the way it should have been.

LOVELL — I didn't see anything big hanging. I saw a lot of stuff straggling out; you know, floating in the breeze.

SWIGERT — I guess the noise at SM SEP was what I expected from what I heard of on previous flights.

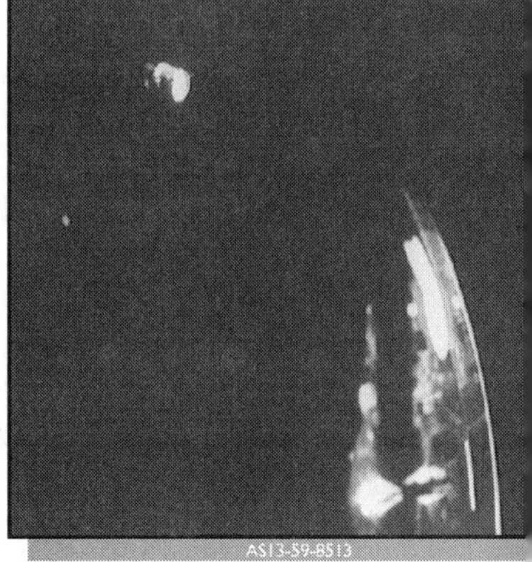

AS13-59-8513

LOVELL — At about that time, we had a discussion with Houston about controllability. I went to a PGNS ATT HOLD mode and used PGNS pulse to fly. They wanted to go to AGS. I disagreed with them and I finally went back again to AGS. However, the pulse in AGS just wasn't where I liked to fly it, and it was adequate in PGNS. The CM and the LM together made a very comfortable mode of flying. I talked to Charlie and I assume they had the fuel computed in AGS and that's why they wanted me to fly it in AGS.

SWIGERT — There was a very large reflection off the LM sublimator and off one of the LM quads. Also, at that time, we were venting something. It appeared to come from the umbilical region, and I surmised that perhaps one of the cutters didn't cut through one of the water tubes or something like that. We were losing some sort of fluid. I asked Jim to come down because I couldn't distinguish any stars because of the stuff that was venting out. I asked Jim to come down and he couldn't distinguish any stars either.

LOVELL — No, I couldn't see any stars through the sextant at all. Essentially, I was keeping the CSM SEP attitude with the LM. Finally, we had to go back to that 91 degrees. They gave us four stars that they thought we could pick up, so we held it there and I held that attitude until 2½ hours.

SWIGERT — I started the power when Jim gave me a countdown.

LOVELL — The next thing, of course, was the alignment. We just waited while the power was up and all squared away.

SWIGERT — It took Houston a long time to lock up on telemetry and it turned out that our attitude was bad. They were trying to transmit through the LM. It took a long time for Houston to get locked up so they could give us the uplink.

SLAYTON — No, there was a little confusion there, I think. They wanted you to use AGS. They didn't care what you used to maneuver, it was what you used after you got into attitude they were concerned about. They were confused about that one.

LOVELL — Oh, maybe that was it. Anyway, I wanted to use PGNS. The ACA PGNS maneuvering with the DAPs load is sufficient to control the spacecraft, that's all. The power-up wasn't until 2½ hours. We already had the rough alignment with the LM. We already had most of the equipment into the LM, so there wasn't much there.

SWIGERT — We really didn't have much to do; we were kind of sitting and waiting.

LOVELL — I went to the SM SEP attitude because that was a good attitude.

SWIGERT — We were looking to see whether we could see stars.

LOVELL — That's right. We went back and forth to see whether we could see stars, and we actually maneuvered from 91 degrees to about 115 degrees to see if that was a better place to see stars. We debated and oscillated back and forth with various angles and pitch.

SWIGERT — I was sitting there just chomping at the bit to get those updates, because I couldn't get my alignment until they got the updates done. We were behind on the time line. I kept looking where we should have been. We were about 5 or 10 minutes behind by the time they finished their alignment. I set the clock and the mission event timer, we got the coarse align angles in, and I went into P52.

LOVELL — There was something wrong before that, though. Why couldn't we get the computer on the line?

SWIGERT — When we powered up, the IMU circuit breaker, the heater circuit breakers, were punched on from the LM power. During LM power down, I was standing by. I pulled the LM circuit breakers as soon as I got word from Fred. That put us all on CSM power. The first thing in powering up the computer is PROGRAM 06 with the flashing 37. They said to proceed and I would not get the STANDBY light and the DSKY would blank immediately. I tried proceeding, but I wasn't holding it long enough. I slithered back up into the LM and talked to them. They said go down and hold it. I did, and then the computer came up. We got both of those things resolved and they didn't cost us any time on the time line at all.

LOVELL — Why didn't we have COMM in the CM, or was that later on in you checklist?

SWIGERT — Yes, that came later on.

LOVELL — Jack had to come on down to the tunnel and put on a headset to talk wit Houston.

SWIGERT — This was at a time when we were still using LM power. We didn't have th CSM powered up. But those were just minor problems. We finally got them square away and I got the coarse-align angles in and immediately started a P52 and got PROGRAM ALARM. I knew right away what I'd done. I hadn't set the REFSMMAT fla and drift flag, and I had to reset. I set the REFSMMAT flag and went into P52 and, let i PICAPAR, Rasalhague, and I let it drive to that, I couldn't see it. The next thing I did wa pick star 36, Altair, and let it drive that. The two stars were Altair or Vega. At that time Jim was saying to hurry up because the Earth was getting bigger. He was chomping t get out of the LM. I did pick up a star in the telescope nearby. I put it in the sextant marked on it, put it in star 40 and let it drive to that, put it in the sextant, marked on i and got a star-angle difference. It was five balls. I proceeded and asked MSFN if I shoul torque.

LOVELL — I don't know what we would have done if it wasn't. I guess we would hav just held off jettisoning the LM and tried to get the hatch back again to get a better sea

SWIGERT — We did have some time that we could have put on the suits, although i would have been pushing it. We actually separated early. We asked if we could SEP early Jim was maneuvering to LM SEP attitude.

LOVELL — I held CSM SEP; it was all squared away. Then they said go to LM SE attitude, and I got it right here. That's where I said that was a lousy attitude. I found ou by talking to John Young that he had tried it and had the same problem. Here it is righ here; roll 130, pitch 125, and yaw 12.4 degrees. So I started going there and I kept gettin stop, because of gimbal lock in the CM. I wondered how to get to the attitude in the L without going through gimbal lock in the CM. We had to go way around.

SWIGERT — That's exactly what we did. I would tell Jim, and he would get a pitch rat started, and then he would get us away from gimbal lock. I would say to roll a little, the he would roll a little bit. Then, we continued to pitch. We just GCA'ed into the thing.

LOVELL — That's what used up much of the gas in the LM. That's where I thought w also had a discussion on whether we should be using AGS or PGNS. I preferred PGN right there. Anyway, we got to that attitude.

SWIGERT — It put us about 65 degrees in yaw on our CM gimbal.

LOVELL — That was uncomfortable.

SWIGERT — It was!

LOVELL — We were very close to gimbal lock. I questioned whether that LM SE attitude is that critical. Was it so critical to be at that attitude, or would it have bee better to stay away from gimbal lock in the CM? At the time, we didn't have a backup We didn't have the BMAGs powered up. If we had gone into gimbal lock, we would hav had to start from scratch again.

SWIGERT — We had one BMAG powered up at that time, and we only had one FDAI powered up. I had the GDC powered up, but, of course, with only one FDAI, it had to switch back and forth. I would recommend that if we had to do this again we stay away from the CM gimbal lock region. When we did SEP, we got a continuous pitch-up in the CM. I was in MINIMUM IMPULSE. I had my MANUAL ATTITUDE switches in MINIMUM IMPULSE, but I had my DIRECT RCS switches ON. When Jim said I was getting near gimbal lock, I just gave it a quick beat down with the DIRECT switches and started pitching down. No sooner had we stabilized then we started to pitch up again. We had a continuous pitch-up in this CM all the way through this thing.

LOVELL — The first thing we had to do was to maneuver away from the gimbal lock attitude, get on the bellyband, and get set up for entry attitude.

SWIGERT — We went to entry attitude, and I got it as close as I could stabilized. Jim did a secondary star check and a star path. We had a lot of confidence.

LOVELL — As I look back on it now, I am trying to see what we would have done if we had of gotten the gimbal lock and lost our alignment. Houston told us about the Moon, and it was a perfect body. The only thing that we could have done would have been to maneuver around to the horizon and find the Moon. We probably could have gotten there that way, but it would have been difficult.

SWIGERT — That was a little too close to gimbal lock.

LOVELL — We had a discussion about that. My ball was aligned. with 3, 3, zero. The only thing that I was worried about was roll. I thought yaw was good, but I had a hard time getting there because of his gimbal lock, and, of course, I didn't want to go into gimbal lock either.

SWIGERT — I thought our coordination there was good.

LOVELL — It's a gas user. One time when we were getting close, I just went to ATT HOLD. I could hear those thrusters firing.

SWIGERT — We had a large rate.

LOVELL — I'll have to take a check on yaw.

SWIGERT — Then I said, "Start up yaw."

LOVELL — Yaw was 360 degrees. I thought sure we had it that way. It took me a long time to get it around that way. Maybe for some reason when I finally got it out of gimbal lock, I went to the other direction.

SWIGERT — The LM/CM DELTA-P was 3.5.

LOVELL — Yes, we bled it down so we wouldn't have too much pressure in the tunnel when we separated.

11.8 CM/LM SEPARATION

SWIGERT — On the CM/LM SEP, the LM moved smartly away and the noise didn't appear to be excessive. Immediately, I noticed a pitch-up in the CM about the same time that Jim called a GIMBAL LOCK. We were sitting right near the gimbal lock limits, and he had a GIMBAL LOCK light on the DSKY status lights. I pitched down, using the DIRECT RCS switches. I came off then with the three MANUAL ATTITUDE switches. We did get out of the gimbal lock region and stabilized. All the time I noticed a continual pitch-up rate. However minor, it would definitely affect the attitudes, continually pitched up. We went from the separation attitude down to the entry attitude, and Jim performed the sextant star check. Our maneuver from that was to the Moon-check attitude, and we maintained this attitude in a kind of wide deadband fashion until our Moon-check time.

HAISE — Maybe it was because of your being busy with the test, whereas I was just an innocent bystander sitting over there, but the LM SEP impressed me as being the loudest pyro event that I heard from stem to stern during the mission.

LOVELL — It was encouraging, I know.

HAISE — It was very close, and it impressed the heck out of me, I know. It actually rocked me off my seat toward the window when it let go.

SWIGERT — I didn't notice that, perhaps because I had both hands at the controls.

LOVELL — Now, my only comment on the Moon thing is that the Moon was a perfect alignment factor right through the center hatch. I could see it work its way right on down. I think that you could use that Moon as an entry point if you had the horizon also.

SWIGERT — Yes.

HAISE — The Moon isn't always going to be there, though.

LOVELL — I know. It was there on Apollo 10; it was there on Apollo 12, and it was there for this one.

SWIGERT — This particular time, it was the Moon that occulted at the correct time.

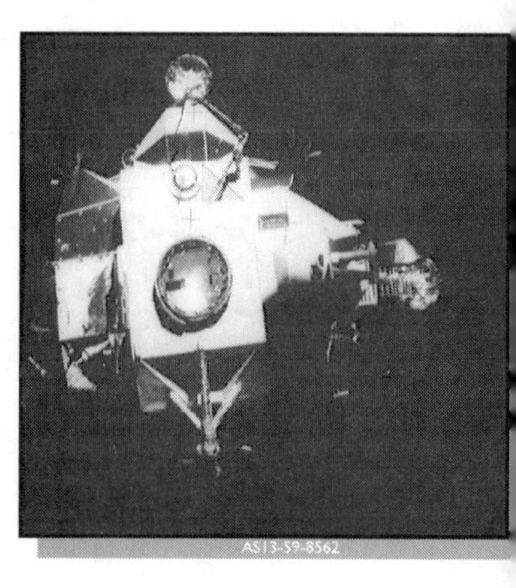
AS13-59-8562

LOVELL — Yes.

SWIGERT — Remember, we counted down to it; and, blink, it went out.

LOVELL — Yes, they were pretty accurate on that.

SWIGERT — They were. After the Moon occulted, we pitched down to entry attitude again and stood by for .05g. We hit our RT, and counted up to 28 seconds, which was .05g time according to the pad. We got the change of displays and the computer. However, the EMS did not start within 3 seconds, and I initiated the EMS start manually, by going to the backup on the EMS. It was apparent when we hit the 4000 ft/sec on the V-axis drive that the EMS was slightly behind in range to go over what it normally is because of the late start. The corridor checks came out okay; it gave us lift vector up.

LOVELL — That corridor was fine.

SWIGERT — The computer drove it throughout the entry and responded well.

LOVELL — Yes. The computer was running right with us.

SWIGERT — Yes.

LOVELL — The control was right with the guidance.

SWIGERT — The g-meter, the EMS g value, and the CMC g values all checked very closely.

LOVELL — And they were close to pad values.

SWIGERT — It was a very quiet entry, I thought.

LOVELL — Not noisy at all, was it?

SWIGERT — No. I've never been through any other entry, but I was quite impressed with it. Of course, Jim kept briefing us on what to expect, and we did get the small bit of ionization just before .05g. We got just a little bit of glow.

LOVELL — Yes. We started getting a glow; in fact, we were all lit up before we started getting any g's.

SWIGERT — The CMC control mode was quite effective. We made a single-ring entry on ring 1, and we had plenty of RCS fuel.

LOVELL — I saw nothing of that whole entry that was off nominal. Everything worked the way it should have worked. We had automatic apex cover JETT and drogue deployment. You verified the drogues, right? You saw the drogues?

SWIGERT — Yes. I got two good drogues. In fact, I called that going to 18,000 feet with two good drogues. Did Houston hear that?

LOVELL — They made one call after blackout before we put on anything, I think.

SWIGERT — Yes.

LOVELL — And Houston confirmed that, too. The COMM, again, was beautiful.

SWIGERT — Yes.

LOVELL — It was okay except for the blackout,

SWIGERT — Yes. And there didn't appear to be any unusual oscillations on the drogues. The drogue had us damped out pretty good.

HAISE — There was another loud metallic sort of noise when the pyros went on the apex cover. When that thing went, there was a clang. Again, that would be my second-order number.

LOVELL — All that is sort of happening above your head there.

SWIGERT — Just above 10,000, we got main chute deployment with three good chutes in a reefed position, and they dereefed in just about the proper 8-second interval. We could hear the recovery choppers calling us. The communications were good during the whole descent.

LOVELL — The last thing that slipped out was the main chute after cold soaking for that time. After that, Fred fell asleep.

SWIGERT — Jim read the checklists. We proceeded down the checklist. We burned RCS and purged. It was a brownish purge and it left a film on the side windows and the rendezvous windows.

HAISE — On both side windows.

SLAYTON — I saw that on TV.

SWIGERT — That's the purge rather than the burn. I never realized that before. I always thought it was the burn that did that.

S70-17646

REEDER — Too bad you missed the recovery on TV. It was the best one yet.

LOVELL — That's great. I'm glad the TV worked out for that part of it. I'm glad you had a nice sight, instead of hearing something whistling through the canvas.

LOVELL — Okay. Visual sightings and oscillations; all that was exactly like cake, even better.

12.0 LANDING AND RECOVERY

12.1 TOUCHDOWN - IMPACT

LOVELL — The impact was as designed because the sea state was slow and we knifed in. It was less impact than Apollo 8 and we stayed stable I. Fred cut in on two circuit breakers and Jack jettisoned the chutes.

S70-35652

HAISE — I think I had only one in at the time he hit the button, but that's all it takes.

LOVELL — Everything worked exactly like the checklist worked.

SWIGERT — We just went right down the checklist, item by item.

LOVELL — The only thing we forgot to do - I guess I forgot to punch to get lat-long out of the computer.

S70-35638

HAISE — The last time I saw, we had miss distance of 0.8 mile. The choppers asked us if we had lat-long laid out and, at that time, we didn't have. It might be of interest to point out that, after we hit and had gone through this smoke and entry, we were all three sitting there on the couches, laying in that 81-degree water, blowing frosty smoke out of our mouths. It was still icy cold in the CM.

LOVELL — I don't think we ever got swimmer communications, did we?

SWIGERT — No, no swimmer communications.

Astronaut James A. Lovell, Jr., Commander for Apollo 13 the third manned lunar landing mission. (above left)
Astronaut Fred W. Haise Jr., lunar module pilot for Apollo 13. (above right)
Astronaut Thomas K. Mattingly II Prime Crew Command Module Pilot for Apollo 13. Mattingly was removed from the prime crew at the last minute due to an accidental exposure to rubella against which it was determined he had no immunity. (below left)
Astronaut John L. Swigert replacement Command Module Pilot for Apollo 13. (below right)

February 1970, Jim Lovell during a moon walk practice at Kennedy Space Center Florida. (above)

Lovell and Haise during egress training (top right).

Fred Haise practices deploying the Solar Wind Experiment at the KSC Flight Crew Training Center February 4th 1970 (right).

The Apollo 13 Saturn V space vehicle undergoes checkout for launch at the Kennedy Space Center's Launch Complex 39. (right)

Preparations are made to remove the Mobile Service Structure during the terminal phase of the Countdown Demonstration Test (CDDT). March 24th 1970. The ill-fated Bay #4 is in the foreground to the left of the reaction control thruster assembly.

Jim Lovell pilots the Lunar Landing Training Vehicle at Ellington Air Force Base (below)

The Apollo 13 (Spacecraft 109/Lunar Module 7/Saturn 508) space vehicle is launched from Pad A, Launch Complex 39, Kennedy Space Center at 2:13 p.m. (EST) April 11, 1970.

Inside Service Module Bay #4. Oxygen tank #2 is at left center, two fuel cells can be seen above, while below is the top of the Liquid Hydrogen tank. (right)

The Earth from Apollo 13 showing the Baja California peninsula at center. (left)

These views of the severely damaged Apollo 13 Service Module (SM) were photographed from the Lunar Module/Command Module (LM/CM) following SM jettisoning. An entire SM panel was blown away by the apparent explosion of oxygen tank number two. The apparent rupture of the oxygen tank caused the Apollo 13 crew members to use the LM as a "lifeboat." The LM was jettisoned just prior to Earth re-entry by the CM.

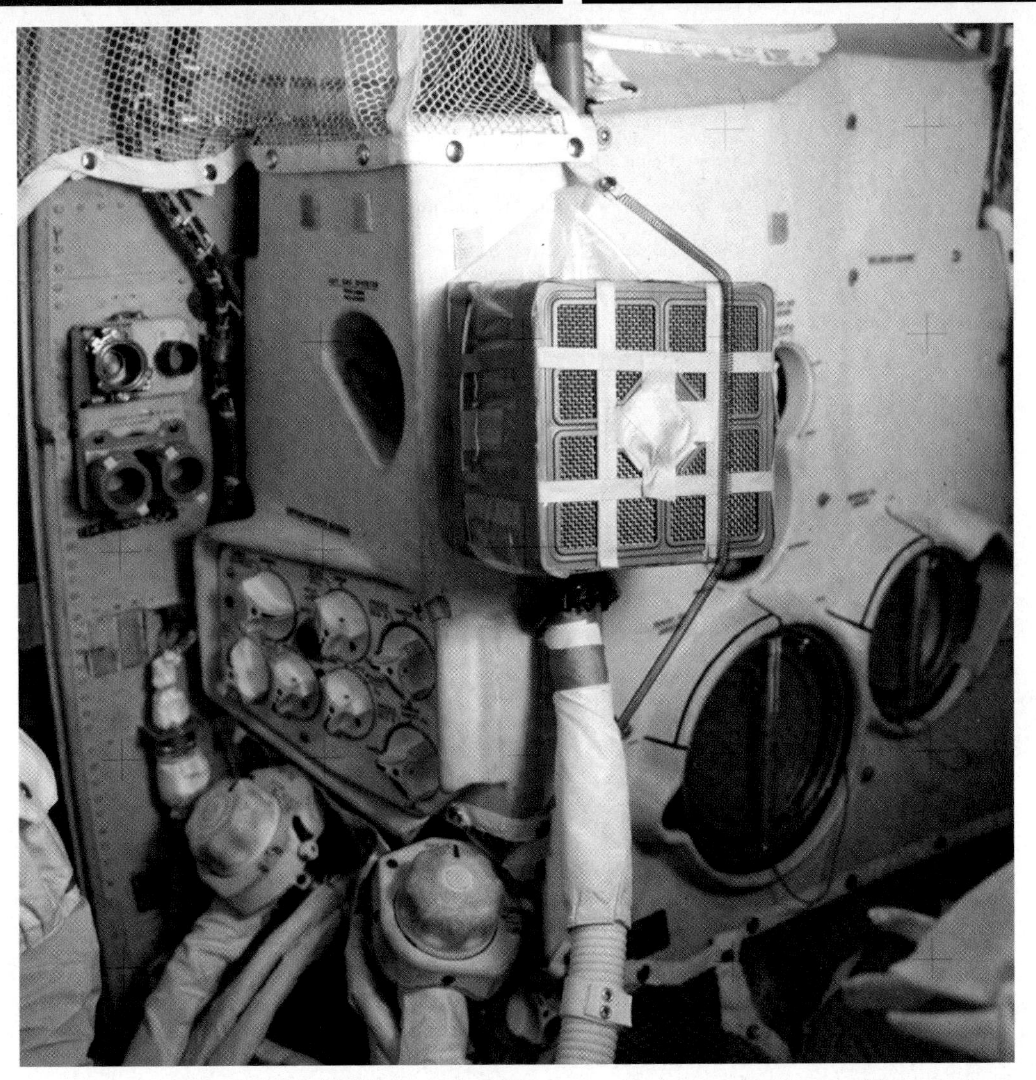

Interior view of the Apollo 13 Lunar Module showing the "mail box", a jerry-rigged arrangement which the astronauts built to use the Command Module Lithium Hydroxide canisters to purge carbon dioxide from the spacecraft's atmosphere.

Tsiolkovsky Crater as seen by Apollo 13. (above)

A picture perfect splashdown four miles from the recovery ship U.S.S. Iwo Jima (below)

Mrs Marilyn Lovell and three of her children greet photographers and newsmen outside the Lovell home in Timber Cove following the successful recovery of Apollo 13 in the South Pacific Ocean. (above right)

Jim Lovell reads the score on board Iwo Jima (right)

The crewmen of Apollo 13 step aboard the U.S.S. Iwo Jima, prime recovery ship for the mission. April 17th 1970. (below)

Crew men aboard the U.S.S. Iwo Jima, prime recovery ship for the Apollo 13 mission, guide the Command Module atop a dolly on board the ship. The CM is connected by strong cable to a hoist on the vessel. The crew were already aboard the Iwo Jima when this photograph was taken. Splashdown occurred at 12:07:44 p.m. (CST) April 17 1970.

Jim Lovell's Apollo 13 EVA glove, checklist and helmet on display at the Chicago Museum of Science & Industry (below)

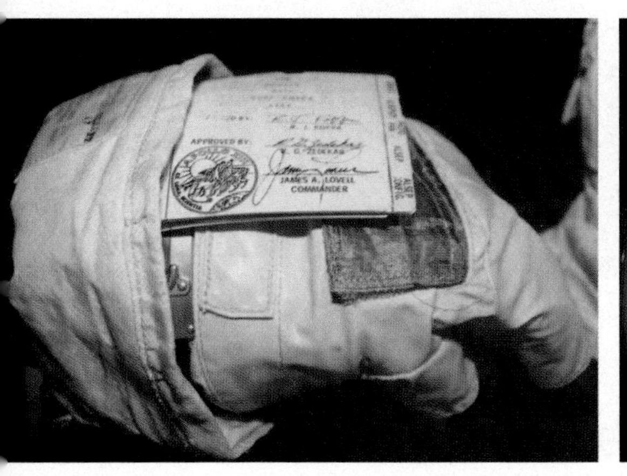

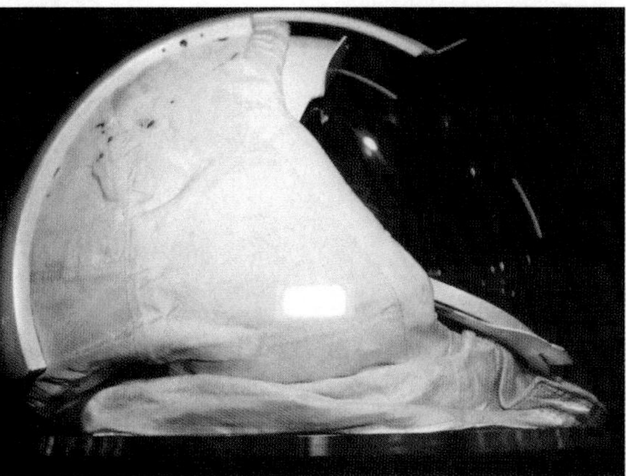

Fred Haise, Jim Lovell, President Nixon and Jack Swigert at Hickam Air Force Base Hawaii 18th April 1970. (above)

President Nixon awards the Medal Of Freedom to the Apollo 13 flight controllers, to the President's left Glynn S. Lunney, Eugene F. Kranz, Gerald Griffin, Milton L. Windler, and Sigurd A. Sjoberg, for their services in guiding the Apollo 13 spacecraft to a successful splashdown and recovery. (below) — Jim Lovell's Apollo 13 flight suit on display at the Kennedy Space Center Florida (inset below)

LOVELL — But he was out. We could see him and we got word from the choppers about what was going on. The swimmer got up and looked at the window and we were going to open up the hatch. Then we got the new life vests which I think are pretty good. Jerry wanted me to make a comment on that. We decided to go with these new marine life vests. Of course, if we had known about no quarantine, we could have used our old ones and never even bothered to open up the hatch, but it was already in work and so we used them and it was okay.

SWIGERT — We put the postlanding vent on low to get some air.

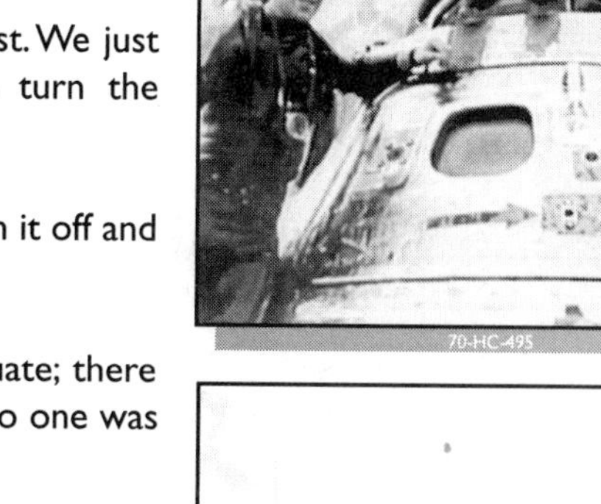

70-HC-495

LOVELL — We used the beacon, which they wanted to turn off.

HAISE — But that was on the checklist. We just went down the checklist. It said to turn the beacon on and we turned it on.

SWIGERT — And they asked us to turn it off and we obliged them.

LOVELL — The ventilation was adequate; there was very little rocking in the boat, so no one was sick.

S70-35610

12.8 COUCH POSITION

LOVELL — I put down my couch because I went down to 250 to open up the circuit breaker.

12.11 RECOVERY OPERATIONS

LOVELL — Recovery operations were very smooth. They got that down to a gnat's eyebrow. Of course, they had good weather to do it and you saw all the recovery operations.

12.12 SPACECRAFT POWERDOWN

LOVELL — We didn't have much powered up, actually.

HAISE — We just yanked the breakers on 250 and that did it.

SWIGERT — That powered us down.

S70-35651

LOVELL — Egress was okay and we had a good crew pickup.

3.0 COMMAND AND SERVICE MODULE SYSTEMS OPERATIONS

3.1 GUIDANCE AND NAVIGATION

13.1.2 Optical Subsystems

SWIGERT — Houston did call me at one time to say they noticed a shaft glitching. I took the optics out of ZERO to call up 16 91 and read the shaft angle and I called them and told them what the TPAC was doing and they could rig the CMC. Their advice was that this was something that they had noticed previously on Apollo 12 and they asked me to turn my OPTICS POWER switch OFF, and I could turn it back on any-time I needed the optics. Of course, this confused me because I went down to my first sextant star check and the optics wouldn't move.

HAISE — Optics power was off manually. I forgot to tell you about that. That was the only anomaly we had in the optical subsystem.

13.2 STABILIZATION AND CONTROL SYSTEM

SWIGERT — Of course, we didn't use any SCS thrust vector control. Minimum impulse was okay.

13.3 SERVICE PROPULSION SYSTEM

SWIGERT — We never used the DIRECT ULLAGE button, We never used the THRUST ON button. We made one burn, the G&N burn, that was completely nominal.

13.4 REACTION CONTROL SYSTEM

SWIGERT — There is some speculation as to exactly how many thrusters we had after the incident and this is something I don't think that we can resolve thoroughly.

LOVELL — They were working normally before the accident.

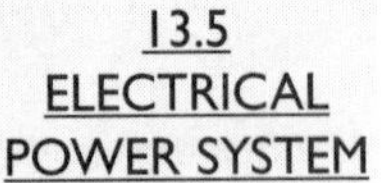

13.5 ELECTRICAL POWER SYSTEM

SWIGERT — I think we've already talked about the fuel cells.

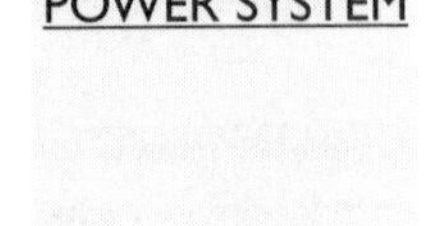

HAISE — I've got one thing to add on the batteries, mainly with respect to simulators. I noted that, after we used the batteries to support the gimbal motors operation during my midcourse burn, when I flipped the bus ties off, rather than as in the simulator with the batteries going immediately back to 35 to 36 volts, they hovered around 32 volts. It would take them a long while to increase to maximum voltage. This is a very small point but something that was a little bit different from what I'd seen in the simulators.

S70-35625

S70-35645

13.5.9 Cryogenic System

SWIGERT — The only anomaly we had in the cryogenic system was a continual unbalance between the two H2 tanks which we were endeavoring to adjust manually. We had the failure of the sensor in the O2 tank 2.

LOVELL — Yes. That might be related somehow. I think what ever caused the O2 transducer to fail might have also been the cause of the catastrophe.

SWIGERT — Particularly, when it occurred. They said it occurred at the time we turned on the fan. Remember, the exact time they said we turned on the fan, they said at that instant this tank quantity sensor pegged full-scale high. Houston came back and they said that because this occurred when we turned on the fans, they' like us to recycle the fans again to see if perhaps we could jar it into operation.

70-H-677

LOVELL — I wonder, if we had purged the fuel cell or something like that.

SWIGERT — The way Pete was talking, even the relief valve couldn't handle th particular flow. The heat source had to generate something like 8 to 10 thousand Btu' an hour to exceed the relief valve. The surprising thing to me is that I did not get a CRYO PRESS light. I don't understand that.

LOVELL — We were getting a CRYO PRESS light on the hydrogen, you know.

SWIGERT — On the low end. We got it on oxygen, also, when we first cycled the fans. Remember I was telling that there was an indication of stratification, but we never did get anything on the high end.

13.6 ENVIRONMENTAL CONTROL SYSTEM

LOVELL — We had no problem up to the time we had the accident.

SWIGERT — We had no problem at all.

13.6.7 Waste Management System

SWIGERT — We found that, during the initial first 8 hours, when the waste storage valve was open, when we went to dump urine and used the urine dump.

LOVELL — We had two vents open.

SWIGERT — We had two vents open but, for some reason, it didn't seem as if we were evacuating that urine.

LOVELL — Very, very, slowly.

70-HC-467

SWIGERT — Very, very slowly. And then, after we got through purging the cabin, we turned the waste stowage valve off, the efficiency of the waste management system seemed to improve.

LOVELL — There is a technique to operating that thing. You've got to sort of push that stuff down by gluing it, somehow; otherwise it all sticks around the honeycomb and the next guy who comes to open it up finds a nice big glob of urine sitting there.

SWIGERT — So we developed a technique, before using it, of turning the vent on, tapping it a couple of times, and raising the cover up and down a couple of times. Then when you opened it up, it was fairly clean.

LOVELL — That urinator requires the same technique.

HAISE — There's no question about your impression about the vents. In fact, the first few times I went, I filled it up to the brim, with liquid. You just sit there and watch it slowly go down. It held its meniscus and didn't break out. Then later on it would go right on down.

13.7 Telecommunications

SWIGERT — We did not use VHF.

LOVELL — We had a little difficulty locking up sometimes. What was causing that?

HAISE — I never had trouble locking up. The problem I had was I just had the switch in the wrong position MANUAL and REACQ. So when Jack maneuvered, it didn't track.

SWIGERT — I didn't have any problem with communications. Some items like USB emergency keying were not used. The ground operated the DSE. Our tape recorder worked adequately. We probably had the minimum tape recorder usage of any flight that's ever flown.

13.8 MECHANICAL

LOVELL — Did you check the Y strut on entry?

HAISE — Yes, when Jack was up there, it was unlatched. Jack tied it up.

SWIGERT — We had the CM in good shape before leaving the LM. Fred kind of went around and double checked me.

13.9 MISCELLANEOUS

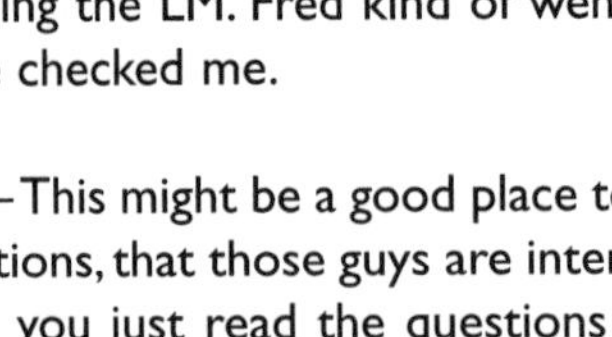

REEDER — This might be a good place to answer these questions, that those guys are interested in. Why don't you just read the questions and give an answer?

LOVELL — Technical Crew Debriefing Questions Generated from Data and Photo Review. Was the oxygen tank on the SM gone? I cannot tell. I don't know.

70-H-641

SWIGERT — I could not tell either; because, by the time I got down there, it had gone.

LOVELL — As long as I looked at it before grabbing the camera, I didn't have a good enough look to see. It was all being reflected back in. Was the hydrogen tank canted? I can't answer that question either. Was the panel cleanly removed? Yes. It was just completely removed away from where it normally swings out and no pieces of panel were still attached that I could see. Did you see anything of those panels?

HAISE — No.

LOVELL — They were just blown out.

SWIGERT — Were all bolts sheared uniformly?

S70-35614

LOVELL — We don't know, we weren't that close. Was there any positive damage to the SPS nozzle? Not to our knowledge. One side of it was stained with something, but we didn't see any bent part of it, pushed in or dented or cracked or anything like that that we could see. Was there any indication of where explosive flow could have exited, streaks, et cetera? Except for the streaks on the engine nozzle and the fact that it looked like it was more damaged back by th high gain antenna than up forward, that's the only indication that I have. It looked lik the damage occurred back in the high gain antenna because it was messed up inside.

HAISE — Yes, it appeared that there was some material like insulation or something lik that still attached to the high gain antenna.

LOVELL — Do you have any knowledge of damage to radial beams, size of hole, shape, et cetera? I have none.

HAISE — No.

LOVELL — Was there any other noticeable deformation on the bays? It looked like the basic structure itself was still intact. It wasn't warped or anything like that. It was just that the panel was missing.

S70-35148

SWIGERT — It appeared that way to me, too.

LOVELL — Did you see anything different?

HAISE — No.

LOVELL — The last question is how many bangs? Was there a second bang? To the best of my knowledge, there was only one explosion.

SWIGERT — I agree.

14.0 LUNAR MODULE SYSTEMS OPERATIONS

14.1 PRIMARY GUIDANCE AND NAVIGATION SYSTEM

LOVELL — In the PGNS INERTIAL, I saw nothing. Optical; nothing there except the use of the Sun filter was a little more difficult than I anticipated because of not being able to see the reticle.

S70-35606

HAISE — Something peculiar to the configuration of the AOT is that, out the front detent, the CM docking light hung down on its staff almost to the center of the AOT field of view.

LOVELL — For Alignments, the CM probe (the docking light) hung way into the middle of the AOT. And if the Sun is behind you and you have a bright sky, you can see the stars and can get an alignment. The Sun reflects right off this thing, so it's like having a light staring at you. We used the rendezvous radar only one time and that was to move it with the PGNS, and it worked. We never used the landing radar. The computer subsystem worked as advertised. G&N controls in space were okay. Procedural data we went around a lot of the procedural data, but what we had was good.

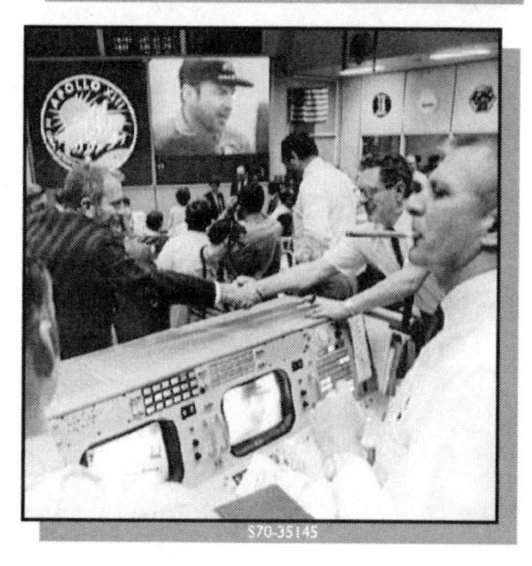

S70-35145

14.2 ABORT GUIDANCE SYSTEM

HAISE — In the AGS modes of operation, all we used were the body axis align and the zero and three accelerometer addresses, but we didn't really use any of that - the normal modes, the external DELTA-V included. Concerning the initialization, we didn't do any. We didn't do any calibrations, we just accepted the ground test data. We didn't do any rendezvous radar, engine commands, or burn programs. We did use some AGS controls and displays; namely, 8-ball and the error needles associated with the AGS; really CES rate needles. They all worked quite well.

14.3 PROPULSION SYSTEM

HAISE — We have talked about the descent system.

LOVELL — The DPS burns worked nominally; we had no problems with them. We never used the ascent engine or pressurized the ascent tanks. We always had an ASCENT PRESSURE light.

14.4 REACTION CONTROL SYSTEM

LOVELL — The attitude control modes were just as I had been briefed, and the operation of the thrusters and the responses to the controls were unusual because we had a different configuration; however, they were manageable. I think we covered that adequately in a previous briefing. The translation control worked as advertised.

HAISE — The only thing I noticed funny was during every pulse the fuel and the OX manifolds, whichever one you were looking at, would get about a 20-psi delta drop. I'm not really sure if that was from some kind of hydraulic shock in the lines or an actual drop. I can't remember reading about that before.

SLAYTON — Did it stabilize there?

HAISE — No, it was momentary. Just very quick and then right back up to normal.

14.5 ELECTRICAL POWER SYSTEM

HAISE — It was great. There was one MASTER ALARM one night with a battery light that quickly turned out to be an obvious sensor problem.

LOVELL — That was another one of those little things that worried me when I first came up.

HAISE — So, basically, the electrical system was flawless.

SWIGERT — It appeared almost impossible to have any problem with the battery, because we were down at 10.3 amps and the total amps didn't change. So, it couldn't have been reverse current. We surmised that it was a temperature sensor problem. It was almost instantaneous.

14.6 ENVIRONMENTAL CONTROL SYSTEM

LOVELL — We never used the suit circuit. The water-glycol was as advertised. The cabin atmospheres were good. The cabin atmospheres were good. Oxygen cabin pressure; no change there. We did "kluge" up the CO2 scrubbing system with the CM canister and that was the only change.

14.7 TELE-COMMUNICATIONS

HAISE — Concerning the monitoring, most of the time we were using the SBA down voice backup mode. I guess we were forced to use that because of the power margin but probably you know more than we do, Deke. I understand it could be pretty noisy and you couldn't hear us very well. But from our end, the uplink was great all the time. We never had any problem hearing Houston, regardless of the mode we were in. Later on, we arbitrarily went to the so-called base band down voice backup mode voice band, which means throwing the BIOMED switch one way or the other — just because I wanted to get off the hot mike for fear, my Commander would come awake and say something he, shouldn't. Then when we had high bit rate, I had the amp in and they wanted high bit rate and all that

HAISE — Then we'd go back to normal voice, which would give you better voice clarity. We never operated the high gain the whole mission. I guess what I am really talking about now is the S-band. We never operated VHF. We had no problems with the audio centers. The volume controls were quite adequate. We never operated the flight recorder. The DSEA was never run.

SWIGERT — One thing, I did want to go back in the CM subsystems in RCS. This was the first time we had ever preheated the CM RCS engines. I think, when we started out, the 3.9 volts was the minimum value I recorded. I reported to Houston what the values were when we first started and it strikes me that 2.8 volts was the lowest engine. We did go a full 20 minutes and at that time there were still several engines below the 3.9 volts. However, after I turned that switch off, I went back down about 5 to 10 minutes later and took a couple of extra readings. There was enough heat soakback that soon we had all the engines over the 3.9-volt minimum. So you might expect the fact that there was some thermal lag in this system and that it might have been 15 minutes. If someone is worried about 20-minute operation of these engines, they might turn them off at 15 minutes, wait 5 minutes, and see how the temperatures have come up - because they increased significantly, maybe half a volt, a significant temperature rise after I turned off the heater. I did want to mention that because no one has ever preheated engines before.

15.0 FLIGHT DATA FILE

15.1 COMMAND AND SERVICE MODULE

SWIGERT — We went right down the launch checklist, and it worked well. We had no problems. The entry checklist from EI minus 19 minutes was the nominal checklist with a few items changed. Prior to that, from EI minus 6½ hours, we rewrote the checklist. We had no problem getting updates. Launch keycards are good. The systems operation G&C checklist was completely adequate. Systems data were good. We read off the malfunction procedures item by item. We used the flight plan right up to 56 hours, and it was completely nominal. We logged things in the flight plan. The solo book, the rescue book, and the star charts weren't used. We used a lot of clips, but that's all.

15.2 LUNAR MODULE

HAISE — The LM was almost as easy. We didn't use the data card book. We lived out of the contingency checklist approximately 60 percent of the time. That was really a base line from which Houston passed up changes that we built on; either erased or added to. About the other 30 percent of the time, we put the TM data in the CM update book, which had a large blank section in the back that was a photo log. That was a convenient place with blank pages to use. The other place we worked was on the activation book. I think we only broke out three of the cue cards; one of them was the DAP card that also had the DPS RCS pressure data on it; another was the BUS loss card; and also we had a DPS card. We used the systems activation checklist, of course, for the first LM entry. We used a few pages of it, and the one addendum page for the tank pressure. Then we used it again, "kluged-up" by Houston, to go to only certain sections to do the first LM activation after we had the problem. Jim may have used the front of the G&N Dictionary. I used every pad I had in the back. We had a lot of P30 pads in the back of the G&N Dictionary, and that's about all I used to write all our pads in. In fact, I ran out of P30 pads. The last pad they gave me was the last blank one I had.

LOVELL — Yes. We used the back of the pads.

HAISE — Do you have any comments concerning the front part of that book?

SWIGERT — No.

HAISE — Charlie ran us in once, and we used the P52 business.

LOVELL — I did the P52 out of here. We also did the Sun check out of here. Everything in here was adequate.

HAISE — I had occasion to use the systems data only to total up my consumables and to start calculating where we stood. I didn't use my function procedures. We never used the Time Line Book.

LOVELL — We never used the star charts.

HAISE — No, we never used the star charts. We did use part of that EVA book, not in its normal usage, on the cartridges.

LOVELL — We used the voice recorder for guitar music.

HAISE — Well, I recorded one whole half hour of LM noise. I decided that I'd save old Aquarius's grinding and moaning, squealing glycol pump, and suit-fan-running noise for posterity. I guess none of us used any charts listed here.

SWIGERT — No. We didn't use any of the CSM monitor charts or any of the orbital science charts.

15.4 GENERAL FLIGHT PLANNING (FDF)

SWIGERT — As far as general flight planning is concerned, I have no comments on the solo phase, because we didn't get that far.

15.5 PREFLIGHT SUPPORT

SWIGERT — I think for the CM that Ken would be more appropriate to answer the preflight support, because, as far as I was concerned, it was entirely adequate.

LOVELL — I thought that we were well ahead of the game, mostly for preparation for the launch. I think that was based on several items. Number 1 is the fact that we've already gone through this landing phase before, and we had that extra month, which I don't recommend, to get the Flight Operating Data File a lot earlier than we have ever had to my knowledge before, either in Gemini or Apollo. We had data on Apollo 13 to train with earlier than we ever had before.

16.0 FLIGHT EQUIPMENT AND GOVERNMENT-FURNISHED EQUIPMENT

SWIGERT — The knob came off a portable timer in the CM. Other than that, our timers and controls were adequate. That one timer is very useful, by the way. It could be used to time fuel-cell flow, purges, and so forth, so it was disturbing that the knob came off.

LOVELL — I don't know how to discuss the clothing and related equipment. Obviously the inflight coveralls we had weren't adequate for the conditions we had. The coveralls are great for a nominal mission. I hate to imply that we ought to carry liners or something like that with us. I sure think we can improve the footwear, though. I know that Grumman, down at the Cape, has for their checkout people a soft boot that is worn in the LM. An insulated soft boot would have been much more adequate than what we had.

HAISE — With the addition of probably just one set of Nomex thermals, we'd have been in good shape. We needed a set of thermal underwear on.

SLAYTON — What did you end up wearing?

LOVELL — Constant wear garments.

HAISE — Two constant wear garments.

LOVELL — Did anybody have any problems with the sensors on the BIOMED harness? I didn't have any problems. We left the sensors on all the time.

SWIGERT — Oh, yes. I got a rash. I got a rash from the sensor paint. I have never been tested for this particular phase, but the doctors attributed this more to the tape reaction than to the paste reaction. I left them on the whole time. They were there when we unsuited.

LOVELL — We had no problems with the pressure garments and connecting equipment, but, of course, the suits came off right after TD&E and we stowed them.

SWIGERT — I tried my suit on once to make sure that I could get it on by myself.

LOVELL — There were certain things that Jack wanted to make sure that he knew how to do. One was to put the suit on by himself. During the quiet period before the accident, he put the suit on by himself. We also mounted all the cameras to make sure that Jack was checked out.

LOVELL — We had no problems with couches that I know of. The restraints were adequate. No problems occurred with the inflight tool sets. We were keeping logs on food and everything else.

HAISE — The only shortage we had was what I mentioned on the air. What we really needed was a big, blank pad of paper for our unusual situation.

LOVELL — We used to carry a crew log on Gemini to put comments on.

HAISE — On the number 2 lunar surface Hasselblad, we had to push the trigger offset to the left to make it work very easily. It was very difficult to work if you pushed it low, center, or to the right side.

17.0 VISUAL SIGHTINGS

17.1 COUNTDOWN

LOVELL — During countdown, we saw the swing arm go back, and that's about all.

17.2 POWERED FLIGHT

LOVELL — We saw the horizon at the proper times. We saw the flash from the separations and some debris go forward.

SWIGERT — Did the BPC hang together in tower JETT? Could you see it?

LOVELL — I just saw it go, I saw a big light, and I went back in.

HAISE — It looked like one big cone. Would that mean that it stayed on?

SWIGERT — We didn't get any moisture on window 5, so it's apparent we didn't have any water under the BPC.

17.3 EARTH ORBIT

LOVELL — There was nothing unusual in Earth orbit.

17.4 TRANSLUNAR AND TRANSEARTH FLIGHT

LOVELL — We saw the S-IVB. We reported the last time we saw it, we saw the SLA panels. I think that Fred was mentioning the fact that during one part of the flight we saw some parts of the SLA panels on the S-IVB close by Post 5. We could see a blinking star that was probably the SLA panel turning.

SLAYTON — Did you see the light flashes in the CM?

LOVELL — Yes, we did. They're right. I didn't see it after the accident. When my eyes were closed, and occasionally, a streak would go through.

HAISE — It's amazing; I didn't see them after the accident either. We never saw them again.

SWIGERT — It is a CM unique phenomenon.

SLAYTON — It would be interesting if you noticed it in the LM or not. Nobody has ever had a chance to do that before.

SWIGERT — I didn't note it at all while I was in the LM.

LOVELL — I won't build a story. I won't say they're there or not. We were so preoccupied after the accident that we weren't looking for something like that.

REEDER — You didn't sleep in the CM any time after that?

SWIGERT — Yes, we did.

REEDER — You still didn't see anything?

LOVELL — I wasn't thinking about it.

HAISE — I think I saw them the very first time after the incident; the very first time I went to bed. I think I saw them then.

LOVELL — I only saw them with my eyes closed.

HAISE — Yes. I have never seen them with my eyes open. There were more directs than there were streaky ones.

LOVELL — Yes. You're right; more pinpoints.

17.5 LUNAR FLYBY

SWIGERT — Tsiolkovsky stuck out.

LOVELL — Our particular orbit around the Moon brought up Tsiolkovsky very nicely.

HAISE — Yes. That was the first actual landmark I saw on the back side that I recognized.

AS13-60-8659

LOVELL — What about the oblong craters that we saw? I'd like to just go back and look at the back-side photography, because you can really see that. The Moon has these oblong craters on the back side. I don't know exactly where we were.

HAISE — I directed a lot of pictures out the right window. Our track was kind of the normal Apollo belt. It might be of interest to somebody, although the Apollo belt has been covered pretty well without pictures. We were starting to gain altitude almost immediately. Fine detail just wasn't really ever there. I mean I never saw anything I could say was a boulder; we never were down that low. The best you could tell was that there was slumping inside some of the craters, on a very large scale. Actually, I disagree with Jim. When we came up on the backside terminator in the CM, the color I saw was a combination black to a reddish-brown to white mantling on some of those features.

SWIGERT — My description would be dirty beach sand.

HAISE — Even right at the terminator?

SWIGERT — That's right. I would say it just looked grayish, a grayish brown, like a white sand that had gotten dirty.

SPEAKER — That's the way I look at it. I thought that was a good description.

17.6 ENTRY

LOVELL — Were there any visual sightings in entry?

SWIGERT — No.

LOVELL — The main chutes.

SWIGERT — Well, I did get main chutes, but there was so much film on that rendezvous window that the drogues and mains did not stand out. If somebody had asked if any panels were missing from the main chutes, I couldn't have told them. I could spot the chute itself. But the purge put enough film on there that distinguishing fine detail was impossible.

HAISE — I didn't get that much out my side apparently. I don't recall ever losing sight of the mains.

SPEAKER — Would you have known if you had panels out?

HAISE — Yes, pretty sure.

SWIGERT — I could see just by the diameters when they reefed.

HAISE — I was watching them when they were going from reef to full.

18.0 PREMISSION PLANNING*

*No debriefing was conducted on this section.

19.0 MISSION CONTROL

LOVELL — We can't give enough praise for what they've done. They were up at all times with consumables, especially after the accident. They kept a pretty handy eye on consumables. Real-time changes were exercised to the utmost. We had a tremendous number of real-time changes, and I think they were handled very adequately. Communications, in general, were good. I thought the LM communications were especially good.

20.0 TRAINING

0.1 COMMAND MODULE SIMULATOR

LOVELL — You might have a comment on the availability because you didn't have priority for both the training simulators.

SWIGERT — I thought the training that the backup crew received was good for the time involved. Ken and I split the simulator time until right near the end. I think that's the way it should be.

LOVELL — You might comment on your reaction on having to replace Ken at the last moment. What did you think about it? Should you have known about it earlier?

SWIGERT — The earlier you know about it, the better off you are. We could have used a session or two together in some of the areas that we didn't have time to run. I had worked with you in one rendezvous before, and I didn't have any misgivings at all about working with you. I thought that we might suffer a little bit on the lunar-orbit activity, but this was a low priority item, so I had no qualms at all about being prepared to do the job.

LOVELL — I think the story here is that, if we have a backup crew and a prime crew, we can replace a prime crew member with a backup crew member if he's had the background training that Jack had. Jack was knowledgeable on the CM to start with. He had worked all the malfunction procedures. He had a lot of good simulator training in our training. The simulator was available so we could give the backup crew 50 percent of the time up to the last 3 weeks of training. In the case where somebody comes aboard new and becomes a CMP, then you're going to have to analyze exactly how muc simulator training he has had and what his background is. We had no problem even wit the minimum amount of training we had with Jack. We were time-line wise and flight plan wise. We were going right ahead of the game, and I had no problem at all. I thin that was a good decision we made to go in April.

HAISE — Part of the thing we were supposed to discuss here on the CMS/LMS was th fidelity of the simulator. I've already broken it out in pieces. I want to discuss them agai in the CMS. I mentioned the voltage/fuel-cell-flow relationship with the gimbal motor and the battery not regaining its voltage as quickly after having the main bus back on.

20.2 LUNAR MODULE SIMULATOR

HAISE — I'd said the thruster noises on the LMS are, at the normal level that they hav for us when we're unsuited, not quite as loud as they really are in the LM itself. I notice that, when things popped, they really popped, particularly when they were the forwar quads that are right outside the window. They really bang. The big things that are reall missing in the LMS, which from a training standpoint I don't consider pertinent at all bu in real life it's something that you have to get used to, such things are the glycol pump and the suit fans running. They make powerful squeaking noises in the LM.

SWIGERT — They change frequency, and they gurgle. I assume the fluid goes fror turbulent to non-turbulent, and it just doesn't sound like it's acting right. That's just the way the LM sounds, and it was pretty rough. Fred told me any time you don't hear those changes, there's something wrong.

S70-34412

HAISE — I'm not really knocking the LMS for that reason, but that was a distinction of what you hear. That's a fairly high-level noise, too, and something that you have to get used to. I've slept for many hours of tests in the LM with all that stuff running, so it didn't bother me much for sleeping.

LOVELL — If you duplicate the noise, I would suggest that you have it such that you can turn it an or off.

HAISE — The only other funny thing I noticed on the LM was the very quick spiking c the RCS pressure when we fired Jets. Everything else seemed high fidelity. We wer doing things with our CM/LM configuration that we'd never looked at in the LMS. I don have any idea what the comparison is there. I thought the firing of the LM stack (the wa we did it, with the translation controllers) was easier for pure pitch and pure roll in re life than it is in the LMS for doing the burns.

LOVELL — It's easier to do the burns that way. One thing I think that we can prove o the CMS and the LMS is using the planet bodies for an alignment. That's one thing tha we didn't do adequately enough using the unit vectors. We don't have the simulation set up where we can get accurate alignments,

SWIGERT — The CMS is limited so that you cannot do planet alignments, or Su alignments. You can make optics calibrations only on selected stars. I couldn't go dow and do a set of P23s because invariably it's only by coincidence that the optic calibration star is one of the stars that you can perform an optics calibration on in th CMS. Also, you cannot get any of the stars that are non-Apollo stars into the sextan All our P23s had one or two non-Apollo stars.

LOVELL — The simulations and the actual operation of what we were doing were excellent. We never expected the amount of work we were going to do with Mission Control after the accident. That was all new. But there's nothing that will ever substitute here.

20.3 CMS/LMS INTEGRATED SIMULATION

LOVELL — We never got to see the areas where we had integrated operations because we never undocked. The only thing we did integrated was trying to get power to one vehicle from the other.

20.4 SIMULATED NETWORK SIMULATIONS

LOVELL — The network simulations are really required. That's the best type of simulation work you can get because it gets you to talk to the guys and see how well you work back and forth. Also, it's good for the CAP COMMs, too.

20.5 DCPS

LOVELL — The DCPS was good. The boost phase was just like our SIMs and just like the DCPS.

20.6 LMPS

LOVELL — We never really got a chance to evaluate the LMPS.

20.7 CMPS

LOVELL — We used the CMPS and the LMPS for rendezvous.

20.8 CENTRIFUGE

LOVELL — Did you get a chance to run the entries in the centrifuge?

SWIGERT — Yes, I did. I didn't do any G&N entries in the centrifuge. I did the EMS entries, and I feel that the guidance was comparable. The control input seemed to increase the g much like I saw. Bank inputs increase and decrease the g during entry.

SLAYTON — Do you think it's necessary?

SWIGERT — Yes. You're used to zero g, so 1 and 2 g's seem like a lot more than one g. You'd think that you could take 1 or 2 transverse g's with no sweat at all, but after zero g, 1 or 2 transverse g's is a significant amount. I think it would be good to have a guy fly centrifuge entry and be able to take over and do it manually.

SLAYTON — Where in the training cycle?

SWIGERT — I'd do it farther along, I'd say a month or two before launch. Just once is enough.

LOVELL — If I were going to fly entry, I think I'd have wanted the centrifuge run just to see how I could do it with respect to the EMS. For G&N entry, the simulator is good enough. It's doing the work, and you're just sitting there monitoring. If you're doing an EMS entry, I'd like to do it.

SWIGERT — I had a number of hours in the entry simulator where they put all sorts of failures in. I didn't have any problem recognizing the failures, taking over, and steering. If I'd had to take over at entry the other day, I felt that it would be much more difficult to do it under a g-load, so I think one centrifuge run would be worthwhile.

20.9 TDS

LOVELL — The TDS is no longer here, and we never used it anyway.

SWIGERT — We used the TDS in the CM side where we put CM moments of inertia in and maneuvered. I did it once to see whether it was worthwhile, and I don't think it's worthwhile.

20.10 NR EVALUATOR & GAEC FMES

LOVELL — I never used either the North American evaluator or the FMES at Grumman.

SWIGERT — I used the North American evaluator.

HAISE — I used the FMES right at the end of Apollo 11.

SWIGERT — It was interesting; you could run some of the rational programs, some c the ropes.

LOVELL — We used FMES only to check out our new AUTO 66. Someone else did th checkout. I think this is the only way you're going to be able to do it. I don't think it wa worthwhile going up there.

20.11 EGRESS TRAINING

LOVELL — Egress training is required for people going through the regular trainin cycle. It's exactly like you're going to see it. You ought to leave it in there. We had n problems with that.

SWIGERT — I guess when John, Charlie, and I went for pad egress training, I had tongu in cheek as to whether it was worthwhile. I'm glad I did go. When you get there for rea you know where to go if you have any problems getting off that booster. I think th backup crew ought to go through it,

LOVELL — Yes. You never can retrace that after you make a change, especially when it that late in the game.

20.12 SPACECRAFT FIRE TRAINING

LOVELL — We had spacecraft fire training on Apollo 8. We didn't do it this time.

SLAYTON — Do you think Jack got it somewhere?

LOVELL — It's good training.

SWIGERT — I had fire training. I did not have the Gulf egress training. I did have th tank training.

LOVELL — You didn't go out in the Gulf?

SWIGERT — No.

LOVELL — No wonder you stumbled over this all the way in.

SWIGERT — We had the frogman there. He said, "When I get the net positioned, yo leap in." So I leaped in, and up I went.

SLAYTON — I was under the impression that you had gotten that. I didn't know yo hadn't.

REEDER — They got it in the tank. Gulf egress would have interfered with some SIM that seemed more important at the time.

LOVELL — That's correct.

SLAYTON — John and Charlie already had Gulf training, didn't they?

REEDER — John did; Charlie didn't.

SLAYTON — I think we will probably want to keep scheduling the training for th backup crews.

SWIGERT — I didn't feel like I was handicapped in any way because I didn't have it.

LOVELL — I think that type of training is good to have just as a part of an astronaut's general training. If you had it for a previous flight, it would carry over for the next flight.

SWIGERT — Yes, I agree.

REEDER — We got into a bind on that training because someone kept wanting to wait until the procedures for quarantine were firm. We ended up doing it too late. At that time, the training would have taken the astronauts away from something we considered to be more valuable, and so we didn't do it.

LOVELL — Right.

20.13 PLANETARIUM

LOVELL — I didn't use the planetarium on this mission because I didn't think it was required.

SWIGERT — I didn't use it either. I did visit the planetarium once for the Apollo 11 mission, but I didn't feel that I needed it at this time.

SPEAKER — It was concluded some time ago that the simulator visuals are good enough. They also give you the field of view you're going to see in flight.

LOVELL — I think the planetarium is for general training, but it was not specifically needed on this mission.

REEDER — Ken did go to Moorehead for half of a day, primarily to familiarize himself with the new stars that he would be needing. 20.14 MIT

LOVELL — MIT did brief us on the changes to our guidance systems. That was mainly for Fred and me and was on the automatic landing program to which they had incorporated a number of changes. We had no training at MIT.

SWIGERT — Ken and I went to MIT once to use the simulator that would duplicate the atmospheric layer of the Earth. We could mark there and get an idea of what kilometric value you were using for your atmospheric layer.

LOVELL — Fred did that on Apollo 8, and it was good for that mission. Of course, you went there for Apollo 13.

SWIGERT — I don't think I would make a special trip up there for that. I would combine it with something else.

LOVELL — They use your MIT results as base-line data to compare with what you do in flight.

SWIGERT — I think that I was pretty near the actual value. I came out with 17 kilometers while I was in flight, and I was using 19 kilometers at MIT. So, it was actually pretty close.

20.15 SYSTEMS BRIEFINGS

HAISE — On our side, we didn't go through any of the CM part at all.

LOVELL — We had deltas and briefings on the last-minute changes and anomalies that can up during the last couple of months.

HAISE — Whatever we did on the LM side was done way early. We had already bee through that once on Apollo 11 and it wasn't particularly needed.

SWIGERT — I didn't feel like I needed it. I did have some systems briefings on systen I thought I could use. They were not very extensive.

20.16 TOPOGRAPHY TRAINING

SWIGERT — We spent an extensive amount of time on topography. This was luna topography.

HAISE — We really needed more.

LOVELL — There were two types of topography training: One was the landing area. We did a lot for this one. Then, there was the area we were to fly around. Ken did more work on that than Jack, Fred, and I did. I still felt that I wasn't up to speed the way I really wanted to be, before launch on orbital geology. We knew some things would have to be cut out, and that was one of them. We never had the time to get it done. Fred and I spent more time learning about the landing site.

70-H-99

20.17 LUNAR SURFACE TRAINING

LOVELL — One thing especially, I think we should change the geology training techniques, and this is basically based on Fred's inputs. It should be more what we would expect to do on the lunar surface. We used walkie-talkie radios and equipment and we kept the time line down to something that was similar to what we were going to do. We learned to be observers and to discuss what we saw. I think that the training along this line is really taking a different curve toward what we're trying to get out of lunar geology.

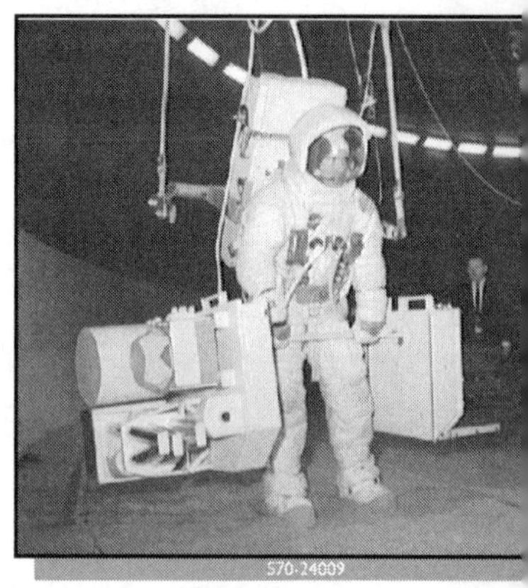

S70-24009

HAISE — We should use only the same scale maps they have from Orbiter pictures of the Moon. In some cases, they laid out known traverses from a known starting point. At other times, we played the game that we didn't know where we landed and we reoriented on the map, built our own traverse, coordinated with the ground station, and operated with CAP COMM and SPAN room people, all the time.

LOVELL — I'm only sorry that we didn't have a chance to exercise our training to determine whether it was adequate. However, I did feel that it was adequate.

S70-24012

HAISE — To me, the KC-135 has the best fidelity of one-sixth g. The POGO seems t help you develop a kind of low running technique, but even it was distorted somewha I was working a lot harder there than I was in the airplane doing the same kind of task When I used the mobile POGO, there was a 30 to 25-knot crosswind. One leg wa completely distorted but the other leg was shielded by the truck. It felt qui comparable to the centrifuge rig. However, both of the simulations distorted my balance; of course, the things I was lugging around were one-g weight, not one-sixth g, and my limbs were one g and not one-sixth g. My heart rate was also saying the same story. It was costing me a heart rate of about 140. On the airplane, I know I got going faster than that, and I bet I could have gone all day without getting above 110.

LOVELL — One g walkthroughs; again we can't prove it, but it seems to me that walking, using those pressure suits and doing that stuff, was the best thing in the world, even though it was horrible. That type of training, according to Pete, is far more difficult than the actual, which makes things easy. I recommend it just to get in condition. We tried to limit the field trips this time to ones we thought were profitable. Early in our training, we went to California. I thought that was profitable, only for the fact that it trained us as observers. The Hawaiian trip was training us for operational time line and for things we thought were peculiar to the lunar surface and the phenomena we should expect on the lunar surface. Although I really didn't think too much of it before we went, I think the last trip out to Flagstaff was very good. I thought Kilbourne Hole was the least interesting, and the least productive. During the training at Flagstaff, they gave us the lunar map that was degraded purposely in the same way that our maps were degraded. They drew holes and it showed this in the traverse. They showed us an actual crater and then showed us what it would look like on the map we had. We could compare that with the Fra Mauro region and get some idea what the crater size would be. That was good training. I think SESL training should be done only as a confidence check concerning your equipment. I recommend that we keep doing this.

S70-27034

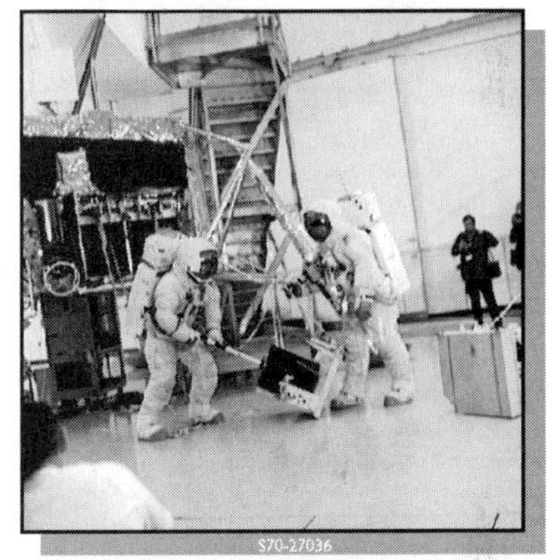
S70-27036

S70-27037

SLAYTON — Were you aware that we are planning to drop this for future training because of budgetary problems?

LOVELL — No, I wasn't.

LOVELL — What are the alternatives?

SLAYTON — We can use the 8 or 11-foot chamber. We need to give E&D guidance on some things we'd like to update there to make it good training.

HAISE — It's the same chamber we were using before on the preliminary runs you mean, the small one?

SLAYTON — Yes.

LOVELL — Well, my point here on the SESL runs is that it gives you confidence that your PLSS and OPS and everything are working right, and that you can actually perform in a vacuum.

SLAYTON — You can do the same thing in the 8-foot chamber.

LOVELL — That's why in Gemini, I was so adamant that we should have one-g chamber run.

SLAYTON — I just wanted to make sure you understood, if you had any strong feeling you should express them here.

LOVELL — Well, I guess what you're saying is the SESL also gave you the thermal environment. I think we've already proved that, but we do need chamber testing and chamber training.

HAISE — You can do it easier in the 8-foot chamber. We had that on Apollo 11, but we didn't have it this time. The 8-foot chamber was used to run off nominals, but they weren't in the SESL. It took a little in-house training to get them to allow you to do anything off-nominal very often.

LOVELL — The briefings. Every time we wanted a briefing, the people were more than happy to respond. We had to work our briefings in with simulator training; so, in most cases the briefer ended up sitting around for an hour.

20.18 CONTINGENCY EVA TRAINING, KC-135, WIF, AND ONE-g WALKTHROUGHS

HAISE — We only had one session of contingency EVA training.

LOVELL — I guess it wasn't adequate. If I had to go out by myself, as in a one-man EVA there was going to be a little talk through. I did train to deploy the ALSEP by myself. I looked at how the drill operated, but I wasn't very proficient at it. There would have to be a lot of talk through.

SLAYTON — This topic was primarily for transfer from the LM to the CSM, if you were not properly docked.

LOVELL — We did the WIF exercise, and my feelings were that if we ever were faced with that in reality we were in deep trouble. As a matter of fact, we came up with a new technique.

HAISE — We had several ways to go. You go through the tunnel or you could go outside. All we determined was that we couldn't make it through the tunnel.

LOVELL — No, but we were trying to determine if we could use the PLSS from the lunar surface. Remember we wanted to leave the PLSSs on instead of taking them off.
HAISE — We never had a PLSS on, in the water tank.

SLAYTON — It should be OPS.

LOVELL — Remember that late in the game we were talking about using a PLSS with John down at the Cape in the one-g mockup. I'm trying to see what the situation was that set us up so we could use that. We said instead of taking the PLSS off and putting the OPS on...

HAISE — We never did any training for that though. There was some idle conversation about that one day, because the hatch jammed and wouldn't seal. It was stuck in there and we couldn't pressurize the LM.

LOVELL — Yes, but the normal thing was to get rid of the PLSS.

HAISE — Yes. Do a vacuum mate/demate.

LOVELL — I think you would be much better off to leave the PLSS on and do the EVT with the PLSS, because you would have communications and you wouldn't have to do all that vacuum demating and mating and get all that stuff squared away.

HAISE — You never get even one-g, on the ascent stage. Why don't you just lift off with the PLSS on your back, and go into orbit that way?

S70-29673

SLAYTON — It depends entirely on what your failure mode is. Even in that case, if you get docked, you can get repress from the CSM and get back to normal, anyway.

LOVELL — We were looking at the case where we had no LM pressurization, and, we couldn't go to the tunnel. We had to go exterior. We thought that we could even recharge our PLSS with the LM system under vacuum conditions, better than we could take off the PLSS, put on the OPSs, and pressurize. We were willing to launch with the PLSS on our back and transfer that way, because we would have communications that way. It would take a long time, maybe 4 hours to recharge the PLSS. That was the only change we had on that.

S70-29672

20.19 MOCKUPS & STOWAGE TRAINING EQUIPMENT

LOVELL — My training suit is just about gone, but all this equipment is adequate.

20.20 PHOTOGRAPHY AND CAMERA TRAINING EQUIPMENT

LOVELL — We were a little late getting the Hycon camera. We felt that we needed more time on that, although Ken knew it pretty good.

SWIGERT — I had a good briefing on the Hycon camera. I didn't have any qualms about setting it up and working it.

HAISE — We were in pretty good shape on the LM side, all along.

LOVELL — Yes, we had the lunar surface cameras soon enough to practice with.

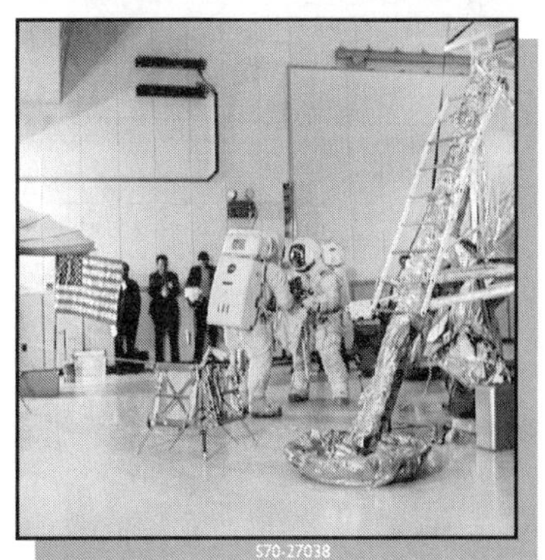

S70-27038

20.21 LUNAR SURFACE EXPERIMENT TRAINING

LOVELL — The equipment on that is getting pretty poor. It's worn out, and we're going to have to start replacing it.

HAISE — A good part of that is replaced just by having a new ALSEP package, Jim. There are no more ALSEPs like that one.

REEDER — There will be a new one coming along for Apollo 14.

HAISE — Yes, a different one though.

LOVELL — The cameras were shot too; all that stuff is worn out.

HAISE — We just brought two back.

REEDER — We just quietly didn't push the backup crew the last couple of weeks, because we didn't want to wear the gear out before your last EVA.

SLAYTON — We'll review that whole subject.

20.22 LUNAR LANDING TRAINING

LOVELL — Well, we can't even talk about that.

HAISE — I have one little thing that isn't listed as an item. Going back to the TV, I felt, by virtue of picking up the old Apollo 12 surface camera and having it to use a few times in conjunction with our surface EVA exercises, that the live TV was pretty well integrated for the lunar party operations. But the TV mockup in the CM doesn't even have the things that turn; the things that force you to an f-stop, or force you to set a range or zoom in or out. You just kind of fake that; at times we'd just play games in the CMS and normally I'd get the mockup out and stuff it in the bracket behind my head and then stick it out the window and then forget about it, and press on about our business. Then all at once that day came and things were a little different. I couldn't stick the handle in that bracket to begin with. I had to hand hold it because I just couldn't hold it straight out. I either had to hunch back or move forward to get the right picture and it was a game of focusing. We also had to be the commentator on the other end of it, trying to tell what you were supposedly showing while doing the other job with the other hand.

SLAYTON — This is CSM specifically.

HAISE — That, to me, was one thing that should be trained for, too. Now it would help if the TV mockup in the CMS could be made so that the end piece was such that you turn to make a change in f-stop and ranges.

LOVELL — It was packed; but quite adequate.

20.23 PLANNING OF TRAINING AND TRAINING PROGRAM

SWIGERT — I think we did some unique things in this flight that hadn't been done before, and we should ask the simulator people, both here and at the Cape, to lay out a training program so that we could accomplish, as far as the CMS goes, all the specific mission phases. We did the same thing in the rendezvous simulator and I'd recommend this plan. When we went over there, their approach was, "What do you want to do?" We came back and said, "Well, if we knew what we wanted to do, we wouldn't be here. You tell us what you think we need." They developed this approach and I'd recommend it.

SLAYTON — We've got a base-line training plan with all the details laid out. What you are saying is that we need to be a little more aggressive in telling you on a specific day to do this rather than leave it up to you?

SWIGERT — That's right.

SLAYTON — Eventually we get, around to that.

SWIGERT — They keep track of things. They say "All right, you've done many mode II's but no mode I's."

LOVELL — I guess when you have a time limit, you have to figure out what you think you're really not proficient at.

SWIGERT — Well, this is, of course, in the latter mission phases. You come down and they say, "Well, now you're getting pretty well near the end," and they'll give you a choice of activities and you will say, "Well, I'd like to run some more time-line work and so forth."

LOVELL — I thought training was very good this time.

SWIGERT — I did too. If it wasn't good, I don't think we'd have been able to switch like we did.

LOVELL — Yes. Maybe that extra month in there was well worth it then.

SWIGERT — We had a good training coordinator.

21.0 HUMAN FACTORS

21.1 PREFLIGHT

LOVELL — I tell you, though, that training is a bottomless pit. You get an extra month and you think you're all set to go; but, before you know it, every day in that extra month is taken up. The LLTV took an awful lot of time, and if it weren't for the slip, I don't think I would have gotten that in.

21.1.1 Preventive Medical Procedures

LOVELL — I'm not going to say a word.

21.1.3 Time for Exercise, Rest, and Sleep

LOVELL — This is a complaint, I guess. I really don't know what to say about it. After you run an EVA and the simulator all day, you don't feel like you can do any exercise. I had a hard time getting Ken to come back to rest, sleep, and things like that. But I really don't know what we should do differently.

SLAYTON — Tell them just to keep working at it.

LOVELL — Yes.

SLAYTON — If you think they're working too hard, just slow them down.

LOVELL — If we hadn't switched CMPs the last few days, we would have had the last few days free.

SLAYTON — Yes.

LOVELL — And we wouldn't have done anything. Having to switch, though, made us do extra work.

21.1.4 Medical Briefings

LOVELL — The medical briefings were adequate.

21.1.5 Eating Habits and Amount of Food Consumption

LOVELL — I think that we're now taking a practical view towards preflight eating. I don't think we had any problems.

SWIGERT — At the Cape, Lou wants to feed you like he thinks you're starving to death. Now, you've been down there, and you know exactly what I'm talking about. I don't need that much food. As a result, I found myself not eating that much, because I just can't take that amount of food.

LOVELL — I guess if we were a little bit stronger, we could probably prevail on Lou to cut down on the food. Maybe I should have told him not to feed us like that.

SLAYTON — I think it's an individual preference. It's up to you to decide how much you want.

LOVELL — Fred, do you have any comments?

HAISE — Well, I guess our training schedule didn't really allow a lot of time for exercise. That was one item that kind of slipped by for me. I normally get a little more exercise than I was getting in about the last 6 weeks of training.

70-H-494

REEDER — EVA didn't give you enough exercise?

HAISE — No, not the right kind. I would have approached the last item a lot differently if I had known the way this mission was going to turn out. If I had known the mission was going to last less than 6 days, I would have gone the route of no bulk foods, and I'd have cavitated the whole system back to the tummy as best I could with enemas and everything else, and I think we'd probably got by the whole time without having to worry about it. As it was, I went three times in 5 days. It's a terrible inconvenience. Jack did the same, and I think that was just because he continued to eat. With a 10-day mission, I don't think you're going to avoid the issue.

LOVELL — That's the way I felt before launch. I'd gone through this low-residue-food clean-yourself-out-good routine at night. Then, I went through the entire flight without going. I said to myself, "One of these days, you're going to face facts that you just can't last 28 days or 56 days without going. You might as well start living normally again." So that's what I did. I went once.

21.2 FLIGHT

21.2.1 Appetite and Food Preference

HAISE — The first day I didn't have much of an appetite. The next morning, I was a bit upset. Then, from there on, I started eating everything that was in each meal.

SWIGERT — We were given no meal A on the first day, but we were given meals B and C on the first day. Of course, our launch wasn't until 2 in the afternoon, which was after meal B.

70-HC-327

LOVELL — We also had snacks stuffed in our pockets.

SWIGERT — As a result, none of us ate either meal B or meal C on the first day, but we did eat the snack. We started off eating regular meals on the second day. We also kept a log every time we ate anything.

LOVELL — I have one general comment on the food. It was good. I thought that the wet packs were a step in the right direction. The bread was good, and the spreads were better, but some of the food was a little difficult to handle. Some of the spreads dried out a little because the water went to the top. They separated, too, so it was hard to spread them because they were hard and because the water floated. That problem can be worked out, though.

LOVELL — I think that the packing of bread packages expand because the pressure goes down. When they pack them, there's 14.7 psi, and there's only 5 Psi in the CM. So, you never can get the bread back in the package. But each package, each complete meal, is in a package by itself and is on a string.

SWIGERT — Yes, that's all right. I think you eat more, because you can't get the food back in its package. You have to do something with it, so you eat it.

LOVELL — We could have packaged the food better. I know we do have more room for food stowage in the CM. The food is too compact. If you pull out one thing, a whole bunch of food comes out. I think we have room enough in the CM that we can devote a little more space to food stowage. They really pack it in there. They must have spent a lot of time packing the food. That's great, except that, in zero g, if you reach in to get one thing, you pull out a bunch of food; boy, it's hard to get the food back in the package. Also, each bread slice is not vacuum packed. That stuff just goes all over the place.

SWIGERT — The orange juice was much better than the orange drink. I'll tell you, those juice drinks really saved us. We used up all of them. In fact, we were even getting into the last meals to get out the juice drinks. But, of course, that was because of our shortage of water.

HAISE — I had a comment on the pantry. The THC cable route interfered with opening the rear door of the pantry. We had to be very careful on the forward door, and I had to be especially careful when I raised the door so that I would not ding the THC cable. I had never noticed that before. I don't know whether or not this is something peculiar to this spacecraft.

SWIGERT — The routing of the THC cables was pretty much standard; so, I think you're going to find that there is a degree of interference. Bat you're right, the door did hit the cable when you raised the door about 60 degrees. The door didn't raise the full 90 degrees.

21.2.3 Food Waste Stowage

LOVELL — We used one temporary stowage bag for the food waste stowage. It's the same old thing, too. The amount of trash really piles up in a hurry, and you have to keep ahead of it. We were keeping ahead of it up until the time of the emergency.

SWIGERT — We didn't even do too badly after the emergency.

HAISE — Is that all it was; that trash bag sitting down with it?

LOVELL — I have been thinking that if we could have jettisoned the stuff in the LM, just gotten rid of it, that would have been fine. Again, if we had a little bit more room in our food stowage area, we could replace the food that we use with the debris that's left over. And we could use that as a stowage spot. But we couldn't do that adequately because of the way the food is packaged now. It's just packaged too tightly. Once you get that food out, you just can't put it back into its package. If you could put it back, it would be much more efficient.

HAISE — The bags of bread needed patches of Velcro on them.

SWIGERT — Did the wetpacks have Velcro on them?

LOVELL — Yes, the wetpacks had Velcro.

HAISE — The ketchup packages did not have Velcro on them.

HAISE — We had one emergency in the LM that you didn't know about.

SWIGERT — Fred went up to get a volts and amps reading, and he said, "Here, hold my frankfurter." When he came back I said, "Fred, I got an emergency here. I squeezed too hard, and it drifted off someplace." I said, "Check around for two loose frankfurters." That broke Fred up for about 5 minutes. I was still holding that empty package when he came back. We had a few laughs over this one.

HAISE — I thought you had eaten them, that's why. You just can't trust these CMPs when you leave them in charge of the LM. The Spoon-bowl packages worked pretty well.

LOVELL — That's a step in the right direction.

21.2.4 Water

LOVELL — We chlorinated the water, and even in the beginning, we saw gas in the water.

SWIGERT — There was gas in the CM potable water all the way through the mission.

LOVELL — The water was hot, and hot water was fine.

SWIGERT — You know, something we all commented on is that none of us noticed a taste of chlorine in the water the next morning.

HAISE — That's right. I thought that it tasted almost as if Jim hadn't chlorinated it.

SWIGERT — There was always some gas in all the juice bags when we filled them, even up to the very end of the mission. Even when the fuel cells weren't working, there was always gas in the water.

LOVELL — I don't know what we're going to do about that situation.

SWIGERT — Jim, you probably can make a comparison between the Apollo 8 potable water and the Apollo 13 potable water. Did you think there was more or less gas?

LOVELL — I thought it was about the same. A good way to tell is by drinking from the juice bag. A lot of gas makes it difficult to drink from the juice bag, because you're drinking air, gas, and juice. I got a little thirsty towards the end of the mission, because we ran out of CM potable water.

21.2.5 Work, Sleep, Rest

LOVELL — I guess we should talk about the worst sleep cycles before and after the emergency.

SWIGERT — The first night I didn't sleep as well as I did the second night. I guess it's just a matter of getting used to sleeping in zero g. The second night I had a good night's sleep.

HAISE — Yes, the same was true for me. I guess I had some sleep the first night on the couch. All I did was set up the couch and fasten my lap belt, and I had a feeling that I was rocking up and down all night. I don't know whether or not this rocking contributed to the severe headache I woke up with in the morning. If I had to do it over again, I think I'd also bring the shoulder harnesses down and latch myself in completely.

LOVELL — Did you wear your COMM carrier?

HAISE — Yes.

LOVELL — On the first night, I wore the COMM carrier. I don't think I would recommend that, though, because of the audible tone. I think we could have taken off the COMM carriers.

HAISE — No, I'm sorry. I didn't wear the COMM carrier. I wore the lightweight headset.

LOVELL — If there's a switch in S-band and OMNI antennas, the antenna will tend to wake you up. Also, we had a third sleep restraint, which I didn't even know we had on board.

HAISE — Yes.

LOVELL — Don't you think the sleep position underneath the couch is better than the sleep position on top of the couch?

HAISE — Yes.

LOVELL — This is true mainly because the sleep restraint under the couch keeps you in position.

SWIGERT — That's right. I agree.

LOVELL — Maybe we ought to think about rigging up a similar sleep restraint on the couch.

SWIGERT — I suspect there are places down in the LEB that a restraint could be tied without any problems. One example is the G&N handhold.

LOVELL — And then, we would have three good sleeping positions. I slept on the couch the second night, and I didn't get as good a sleep as I did the first night for two reasons. First, a MASTER ALARM occurred just about the time I was falling asleep. That really made me jump out of the couch.

SWIGERT — We never did find out what that was.

LOVELL — No, and the second reason was that the hose by my side was rotted. It was blowing cool air on me all night, and that distracted me.

21.2.10 Medical Kits

LOVELL — The medical kits were adequate. We used aspirin.

SWIGERT — I used two Lomotils and one Dexedrine.

LOVELL — I used Dexedrine and we also used quite a bit of aspirin and one Darvon.

HAISE — You used two Lomotils?

SWIGERT — They didn't do any good.

HAISE — Are you sure you got the right compartment?

SWIGERT — I think so.

HAISE — And we used seasickness pills.

LOVELL — No you didn't.

SWIGERT — I took the Marzine.

LOVELL — Dexedrine.

SWIGERT — I took a Dexedrine.

LOVELL — I just took one (Dexedrine).

HAISE — I didn't take any because the seasickness pills had that in it.

LOVELL — Yes. I was a bit concerned about taking too much Dexedrine. I was afraid it might wear off before I got down.

21.2.11 Housekeeping

LOVELL — We kept up with all the debris. We never had any loose packages around. After every meal, we immediately used the pills and the debris went in the garbage bag.

SWIGERT — Even after the incident, we always had a clean house. There were never any odors; never any mold or anything.

21.2.12 Shaving

LOVELL — Fred, why don't you talk about shaving?

HAISE — The problem was one of two things, I guess. Deke's point that it had been done before, and quite successfully, leads me to believe that our selection of the type of cream, the Mennen was not the right one. What happened with all of us was that the shaving cream caked underneath the razor blade on the Techmatic we had and it allowed the blade to skim very neatly over the rest of the face without even touching the whiskers. You really had to apply a lot of pressure and scraping back and forth to get it to dig in a little bit to do any cutting. It was a very long-term, meticulous job to get a decent shave with that apparatus. I guess the next guys should follow more in line with that used previously - which I didn't have any knowledge of - or make use of a more selective sampling in the available creams. We really only looked at two - Gillette and Mennen. I thought if any of them would be good, those two would. I guess that wasn't necessarily true for the environment we were in. Another possibility we thought of before the mission was the benefit of having a razor that you can either remove the head or move the razor to allow cleaning of the blade. With the Techmatic we weren't able to do either.

21.2.13 Radiation Dosimetry

LOVELL — We erred here a little bit. Fred and I took off our suits and left our dosimeters in our suit pockets. Because we already had them stowed and wanted to be very careful with the suits, we were reluctant to unstow them. So we relied on Jack's dosimeter as an overall dosimeter for the flight, after the accident. After the accident we couldn't bother with taking any dosimeter readings.

SWIGERT — I did give them the dosimeter out of my pouch. It was stowed underneath the LEB in one of those little pouches underneath the optics.

LOVELL — The PRDs were not worn throughout the mission. Two of them were stowed in the suit bag and one was stowed down by the LEB.

SWIGERT — We didn't use the radiation survey meter.

LOVELL — We activated it one time when we couldn't find a dosimeter, to see if there was any change. We didn't see anything in excess of the 10-mrad/hr range. It was outside the radiation belt.

21.2.14
Personal Hygiene

LOVELL — I've always thought those wet-wipes were too small, whether they are packaged with the food or packaged with the other. They're very small. I'd much prefer the wet-wipes packaged in the AF inflight lunch kits. At least they smell good. We've had these things from the Gemini days. I don't know what they put on them, but they're awful, and they are small. I guess that's something we can live with.

SLAYTON — Do you know of any reason why we can't improve them?

LOVELL — You know, we're looking toward Skylab and long-duration flights and improving crew comfort.

SLAYTON — I don't know why we can't use the same thing they use on the airlines.

LOVELL — That's right.

SLAYTON — Was potable water used for personal hygiene?

LOVELL — Sure it was. We used it to keep our faces clean, and for shaving. We used the hot water to try to soften the beard.

SWIGERT — As far as tissues go, there was plenty of tissues.

HAISE — Yes, we never ran out of tissues.

22.0
MISCELLANEOUS

22.1 MEDICAL
REQUIREMENTS

LOVELL — Deke could probably better explain it than I could, but it seemed to me that this physical was different from the Apollo physical. This time, we had a physical every day, something which came as a complete surprise to me. I must have missed the briefing somewhere along the line.

SLAYTON — We've been doing that since Apollo 9; the physicals are in terms of a quick look each morning. They are just a kind of nose and throat check to make sure nobody is getting a red throat or something. Before Apollo 9, we weren't conducting physicals each day. But remember that all of a sudden we came along and had a problem. So since then, from T minus 5 days to lift-off, we have been doing physicals each day.

LOVELL — I didn't know about it until Jack Teegen said, "Where do you want to take the physical?" I said, "What physical?" He said, "Well, we have to look at you every day." Another thing I thought of, which you might want to consider, is that the backup crew and the prime crew never follow the same physical regimen.

SLAYTON — That's right; from T minus 30 days to lift-off.

LOVELL — If you have to repeat what we experienced, you might think about changing that technique.

SLAYTON — Again, I think it worked out all right, because from T minus 5 days to lift-off is really the critical period.

22.2 PAO
REQUIREMENTS

LOVELL — There was no interference by PAO requirements with flight preparation.

70-HC-323

70-HC-541

S70-15501

S70-35748

70-HC-564

S70-15526

S70-15506

S70-35600

THE APOLLO 13 ACCIDENT

HEARINGS

BEFORE THE

COMMITTEE ON SCIENCE AND ASTRONAUTICS U.S. HOUSE OF REPRESENTATIVES

NINETY-FIRST CONGRESS

SECOND SESSION

JUNE 16, 1970

[No. 19]

Printed for the use of the
Committee on Science and Astronautics

COMMITTEE ON SCIENCE AND ASTRONAUTICS

GEORGE P. MILLER, California. CHAIRMAN

OLIN E. TEAGUE, TEXAS
JOSEPH E. KARTH, MINNESOTA
KEN HECHLER, West Virginia
EMILIO Q. DADDARIO, Connecticut
JOHN W. DAVIS, Georgia
THOMAS N. DOWNING, Virginia
JOE D. WAGGONNER, Jr., Louisiana
DON FUQUA, Florida
GEORGE E. BROWN, Jr., California
EARLE CABELL, Texas
BERTRAM L. PODELL, Now York
WAYNE N. ASPINALL. Colorado
ROY A. TAYLOR, North Carolina
HENRY HELSTOSKI, New Jersey
MARIO BLAGGI, New York
JAMES W. SYMINGTON Missouri
EDWARD I. KOCH New York

JAMES G. FULTON, Pennsylvania
CHARLES A. MOSHER, Ohio
RICHARD L. ROUDEBUSH, Indiana
ALPHONZO BELL, California
THOMAS M. PELLY, Washington
JOHN W. WYDLER, New York
GUY VANDER JAGT ,Michigan
LARRY WINN, Jr., Kansas
D. E. (BUZ) LUKENS, Ohio
ROBERT PRICE, Texas
LOWELL P. WEICKER. JR., Connecticut
LOUIS FREY JR., Florida
BARRY M. GOLDWATER, JR., California

CHARLES F. DUCANDER Executive Director and Chief Counsel
JOHN A. CARSTARPHEN Jr., Chief Clerk and Counsel
PHILIP B. YEAGER Counsel
FRANK R. HAMMILL Jr.. Counsel
W. H. BOONE Technical Consultant
JAMES E. WILSON Technical Consultant
RICHARD P. HINES, staff Consultant
HAROLD A. GOULD, Technical Consultant
J. THOMAS RATCHFORD Science Consultant
PHILIP P. DICKINSON, Technical Consultant
WILLIAM G. WELLS, Jr., Technical Consultant
K. GUILD NICHOLS. Jr., Staff Consultant
ELIZABETH S. KERNAN, Scientific Research Assistant
FRANK J. GIROUX Clerk
DENIS C. QUIGLEY, Publications Clerk
RICHARD K. SHULLAW Assistant Publications Clerk
JAMES A. ROSE Jr.. Minority Staff

THE APOLLO 13 ACCIDENT

TUESDAY, JUNE 16, 1970

House Of Representatives
Committee On Science and Astronautics,
Washington, D.C.

The committee met, pursuant to notice, at 10:05 a.m. in room 2318, Rayburn House Office Building, Hon. P. Miller, chairman, presiding.

The CHAIRMAN — The committee will be in order. Dr. Paine, Mr. Cortright, members of the Apollo 13 Review Board, we are pleased to welcome you to the committee today for the purpose of presenting findings, determinations, and recommendations of the Apollo 13 review board.

Dr Paine, I would like to commend you for the appointment of a most competent and outstanding board to review the Apollo 13 accident and the circumstances surrounding it.

Mr. Cortright, whom this committee knows well, has distinguished himself not only as an administrator in the NASA Headquarters organization but also as a field center director at Langley Research Center.

The other members of the board are similarly well qualified to have participated in this intensive and searching review.

As I stated at the time of the Apollo 13 accident the committee decided that sufficient time should be allowed for NASA to fully investigate the accident and at such time that this investigation was completed the committee would convene to receive NASA's evaluation of the accident.

Therefore, I have asked you to appear here today even though the board's report was only submitted to you yesterday, Dr. Paine, because I feel it is important that the members of the committee receive a firsthand and timely review of the Apollo 13 accident.

Dr. Paine, I understand you have a short statement and then Mr. Cortright will go into the details of the accident and the board's findings.

I want to give all members an opportunity to ask questions, so will you please proceed.

Before proceeding, I would also like to make a part of the record the fact that Mr. Wilson of the staff of the committee was appointed to act as an observer with the board.

I want to thank you for the courtesies you have shown us and it has given us a new system of liaison. Please proceed, Dr. Paine.

STATEMENT OF DR. THOMAS O. PAINE ADMINISTRATOR, NASA

Dr. PAINE — Mr. Chairman and members of the committee, on April 17, Dr. George Low and I established the Apollo 13 Review Board under the direction of Mr. Edgar M. Cortright, director of the Langley Research Center. The instructions to the board are contained in a memorandum dated April 17, and the membership of the board in a memorandum dated April 20, 1970, which are reproduced in the summary volume of the report you have received.

The past 2 months have involved long hours and very hard work by the review board and supporting elements in NASA and the industrial community. I would like to take this opportunity to extend my thanks to them for the thoroughness of their investigation and their dedication to this arduous assignment.

Since I received the review board report only yesterday, I have not had a chance to review it in detail. Nor have I had the benefit of the independent assessment which is being carried out by the Aerospace Safety Advisory Panel, chaired by Dr. Charles Harrington.

The Office Of Manned Space Flight is also conducting a separate review of the report.

In about 10 days I will receive the results of the safety panel and the manned space flight review. Until I have received and studied these reports. I will obviously not be in a position to give you my evaluation of the board's recommendations or NASA's future actions. Earlier we announced a change in our lunar landing schedule involving a delay of the Apollo 14 launch from October to the December launch window. However, this is subject to review in light of the report of the Apollo 13 Review Board and we will not fly Apollo 14 to the moon until we are confident that we have done everything necessary to eliminate the conditions that caused or contributed to the problems on Apollo 13.

I believe that, as we plan man's future course in space, the preface to this report should be a reminder of the nature of the challenge we have undertaken. Let me quote:

> *The Apollo 13 accident, which aborted man's third mission to explore the surface of the moon, is a harsh reminder of the immense difficulty of this undertaking.*
>
> *The total Apollo system of ground complexes, launch vehicle, and spacecraft constitutes the most ambitious and demanding engineering development ever undertaken by man. For these missions to succeed, both men and equipment must perform to near perfection. That this system has already resulted in two successful lunar surface explorations is a tribute to those men and women who conceived, designed, built, and flew it.*
>
> *Perfection is not only difficult to achieve, but difficult to maintain. The imperfection in Apollo 13 constituted a near disaster, averted only by outstanding performance on the part of the crew and the ground control team which supported them.*
>
> *The Board feels that the nature of the Apollo 13 equipment failure holds important lessons which, when applied to future missions, will contribute to the safety and effectiveness of manned space flight.*

Mr. Chairman, there has been time for me to reach one conclusion on the report of the Apollo 13 Review Board, and that is that the board and their supporting teams have done a magnificent piece of technical detective work that carefully reconstructs the background and the events which took place aboard Apollo 13 200,000 miles from earth.

I would now like to introduce the chairman of the review board, Mr. Edgar Cortright who will briefly discuss the report and respond to you.

CHAIRMAN — Thank you very much, Dr. Paine. We are very happy to have you here, Mr. Cortright.

STATEMENT OF EDGAR M. CORTRIGHT, CHAIRMAN, APOLLO 13 REVIEW BOARD; DIRECTOR, LANGLEY RESEARCH CENTER

Mr. CORTRIGHT — Thank you, Mr. Chairman. I have a prepared statement and, with your permission, I will submit this for the record and attempt to convey to you what the Board has done and what our conclusions have been in a more informal manner.

The CHAIRMAN — Without objection, that will be the manner in which we will proceed.

(The prepared statement of Mr. Cortright is as follows:)

PREPARED STATEMENT OF EDGAR M. CORTRIGHT, CHAIRMAN, APOLLO 13, REVIEW BOARD, NATIONAL AERONAUTICS AND SPACE ADMINISTRATION

Mr. Chairman and Members of the Committee:

I appreciate this opportunity to appear before the Committee to summarize the Report of the Apollo 13 Review Board.

As you know, yesterday I presented this Report on behalf of the Board to the Administrator and Deputy Administrator. Copies of the Report were given to the Members and Staff of the Committee, and the Report was made public yesterday afternoon, at which time Dr. Paine and I held a press conference.

This morning I would like first to outline for the Committee how the Board was established and how it organized itself to review and report on the Apollo 13 accident. Then I will cover in some detail the findings and determinations of the Board regarding the accident, including pre-accident mission events, the events of the accident itself, and the recovery procedures which were implemented to return the crew safely to earth. I will also summarize the Board's findings and determinations regarding the management, design, manufacturing, and test procedures employed in the Apollo Program as they relate specifically to the accident.

Based on its findings and determinations, the Board made a series of detailed recommendations. I will report these to you and be pleased to answer any questions you may have on the Board's work.

ESTABLISHMENT AND HISTORY OF THE BOARD

The Apollo 13 Review Board was established, and I was appointed Chairman, on April 17, 1970. The charter of the Board was set forth in the memorandum which established it. Under this charter the Board was directed to:

(a) Review the circumstances surrounding the accident to the spacecraft which occurred during the flight of Apollo 13 and the subsequent flight and ground actions taken to recover, in order to establish the probable cause or causes of the accident and assess the effectiveness of the recovery actions.

(b) Review all factors relating to the accident and recovery actions the Board determines to be significant and relevant, including studies, findings, recommendations, and other actions that have been or may be undertaken by the program offices, field centers, and contractors involved.

(c) Direct such further specific investigations as may be necessary.

(d) Report as soon as possible its findings relating to the cause or causes of the accident and the effectiveness of the flight and ground recovery actions.

(e) Develop recommendations for corrective or other actions, based upon its findings and determinations or conclusions derived therefrom.

(f) Document its findings, determinations, and recommendations and submit a final report.
The Membership of the Board was established on April 21, 1970. The members are:

Mr. Edgar M. Cortright,	Chairman (Director, Langley Research Center)
Mr. Robert F. Allnutt	(Assistant to the Administrator, NASA Hqs)
Mr. Neil Armstrong	(Astronaut, Manned Spacecraft Center)
Dr. John F. Clark	(Director, Goddard Space Flight Center)
Brig. General Walter R. Hedrick, Jr.	(Director of Space, DCS/R&D, Hqs. USAF)
Mr. Vincent L. Johnson	(Deputy Associate Administrator-Engineering, Office of Space Science and Applications)
Mr. Milton Klein	(Manager, ABC—NASA Space Nuclear Propulsion Office)
Dr. Hans M. Mark	(Director, Ames Research Center)

Legal Counsel to the Board is Mr. George T. Malley, Chief Counsel, Langley Research Center.

Appointed as Observers were:

Mr. William A. Anders	(Executive Secretary, National Aeronautics and Space Council)
Dr. Charles D. Harrington	(Chairman, NASA Aerospace Safety Advisory Panel)
Mr. I. I. Pinkel	(Director, Aerospace Safety Research and Data Institute, NASA Lewis Research Center)
Mr. James E. Wilson, Jr.	(Technical Consultant, House of Representatives, Committee on Science and Astronautics)

The documents establishing the Board and its membership and other relevant documents are included in Chapter I of the Board's Report.

The Review Board convened at the Manned Spacecraft Center (MSC), Houston, Texas, on Tuesday, April 21, 1970. Four Panels of the Board were formed, each under the overview of a member of the Board. Each of the Panels was chaired by a senior official experienced in the area of review assigned to the Panel. In addition, each Panel was manned by a number of experienced specialists to provide in-depth technical competence for the review activity. During the period of the Board's activities, the Chairmen of the four Panels were responsible for the conduct of reviews, evaluations, analyses, and other studies bearing on their Panel assignments and for preparing documented reports for the Board's consideration. Complementing the Panel efforts, each member of the Board assigned specific responsibilities related to the overall review.

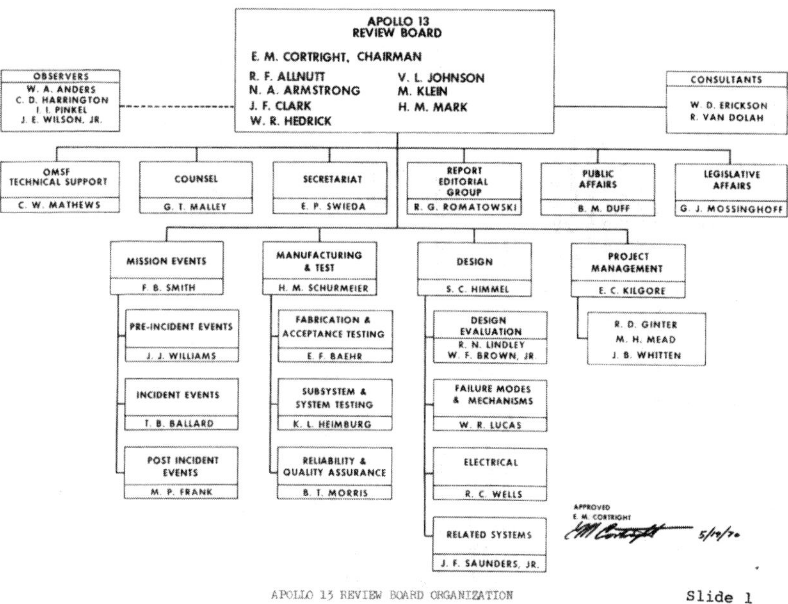

APOLLO 13 REVIEW BOARD ORGANIZATION Slide 1

On Slide 1 is shown a chart depicting the organization of the Board. The four Panels — Mission Events, Manufacturing and Test, Design, and Project Management, are shown along with the sub-panels and the supporting office structure. The membership and responsibilities of each Panel are set forth in the Report.

While the Board's intensive review activities were underway, the Manned Spacecraft Center Apollo 13 Investigation Team, under James A. McDivitt, Director of the MSC Apollo Spacecraft Program Office, was also conducting its own analysis of the Apollo 13 accident. Coordination between the Investigation Team Work and the Apollo 13 Review Board activities was effected through the Manned Space Flight Technical Support official and by maintaining a close and continuing working relationship between the Panel Chairmen and officials of the MSC Investigation Team. In addition, Board members regularly attended daily status meetings of the Manned Spacecraft Center Investigation Team.

In general, the Board relied on Manned Spacecraft Center post-mission evaluation activities to provide the factual database for evaluation, assessment, and analysis efforts. However, the Board, through a regular procedure, also levied specific data collection, reduction, and analysis requirements on MSC. Test support for the Board was provided by MSC, but in addition, the Board established an extensive series of special tests and analyses at other NASA Centers and at contractor facilities. Members of the Board and its Panels also visited contractor facilities to review manufacturing, assembly, and test procedures applicable to Apollo 13 mission equipment.

In this test program, which included nearly 100 separate tests, and which involved several hundred people at its peak, the elements of the inflight accident were reproduced. All indications are that electrically initiated combustion of Teflon insulation in oxygen tank No. 2

in the service module was the cause of the Apollo 13 accident. One series of tests demonstrated electrical ignition of Teflon insulation in supercritical oxygen under zero g and at one g, and provided data on ignition energies and burning rates. Other tests culminating in a complete flight tank combustion test, demonstrated the most probable tank failure mode. Simulated tank rupture tests in a ½ scale service module verified the pressure levels necessary to eject the panel from the service module. Other special tests and analyses clarified how they might have been generated. I have with me a brief film, highlighting these tests, which I would like to show at the conclusion of my statement.

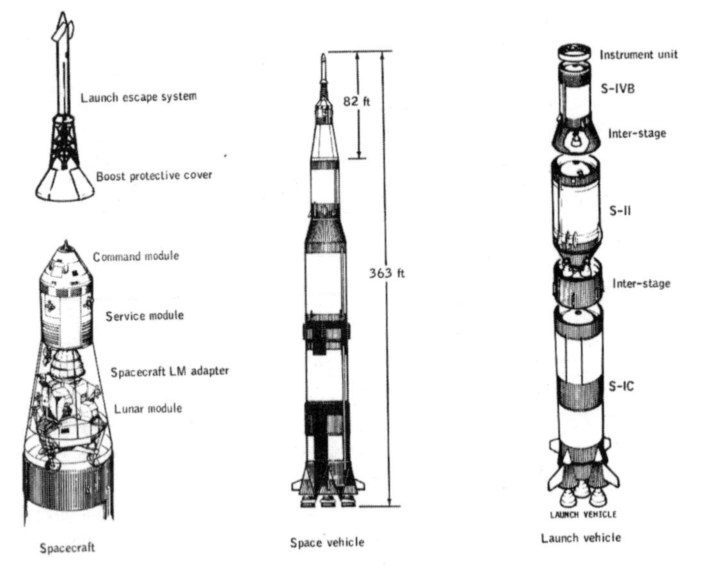

Figure 3-1.- Apollo/Saturn V space vehicle.

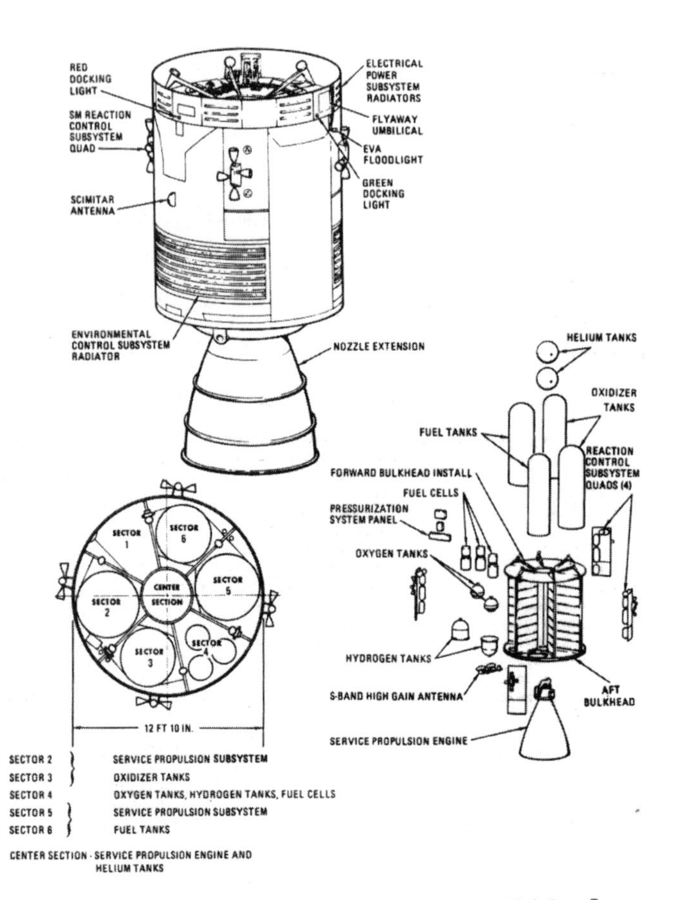

Figure 3-6.- Service module. Slide 3

APOLLO 13 SYSTEMS

Before tracing the analyses which lead to the Board's conclusions — and to place them in proper context — I would like to explain the design and functions of the oxygen tank #2 as a part of the Apollo system. Details of the entire Apollo/ Saturn Space Vehicle are set forth in the Report and its Appendices.

Slide 2 shows the Apollo/Saturn Space Vehicle with which you are all familiar. Slide 3 shows the service module which, as you know, is designed to provide the main spacecraft propulsion and maneuvering capability during a mission. It also contains most of the spacecraft consumables (oxygen, water, propellant, and hydrogen) and supplies electrical power. The service module is divided into six sectors or bays surrounding a center section. The oxygen tank, to which I referred, is located in Bay 4 (shown in more detail on Slide 4), along with another oxygen tank, two hydrogen tanks, three fuel cells and interconnecting lines, and measuring and control equipment.

The tanks supply oxygen to the environmental control system (ECS) for the astronauts to breathe, and oxygen and hydrogen to the fuel cells. The fuel cells generate the electrical power for the command and service modules during a mission. The next slides (Nos. 5, 6 and 7) are photographs of Bay 4 of the Service module for Apollo 13 showing the major elements and their interconnection Slide 7 shows the oxygen tank #2 in place.

As the simplified drawing in Slide 8 indicates each oxygen tank has an outer shell and an inner shell, arranged to provide a vacuum space to reduce heat leak, and a dome enclosing paths into the tank for transmission of fluids, and electrical power and signals.

The space between the shells and the space in the dome are filled with insulating materials. Mounted in the tank are two tubular assemblies. One, called the heater tube, contains two thermostatically protected heater coils and two small fans driven by 1800 RPM motors to stir the tank contents. The other assembly, called the quantity probe, consists of a cylindrical capacitance gage used to measure electrically the quantity of fluid in the tank. The inner cylinder of this probe is connected through the top of the tank to a fill line from the exterior of the SM and serves both as a fill and drain tube and as one plate of the capacitance gage. In addition, a temperature sensor is mounted on the outside of the quantity probe near the head. Wiring for the quantity gage, the temperature sensor, the fan motors, and the heaters passes through the head of the quantity probe, through a conduit in the dome and to a connector to the appropriate external circuits in the CSM. The routing of wires and lines from the tank through the dome is shown in Slide 9.

The oxygen tank, as designed, contained materials, which if ignited will burn in supercritical oxygen. These include Teflon, used for example, to insulate the wiring, and aluminum.

NASA-S-70-512-V

ARRANGEMENT OF FUEL CELLS AND CRYOGENIC SYSTEMS IN BAY 4

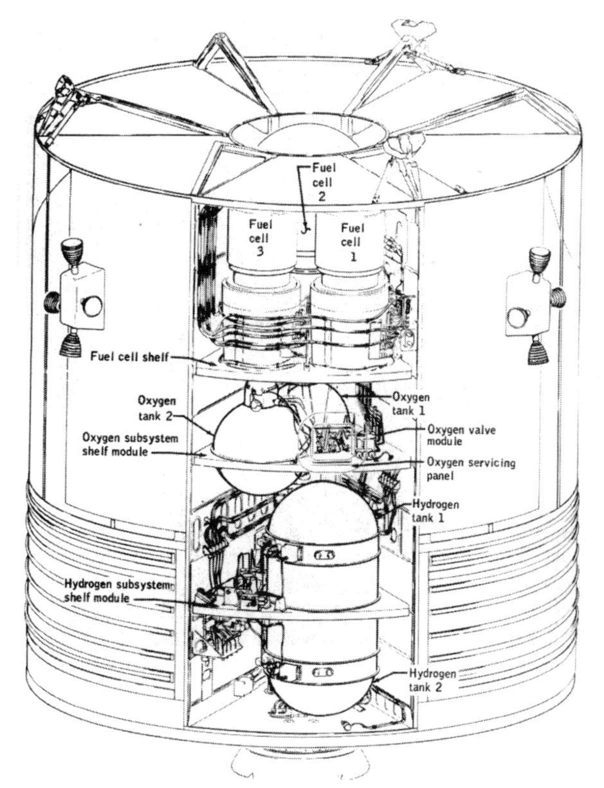

Slide 4

Pressure in the tank is measured by a pressure gage in the supply line, and a pressure switch near this gage is provided to turn on the heaters in the oxygen tank if the pressure drops below a pre-selected value. This periodic addition of heat to the tank maintains the pressure at a sufficient level to satisfy the demand for oxygen as tank quantity decreases during a flight mission.

The oxygen tank is designed for a capacity of 320 pounds of supercritical oxygen at pressures ranging between 865 and 935 pounds per square inch absolute (psia). The tank is initially filled with liquid oxygen at -297°F and operates over the range from -340° to +80°F. The term "supercritical" means that the oxygen is maintained at a temperature and pressure which assures that it is a homogeneous, single-phase fluid.

The burst pressure of the oxygen tank is about 2200 psia at -150°F, over twice the normal operating pressure at that temperature. A relief valve in the supply line leading to the fuel cells and the ECS is designed to relieve pressure in the oxygen tank at a pressure of approximately 1000 psi. The oxygen tank dome is open to the vacuum between the inner and outer tank shell and contains a rupture disc designed to blow out at about 75 psi.

As shown in Slide 9, each heater coil is protected with a thermostatic switch, mounted on the heater tube, which is intended to open the heater circuit when it senses a temperature of 80°F. As I will point out later in tracing the Board's conclusions as to the cause of the accident, when the heaters were powered from a 65 volt DC supply at KSC during an improvised detanking procedure, these thermostatic switches, because they were rated at only 30 V DC, could not prevent an overheating condition of the heaters and the associated wiring. Tests conducted for the Board indicate that the heater tube assembly was probably heated to a temperature of as much as 1000° F during this detanking procedure.

Slide 5

Figure 4-5.- Fuel cells shelf.

Slide 6

THE APOLLO 13 MISSION

/ith this general background, I will now summarize the Apollo 13 mission. This mission, as you know, was designed to perform the third anned lunar landing. The selected site was in the hilly uplands of the Fra Mauro formation. A package of five scientific experiments was anned for emplacement on the lunar surface near the lunar module landing point. Additionally the Apollo 13 landing crew was to gather e third set of selenological samples of the lunar surface for return to earth for extensive scientific analysis. Candidate future landing

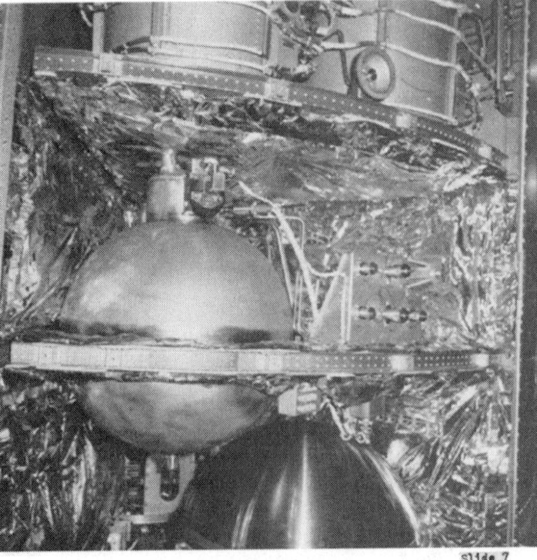

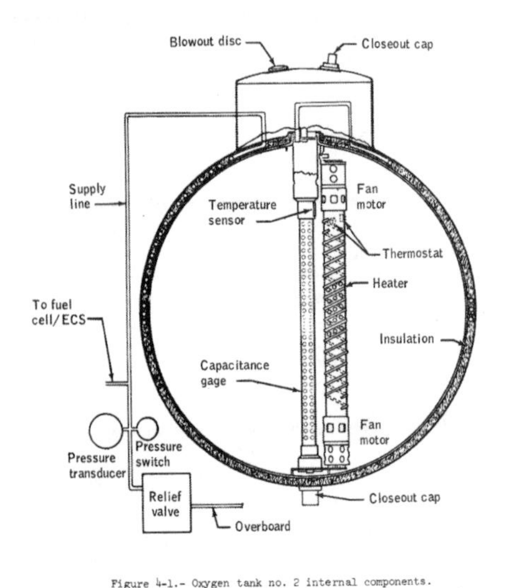

Figure 4-1.- Oxygen tank no. 2 internal components.
Slide 8

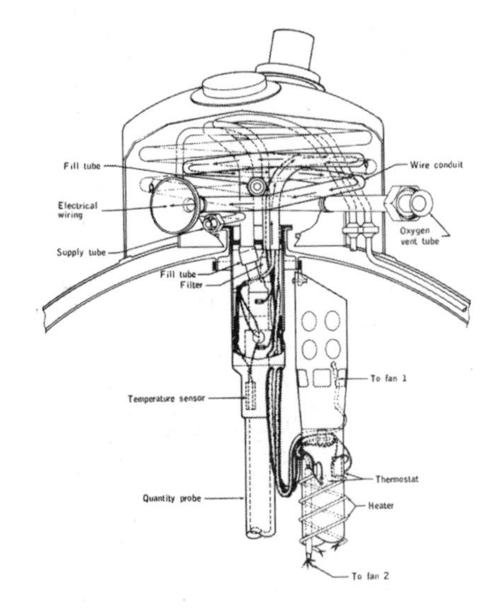

Figure 4-2.- Oxygen tank wiring and lines.
Slide 9

sites were scheduled to be photographed from lunar orbit. The crew consisted of Captain James A. Lovell, Commander, Fred W. Haise, Lunar Module Pilot; and John L. Swigert, Jr., Command Module Pilot, who replaced Thomas K. Mattingly, III, who had been exposed to rubella and, after tests, found not to be immune.

Launch was on time at 2:13 p.m., EST on April 11 from the KSC Launch Complex 39A. The spacecraft was inserted into a 100-nautical mile circular earth orbit. The only significant launch phase anomaly was premature shutdown of the center engine of the S-II second stage. This anomaly, although serious, was not related to the subsequent accident. It is being investigated by the Apollo organization. As a result of this shutdown, the remaining four S-II engines burned 34 seconds longer than planned and the S-IVB third stage engine burned a few seconds longer than planned. At orbital insertion, the velocity was within 1.2 feet per second of the planned velocity. Moreover, an adequate propellant margin was maintained in the S-IVB for the translunar injection burn.

After spacecraft systems checkout in earth orbit, the S-IVB restarted for the translunar injection (TLI) burn, with shutdown coming some six minutes later. After TLI, Apollo 13 was on the planned free-return trajectory with a predicted closest approach to the lunar surface o 210 nautical miles.

The command and service module (CSM) was separated from the S-IVB about three hours into the mission, and after a brief period o station-keeping, the crew maneuvered the CSM into dock with the LM vehicle in the LM adapter atop the S-IVB stage. The S-IVB stage was separated from the docked CSM and LM shortly after four hours into the mission, and placed on a trajectory to ultimately impact the moon near the site of the seismometer emplaced by the Apollo 12 crew.

At 30:40:49 g.e.t. (ground elapsed time) a midcourse correctio maneuver was made using the service module propulsion system This maneuver took Apollo 13 off a free-return trajectory and place it on a non-free return trajectory A similar profile had been flown o Apollo 12. The objective of leaving a free-return trajectory is to control the arrival time at the moon to insure the proper lightin conditions at the landing site. The transfer maneuver lowered th predicted closest approach to the moon, or pericynthion altitude from 210 to 64 nautical miles.

From launch through the first 46 hours of the mission, th performance of the oxygen tank #2 was normal, so far a telemetered data and crew observations indicate. At 46:40:02, th crew turned on the fans in oxygen tank #2 as a routine operatio and the oxygen tank #2 quantity indication changed from a norma reading to an obviously incorrect reading "off scale high" of over 10 percent. Subsequent events indicate that the cause was a shor circuit which was not hazardous in this case.

At 47:54:50 and at 51:07:44 the oxygen tank #2 fans were turned o again, with no apparent adverse effects. The quantity gage continue to read "off scale high."

Following a rest period, the Apollo 13 crew began preparations fo activating and powering up the lunar module for checkout. At abou 53 and one-half hours g.e.t. Astronauts Lovell and Haise were cleare to enter the LM to commence inflight inspection for the LM. Afte this inspection period, the lunar module was powered down an preparations were underway to close the LM hatch and run throug the pre-sleep checklist when the accident in oxygen tank # occurred.

At about 55:53, flight controllers in the Mission Control Center a MSC requested the crew to turn on the cryogenic system fans an heaters, since a master alarm on the CM Caution and Warnir System had indicated a low pressure condition in the cryogen hydrogen tank #1. This tank had reached the low end of its norm operating pressure range several times previously during the fligh Swigert acknowledged the fan cycle request and data indicate th current was applied to the oxygen tank #2 fan motors at 55:53:20.

About 2½ minutes later, at 55:54:53.5, telemetry from the spacecra was lost almost totally for 1.8 seconds. During the period of data los the Caution and Warning System alerted the crew to a low voltag condition on DC Main Bus B, one of the two main buses which supp electrical power for the command module. At about the same tim the crew heard a loud "bang" and realized that a problem existed the spacecraft. It is now clear that oxygen tank #2 ,or its associate tubing lost pressure integrity because of combustion within the tan

NASA AS13-58-8464 Slide 10

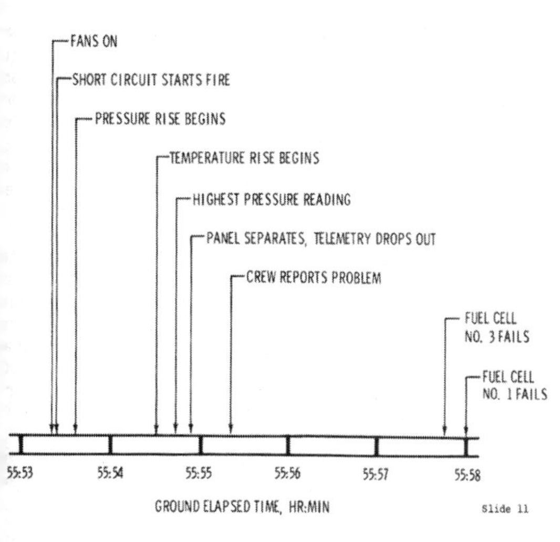

Slide 11

and that the effects of oxygen escaping from the tank, caused the removal of the panel covering Bay 4 and a relatively slow leak in oxygen tank #1 or its lines or valves. Photographs of the service module taken by the crew later in the mission (Slide 10) show the panel missing, the fuel cells on the shelf above the oxygen shelf tilted, and the high gain antenna damaged.

The resultant loss of oxygen made the fuel cells inoperative, leaving the CM with batteries normally used only during reentry as the sole power source and with only that oxygen contained in a surge tank and repressurization packages. The lunar module, therefore, became the only source of sufficient battery power and oxygen to permit safe return of the crew to earth.

SUMMARY ANALYSIS OF THE ACCIDENT

The Board determined that combustion in oxygen tank #2 led to failure of that tank, damage to oxygen tank #1 or its lines or valves adjacent to tank #2, removal of the Bay 4 panel and, through the resultant loss of all three fuel cells, to the decision to abort the Apollo 13 mission. In the attempt to determine the cause of ignition in oxygen tank #2, the course of propagation of the combustion, the mode of tank failure, and the way in which subsequent damage occurred, the Board has carefully sifted through all available evidence and examined the results of nearly 100 special tests and analyses conducted by the Apollo organization and by or for the Board after the accident.

Although tests and analyses are continuing, sufficient information is now available to provide a clear picture of the nature of the accident and the events which led up to it. It is now apparent that the extended heater operation at KSC damaged the insulation on wiring in the tank and that this set the stage for the electrical short circuits which initiated combustion within the tank. While the exact point of initiation of combustion and the specific propagation path involved may never be known with certainty, the nature of the occurrence is sufficiently well understood to permit taking corrective steps to prevent its recurrence.

The Board has identified the most probable failure mode.

The following discussion treats the accident in its key phases: initiation, propagation and energy release, loss of oxygen tank #2 system integrity, and loss of oxygen tank #1 system integrity. Slide 11 shows the key events in the sequence.

Initiation

The evidence points strongly to an electrical short circuit with arcing as the initiating event. Near the end of the 55th hour of flight, about 2.7 seconds after the fans were turned on in the SM oxygen tanks, an 11.1 ampere current spike and simultaneously a voltage drop spike were recorded in the spacecraft electrical system. Immediately thereafter current drawn from the fuel cells decreased by an amount consistent with the loss of power to one fan. No other changes in spacecraft power were being made at the time. No power was on the heaters in the tanks at the time and the quantity gage and temperature sensor are very low power devices. The next anomalous event recorded was the beginning of a pressure rise in oxygen tank # 2, 13 seconds later. Such a time lag is possible with low level combustion at the time. These facts point to the likelihood that an electrical short circuit with arcing occurred in the fan motor on its leads to initiate ᵉe accident sequence. The energy available from the short circuit is estimated to have been at least 10 to 20 joules. Tests conducted ʲring this investigation have shown that this energy is more than adequate to ignite Teflon wire insulation of the type contained within ᵉe tank.

his likelihood of electrical initiation is enhanced by the high probability that the electrical wires within the tank were damaged during ᵉe abnormal detanking operation at KSC prior to launch. The likelihood of damage and the possibility of electrical ignition have been ᵉrified by tests.

ropagation

/hile there is enough electrical power in the tank to cause ignition in the event of an arcing short circuit in defective wire, there is not ᶠfficient electric power to account for all of the energy required to produce the observed pressure rise.

ıere are materials within the tank that can, if ignited in the presence of supercritical oxygen, react chemically with the oxygen in ᵉat-producing chemical reactions. The most readily reactive is Teflon, used for electrical insulation in the tank. Also potentially reactive ᵉ aluminum and solder. Our analyses indicate that there is more than sufficient Teflon in the tank, if reacted with oxygen, to account ʳ the pressure and temperature increases recorded. Furthermore, the pressure rise took place over a period of more than 69 seconds, relatively long period, and one which would be more likely characteristic of Teflon combustion than metal-oxygen reactions.

ıus, the Board concluded that combustion caused the pressure and temperature increases recorded in oxygen tank #2. The pressure ading for oxygen tank #2 began to increase about 13 seconds after the first electrical spike and about 55 seconds later the temperature began to increase. The temperature sensor reads local temperature, which need not represent bulk fluid temperature. Since the rate of pressure rise in the tank indicates a relatively slow propagation of burning along the wiring, it is likely that the region immediately around the temperature sensor did not become heated until this time.

The data on the combustion of Teflon in supercritical oxygen in zero gravity, developed in special tests in support of the Board, indicate that the rate of combustion is generally consistent with these observations.

Loss of Oxygen Tank #2 System Integrity

After the relatively slow propagation process described above took place, there was a relatively abrupt loss of oxygen tank #2 integrity. About 69 seconds after the pressure began to rise, it reached the peak recorded, 1008 psia, the pressure at which the cryogenic oxygen tank relief valve is designed to be fully open. Pressure began a decrease for 8 seconds, dropping to 996 psia before reading's were lost. About 1.85 seconds after the last presumably valid reading from within the tank (a temperature reading) and .8 seconds after the last presumably valid pressure reading (which may or may not reflect the pressure within the tank itself since the pressure transducer is about 20 feet of tubing length distant), virtually all signal from the spacecraft was lost. Abnormal spacecraft accelerations were recorded approximately .42 seconds after the last pressure reading and approximately 38 seconds before the loss of signal. These facts all point to a relatively sudden loss of integrity. At about this time, several solenoid valves, including the oxygen valves feeding two of the three fuel cells, were shocked to the closed position. The "bang" reported by the crew also occurred in this time period. Telemetry signal from Apollo 13 were lost for a period of 1.8 seconds. When signal was reacquired, all instrument indicators from oxygen tank #2 were off-scale high or low. Temperatures recorded by sensors in several different locations in the service module showed slight increases in the several seconds following reacquisition of signal.

Data are not adequate to determine precisely the way in which the oxygen tank #2 system failed. However, available information, analyses, and tests performed during this investigation indicate that the combustion within the pressure vessel ultimately led to localized heating and failure at the pressure vessel closure. It is at this point, the upper end of the quantity probe, that the 1/2-inch Inconel conduit is located, through which the Teflon insulated wires enter the pressure vessel. It is likely that the combustion progressed along the wire insulation and reached this location where all of the wires come together. This, possibly augmented by ignition of other Teflon parts and even metal in the upper end of the probe, led to weakening and failure of the closure or the conduit or both.

Failure at this point would release the nearly 1000 psi pressure in the tank into the tank dome, which is equipped with a rupture disc rated at 75 psi. Rupture of this disc or of the entire dome would then release oxygen, accompanied by combustion products, into Bay 4. The accelerations recorded were probably caused by this release.

Release of the oxygen then began to rapidly pressurize the oxygen shelf space of Bay 4. If the hole formed in the pressure vessel were large enough and formed rapidly enough, the escaping oxygen alone would be adequate to blow off the Bay 4 panel. However, it is also quite possible that the escape of oxygen was accompanied by combustion of Mylar and Kapton (used extensively as thermal insulation in the oxygen shelf compartment and in the tank dome) which would augment the pressure caused by the oxygen itself. The slight temperature increases recorded at various locations in the service module indicate that combustion external to the tank probably took place. The ejected Bay 4 panel then struck the high gain antenna, disrupting communications from the spacecraft for the 1.8 seconds.

Loss of Oxygen Tank #1 Integrity

There is no clear evidence of abnormal behavior associated with oxygen tank #1 prior to loss of signal, although the one data bit (4 psi) drop in pressure in the last tank #1 pressure reading prior to loss of signal may indicate that a problem was beginning. Immediately after signal strength was regained, data show that the tank #1 system had lost its integrity. Pressure decreases were recorded over a period of approximately 130 minutes, indicating that a relatively slow leak had developed in the tank #1 system. Analysis has indicated that the leak rate is less than that which would result from a completely ruptured line, but could be consistent with a partial line rupture or a leaking check valve or relief valve.

Since there is no evidence that there were any anomalous conditions arising within oxygen tank #1, it is presumed that the loss of oxygen tank #1 integrity resulted from the oxygen tank #2 system failure. The relatively sudden, and possibly violent, event associated with the failure of the oxygen tank #2 system could have ruptured a line to oxygen tank #1, or have caused a valve to leak because of mechanical shock.

APOLLO 13 RECOVERY

Understanding the Problem

In the period immediately following the Caution and Warning Alarm for Main Bus B under-voltage, and the associated "bang" reported by the crew, the cause of the difficulty and the degree of its seriousness were not apparent.

The 1.8 second loss of telemetered data accompanied by the switching of the CSM high gain antenna mounted on the SM adjacent to Bay 4 from narrow beam width to wide beam width. The high gain antenna (HGA) does this automatically 200 milliseconds after its directional lock on the ground signal has been lost.

A confusing factor was the repeated firings of various SM attitude control thrusters during the period after data loss. In all probability these thrusters were being fired to overcome the effects that oxygen venting and panel blow-off were having on spacecraft attitude, but it was believed for a time that perhaps the thrusters were malfunctioning.

The failure of oxygen tank #2 and consequent removal of the Bay 4 panel produced a shock which closed valves in the oxygen supply lines to fuel cells 1 and 3. These fuel cells ceased to provide power in about three minutes, when the supply of oxygen between the closed valves and the cells was depleted.

The crew was not alerted to closure of the oxygen feed valves to fuel cells 1 and 3 because the valve position indicators in the CM were arranged to give warning only if both the oxygen and hydrogen valves closed. The hydrogen valves remained open. The crew had not been alerted to the oxygen tank #2 pressure rise or to its subsequent drop because a hydrogen tank low pressure warning had blocked the cryogenic subsystem portion of the Caution and Warning System several minutes before the accident. A limit sense light presumably came on in Mission Control during the brief period of tank overpressure, but was not noticed.

When the crew heard the "bang" and got the master alarm for low DC Main Bus B voltage, Lovell was in the lower equipment bay of the command module, stowing a television camera which had just been in use. Haise was in the tunnel between the CSM and the LM, returning to the CSM. Swigert was in the left hand couch, monitoring spacecraft performance. Because of the master alarm indication low voltage, Swigert moved across to the right hand couch where CSM voltages can be observed. He reported that voltages were "looking good" at 55:56:10. At this time, voltage on Main Bus B had returned to normal levels and fuel cells 1 and 3 did not fail for another 1½ to 2 minutes. He also reported fluctuations in the oxygen tank #2 quantity, followed by a return to the off-scale high position.

When fuel cells 1 and 3 electrical output readings went to zero, the ground controllers could not be certain that the cells had not somehow been disconnected from their respective buses and were not otherwise all right. Consequently about five minutes after the accident, controllers asked the crew to connect fuel cell 3 to DC Main Bus B in order to be sure that the configuration was known. When it was realized that fuel cells 1 and 3 were not functioning, the crew was directed to perform an emergency power-down to reduce the load on the remaining fuel cell. Observing the rapid decay in oxygen tank #1 pressure, controllers asked the crew to re-power instrumentation in oxygen tank #2. When this was done, and it was realized that oxygen tank #2 had failed, the extreme seriousness of the situation became clear.

During the succeeding period, efforts were made to save the remaining oxygen in the oxygen tank #1. Several attempts were made, but had no effect The pressure continued to decrease.

It was obvious by about one and one-half hours after the accident that the oxygen tank #1 leak could not be stopped and that it would soon become necessary to use the LM as a "lifeboat" for the remainder of the mission.

By 58:40, the LM had been activated, the inertial guidance reference transferred from the CSM guidance system to the LM guidance system, and the CSM system were turned off.

Return to Earth

The remainder of the mission was characterized by two main activities — planning and conducting the necessary propulsion maneuvers to return the spacecraft to earth, and managing the use of consumables in such a way that the LM, which is designed for a basic mission with two crewman for a relatively short duration, could support three men and serve as the control vehicle for the time required.

One significant anomaly was noted during the remainder of the mission. At about 97 hours 14 minutes into the mission, Haise reported hearing a "thump" and observing venting from the LM Subsequent data review shows that the LM electrical power system experienced a brief but major abnormal current flow at that time. There is no evidence that this anomaly was related to the accident. Analysis by the Apollo organization is continuing.

A number of propulsion options were developed and considered. It was necessary to return the spacecraft to a free-return trajectory and to make any required midcourse corrections. Normally, the Service Propulsion System (SPS) in the SM would be used for such maneuvers. However, because of the high electrical power requirements for using that engine, and in view of its uncertain condition and the uncertain nature of the structure of the SM after the accident, it was decided to use the LM descent engine if possible.

The minimum practical return time was 133 hours to the Atlantic Ocean. and the maximum was 152 hours to the Indian Ocean. Recovery forces were deployed in the Pacific. The return path selected was for splashdown in the Pacific Ocean at 142:40 g.e.t. This required a minimum of two burns of the LM descent engine. A third burn was subsequently made to correct the normal maneuver execution variations in the first two burns. One small velocity adjustment was also made with reaction control system thrusters. All burns were satisfactory. Slides 12 and 13 depict the flight plan followed from the time of the accident to splashdown.

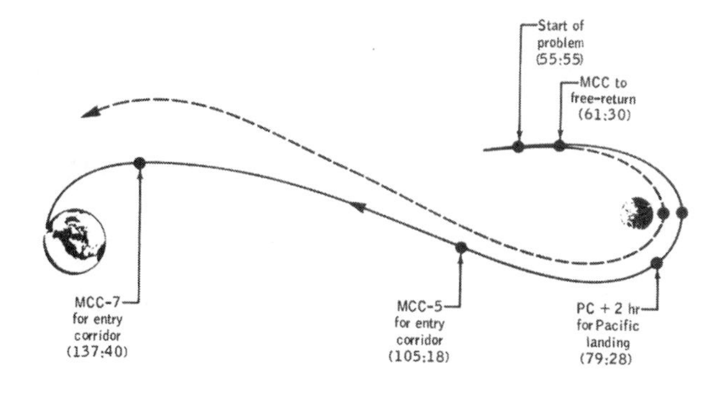

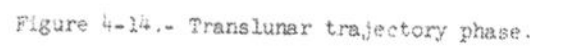
Figure 4-14.- Translunar trajectory phase. Slide 12

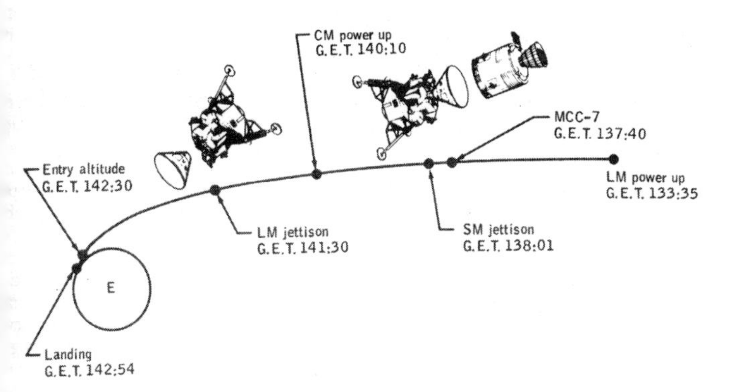

Figure 4-15.- Final trajectory phase. Slide 13

The most critical consumables were water, used to cool the CSM and LM systems during use; CSM and LM battery power, the CSM batteries being for use during reentry and the LM batteries being needed for the rest of the mission: LM oxygen for breathing; and lithium hydroxide (LiOH) filter canisters used to remove carbon dioxide from the spacecraft cabin atmosphere. These consumables, and in particular the water and LiOH canisters, appeared to be extremely marginal in quantity shortly after the accident, but once the LM was powered down to conserve electric power and to generate less heat and thus use less water, the situation greatly improved. Engineers at MSC developed a method which allowed the crew to use materials onboard to fashion a device allowing the use of the CM LiOH canisters in the LM cabin atmosphere cleaning system. At splashdown time, many hours of each consumable remained available.

With respect to the steps taken after the accident, Mission Control and the crew worked, under trying circumstances, as well as was humanly possible, which was very well indeed.

The Board's conclusion that the Apollo 13 accident resulted from an unusual combination of mistakes, coupled with a somewhat deficient and unforgiving design, is based on the Board's in-depth analysis of the oxygen tank, its design, manufacturing, test, handling, checkout, use, failure mode, and eventual effects on the rest of the spacecraft.

OXYGEN TANK #2 HISTORY

On February 26, 1966, the North American Aviation Corporation, now, North American Rockwell (NR), prime contractor for the Apollo command and service modules (CSM), awarded a subcontract to the Beech Aircraft Corporation (Beech) to design, develop, fabricate, assemble, test, and deliver the Block II Apollo cryogenic gas storage subsystem. This was a follow-on to an earlier subcontract under which the somewhat different Block I subsystem was procured.

Manufacture

The manufacture of oxygen tank #2 began in 1966. In its review, the Board noted that the design inherently requires during assembly a substantial amount of wire movement inside the tank, where movement cannot be readily observed, and where possible damage to wire insulation by scraping or flexing cannot be easily detected before the tank is capped off and welded closed. It does not appear, however, that these design deficiencies played any part in the accident.

Several minor manufacturing flaws were discovered in the oxygen tank #2 in the course of testing. A porosity in a weld on the lower half of the outer shell necessitated grinding and re-welding. Rewelding was also required when it was determined that incorrect welding wire had been inadvertently used for a small weld on a vacuum pump mounted on the outside tank dome. The upper fan motor originally installed was noisy and drew excessive current. The tank was disassembled and the heater assembly, fans, and heaters were replaced.

Following acceptance testing at Beech, during which the tank was filled and detanked without apparent difficulty, oxygen tank #2 was shipped to NR on May 3, 1967, for installation, which was completed on March 11, 1968, on a shelf to be installed in Service module 106 for flight in the Apollo 10 mission.

From April 27 to May 29, 1968, the assembled oxygen shelf underwent standard proof pressure, leak, and functional checks. One valve on the shelf leaked and was repaired, but no anomalies were noted with regard to oxygen tank #2, and therefore no rework of oxygen tank #2 was required.

On June 4, 1968, the shelf was installed in SM 106.

Between August 3 and August 8, 1968, testing of the shelf in the SM was conducted, including operation of the heater controls and fan motors. No anomalies were noted.

Due to electromagnetic interference problems with the vacuum pumps on cryogenic tank domes in earlier Apollo spacecraft, a modification was introduced and a decision was made to replace the complete oxygen shelf in SM 106. An oxygen shelf with approved modifications was prepared for installation in SM 106. On October 21, 1968, the oxygen shelf was removed from SM 106 for the required modification and installation in a later spacecraft.

During the initial attempt to remove the shelf, one shelf bolt was mistakenly left in place; and as a consequence, after the shelf was raised about two inches, the lifting support broke, allowing the shelf to drop back into place. At the time, it was believed that the oxygen shelf had simply dropped back into place, and an analysis was performed to calculate the forces resulting from a drop of two inches. It now seems likely that the shelf was first accelerated upward and then dropped.

The remaining bolt was then removed, the incident recorded, and the oxygen shelf was removed without further difficulty. Following removal, the oxygen shelf was re-tested to check shelf integrity, including proof pressure tests, leak tests, and fan and heater operation. Visual inspection revealed no problem. These tests would have disclosed external leakage or serious internal malfunctions of most types but would not disclose fill line leakage within oxygen tank #2. Further calculations and tests conducted during this investigation have indicated that the forces experienced by the shelf were probably close to those originally calculated, assuming a 2-inch drop only. The probability of tank damage from this incident, therefore, is now considered to be rather low, although it is possible that a loosely fitting fill tube assembly could have been displaced by the event.

The shelf passed these tests and was installed in SM 109, the Apollo 13 service module, on November 22, 1968. The shelf tests accomplished earlier in SM 106 were repeated in SM 109 in late December and early January, with no significant problems, and SM 109 was shipped to KSC in June of 1969 for further testing, assembly on the launch vehicle, and launch.

Testing at KSC

At the Kennedy Space Center the CM and the SM were mated, checked, assembled on the Saturn V launch vehicle, and the total vehicle was moved to the launch pad.

The Countdown Demonstration Test (CDDT) began on March 16, 1970. Up to this point nothing unusual about oxygen tank #2 had been noted during the extensive testing at KSC. Cryogenic oxygen loading and tank pressurization to 331 psi was completed without abnormalities. At the time during CDDT when the oxygen tanks are normally vented down to about 50 percent of capacity, oxygen tank #1 behaved normally, but oxygen tank #2 only went down to 92 percent of its capacity. The normal procedure during CDDT to reduce the quantity in the tank is to apply gaseous oxygen at 80 psi through the vent line and to open the fill line. When this procedure failed it was decided to proceed with the CDDT until completion and then look at the oxygen detanking problem in detail.

On Friday, March 27, 1970, detanking operations were resumed, after discussions of the problem had been held with KSC, MSC, NR and Beech personnel participating, either personally or by telephone. As a first step, oxygen tank #2, which had self-pressurized to 17[illegible] psi and was about 83 percent full, was vented through its fill line. The quantity decreased to 65 percent. Further discussions between KSC, MSC, NR, and Beech personnel considered that the problem might be due to a leak in the path between the fill line and the quantity probe due to loose fit in the sleeves and tube. Such a leak would allow the gaseous oxygen being supplied to the vent line to leak directly to the fill line without forcing any significant amount of LOX out of the tank. At this point, a Discrepancy Report against the spacecraft system was written.

A "normal" detanking procedure was then conducted on both oxygen tanks, pressurizing through the vent line and opening the fill lines. Tank #1 emptied in a few minutes; tank #2 did not. Additional attempts were made with higher pressures without effect, and a decision was made to try to "boil off" the remaining oxygen in tank #2 by use of the tank heaters. The heaters were energized with the 65 volt DC GSE power supply and, about 1½ hours later, the fans were turned on to add more heat and mixing. After 6 hours of heater operation, the quantity had only decreased to 35 percent, and it was decided to attempt a pressure cycling technique. With the heaters and fans still energized, the tank was pressurized to about 300 psi, held for a few minutes, and then vented through the fill line. The first cycle produced a 7 percent quantity decrease, and the process was continued, with the tank emptied after five pressure/vent cycles. The fans and heaters were turned off after 8 hours of heater operation.

Slide 14

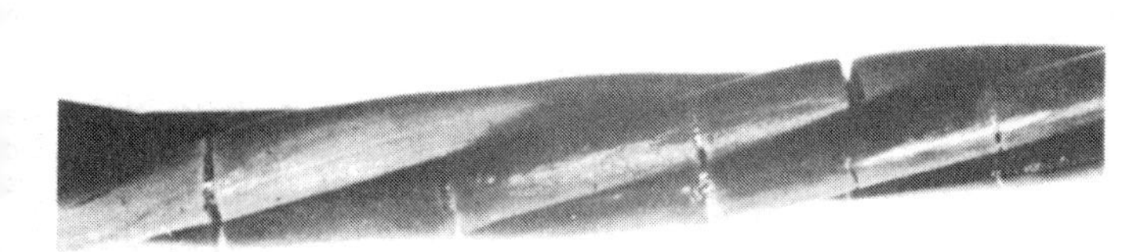

Slide 15

Slide 16

Suspecting the loosely fitting fill line connection to the quantity probe inner cylinder, KSC personnel consulted with cognizant personnel at MSC and at NR. It was decided that if the tank could be filled, the leak in the fill line would not be a problem in flight, since it was felt that even a loose tube resulting in an electrical short between the capacitance plates of the quantity gage would result in an energy level too low to cause any other damage. Replacement of the oxygen shelf in the CM would have been difficult and would have taken at least 45 hours. In addition, shelf replacement would have had the potential of damaging or degrading other elements of the service module in the course of replacement activity. Therefore, the decision was made to test the ability to fill oxygen tank #2 on March 30, 1970, 12 days prior to the scheduled Saturday, April 11, launch, so as to be in a position to decide on shelf replacement well before the launch date.

Flow tests were first made with gaseous oxygen on oxygen tank #2 and on oxygen tank #1 for comparison. No problems were encountered, and the flow rates in the two tanks were similar. In addition, Beech was asked to test the electrical energy level reached in the event of a short circuit between plates of the quantity probe capacitance gage. This test showed that very low energy levels would result. Then, oxygen tanks #1 and #2 were filled with LOX to about 20 percent of capacity on March 30 with no difficulty Tank #1 emptied in the normal manner, but emptying oxygen tank #2 again required pressure cycling with the heaters turned on. As the launch date approached, the oxygen tank #2 detanking problem was considered by the Apollo organization. At this point, the "shelf drop" incident on October 21, 1968, at NR was not considered and it was felt that the apparently normal detanking which had occurred in 1967 at Beech was not pertinent because it was believed that a different procedure was used by Beech. In fact, however, the last portion of the procedure was quite similar, although at a slightly lower pressure.

Throughout these considerations, which involved technical and management personnel of KSC, MSC, NR, Beech, and NASA Headquarters, emphasis was directed toward the possibility and consequence of a loose fill tube; very little attention was paid to the extended heater and fan operation, except to note that they operated during and after the detanking sequences.

Many of the principals in the discussions were not aware of the extended heater operations. Those that did know the details of the procedure did not consider the possibility of damage due to excessive heat within the tank, and therefore did not advise management officials of any possible consequences of the unusually long heater operations.

As I noted earlier each heater is protected with a thermostatic switch, mounted on the heater tube, which is intended to open the heater circuit when it senses a temperature of about 80°F. In tests conducted since the accident, however, it was found that the switches failed to open when the heaters were powered from a 65 volt DC supply similar to the power used at KSC during the detanking sequence. Subsequent investigations have shown that the thermostatic switches used, while rated as satisfactory for the 28 volt DC spacecraft power supply, could not open properly at 65 volts DC with 6-7 amps of current. A review of the voltage recordings made during the detanking at KSC indicates that, in fact, the switches did not open when the temperature of the switches rose past 80° F. Slide 14 shows a thermostatic switch welded closed after application of 1½ amperes of 65 volts DC.

Further tests have shown that the temperatures on the heater tube subsequent to the switch failures may have reached as much 1000° F. during the detanking. This temperature can cause serious damage to adjacent Teflon insulation, and such damage almost certainly occurred. Slides 15 and 16 show the condition of wires, such as those used in the fan motor circuit, after they have been subjected to temperatures of about 1000° F.

None of the above, however, was known at the time and, after extensive consideration was given to the possibilities of damage from a loose fill tube, it was decided to leave the oxygen shelf and oxygen tank #2 in the SM and to proceed with preparations for the launch of Apollo 13. In fact, following the special detanking the oxygen tank #2 was in hazardous condition whenever it contained oxygen and was electrically energized. This condition caused the Apollo 13 accident, which was nearly catastrophic. Only the outstanding performance on the part of the crew, Mission Control, and other members of the team which supported the operations, successfully returned the crew to earth.

In investigating the Apollo 13 accident, the Board attempted to identify those additional technical and management lessons which can be applied to help assure the success of future spaceflight missions. Several recommendations of this nature are included.

RECOMMENDATIONS

Before reading the Board's recommendations. I would like to point out that each Member of the Board concurs in each finding determination, and recommendation.

The Board's recommendations are as follows:

1. The Cryogenic oxygen storage system in the service module should be modified to:

a). Remove from contact with the oxygen all wiring, and the unsealed motors, which can potentially short circuit and ignite adjacent materials; or otherwise insure against a catastrophic electrically induced fire in the tank.

b). Minimize the use of Teflon, aluminum, and other relatively combustible materials in the presence of the oxygen and potential ignition sources.

2. The modified cryogenic oxygen storage system should be subjected to a rigorous requalification program, including careful attention to potential operational problems.

3. The warning systems onboard the Apollo spacecraft and in the Mission Control Center should be carefully reviewed and modified when appropriate, with specific attention to the following:

a). Increasing the differential between master alarm trip levels and expected normal operating ranges to avoid unnecessary alarms.

b). Changing the caution and warning system logic to prevent an out-of-limits alarm from blocking another alarm when a second quantity in the same subsystem goes out of limits.

c). Establishing a second level of limit sensing in Mission Control on critical quantities with a visual or audible alarm which cannot be easily overlooked.

d). Providing independent talkback indicators for each of the six fuel cell reactant valves plus a master alarm when any valve closes

4. Consumables and emergency equipment in the LM and the CM should be reviewed to determine whether steps should be taken to enhance their potential for use in a "lifeboat" mode.

5. The Manned Spacecraft Center should complete the special tests and analyses now underway in order to understand more completely the details of the Apollo 13 accident. In addition, the lunar module power system anomalies should receive careful attention. Other NASA Centers should continue their support to MSC in the areas of analysis and test.

6. Whenever significant anomalies occur in critical subsystems during final preparation for launch, standard procedures should require presentation of all prior anomalies on that particular piece of equipment, including those which have previously been corrected or explained. Furthermore, critical decisions involving the flightworthiness of subsystems should require the presence and full participation of an expert who is intimately familiar with the details of that subsystem.

7. NASA should conduct a thorough reexamination of all of its spacecraft, launch vehicle, and ground systems which contain high-density oxygen, or other strong oxidizers, to identify and evaluate potential combustion hazards in the light of information developed in this investigation.

8. NASA should conduct additional research on materials compatibility, ignition, and combustion in strong oxidizers at various g levels and on the characteristics of supercritical fluids. Where appropriate, new NASA design standards should be developed.

9. The Manned Spacecraft Center should reassess all Apollo spacecraft subsystems, and the engineering organizations responsible for them at MSC and at its prime contractors, to insure adequate understanding and control of the engineering and manufacturing details of these subsystems at the subcontractor and vendor level. Where necessary, organizational elements should be strengthened and in-depth reviews conducted on selected subsystems with emphasis on soundness of design, quality of manufacturing, adequacy of test and operational experience.

CONCLUSION

In concluding, I would stress two points.

The first is that in this statement I have attempted to summarize the Board's Report. This Report and its appendices are the result of more than seven weeks of intensive work by the Board, its Panels, and staff, supported by the NASA and contractor organizations. In the interest of time, I have not included many supporting findings and determinations which are set forth in the Report.

The Second point I wish to make is this:
The Apollo 13 accident, which aborted man's third mission to explore the surface of the moon, is a harsh reminder of the immense

ifficulty of this undertaking.

he total Apollo system of ground complexes, launch vehicle, and spacecraft constitutes the most ambitious and demanding engineering evelopment ever undertaken by man. For these missions to succeed, both men and equipment must perform to near perfection. That his system has already resulted in two successful lunar surface explorations is a tribute to those men and women who conceived, esigned, built, and flew it.

erfection is not only difficult to achieve, but difficult to maintain. The imperfection in Apollo 13 constituted a near disaster, averted only y outstanding performance on the part of the crew and the ground control team which supported them.

he Board feels that the Apollo 13 accident holds important lessons which, when applied to future missions, will contribute to the safety nd effectiveness of manned space flight.

1r. Chairman, this concludes my prepared statement.

(Editor's note: Slide numbers in this document have been renumbered to avoid any repetition of images.)

Ir. CORTRIGHT — The prepared statement is a summary of the Board report, and I will give you a summary of the statement.

s Dr. Paine pointed out, the Review Board was created about April 17, and we first set about to put together a technical and anagement team which could tackle the problem of understanding this particular accident. Our charter, in essence, was to find out hat happened, why it happened, and what do we do about it.

Iow, the first slide indicates the Review Board that was formed. (Slide 1.)

was the Chairman, Robert Allnutt, who is sitting at the end of the table here on my left, was Assistant to the Administrator; Neil rmstrong, whom I hope will join us shortly; Dr. John Clark, seated second from my left, the Director of the Goddard Space Flight enter; General Hedrick, Director of Space, DCS/RVD, Headquarters, USAF, is sitting second from the end on the right; Mr. Vince ohnson, seated at my left here, Deputy Associate Administrator (Engineering), Office of Space Science and Applications; and Mr. Milton lein, seated on the other side of Dr. Paine, is Manager of the AEC-NASA Space Nuclear Propulsion Office.

r. Hans Mark, who was unable to be here today, is Director of the Ames Research Center.

s you can see from the organization chart, we put together four major panels: Mission events, manufacturing and test, design, and roject management. And in addition, we have a number of supporting staff functions.

he mission events panel was assigned the job of reviewing in meticulous detail everything that telemetry and any other type of records now as to what happened in the hours preceding the accident, actually from liftoff to the accident — that was the pre-incident events ortion — then the actual few minutes of the accident, and then the post-incident events, which cover the question of how well did we ecover from this problem and get the astronauts back to earth. Frank Smith from NASA Headquarters headed up mission events work.

lanufacturing and Test, was headed up by Mr. Schurmeier, whom you may remember from his Ranger and Mariner days. He had three roups under him, Fabrication and Acceptance Testing, Subsystem and System Testing, and Reliability and Quality Assurance.

the area of Design, we brought in Mr. Himmel, who managed the Agena and Centaur programs. He had four groups: Design Evaluation, ailure Modes and Mechanisms, Electrical, and Related Systems.

nd then we put together a somewhat smaller group, headed by Mr. Kilgore, to look into the project management aspects of this roblem.

Iow, the manner in which we worked at the Manned Spacecraft Center is something like this: we relied heavily on the technical team the Manned Spacecraft Center to generate basic factual data for us. In addition, we levied on this group special requirements for nalyses, additional data, and special tests.

o do this and to make this a smooth working relationship, we had Mr. Charles Mathews of the Office of Manned Space Flight serve as liaison representative and in addition, we established working relationships which were quite effective and minimized, if not eliminated, ny duplication of effort. I think this worked quite effectively.

addition, we levied special testing requirements on the Manned Spacecraft Center and brought to bear a rather wide NASA effort of nalyses and special tests.

or example, at the peak of our testing, and I will go into that a little bit more later, there were nearly 100 special tests run and several undred people running tests to attempt to duplicate on the ground most of the elements of this accident as we think we understand

Iow, let me switch for a moment to the problem as it confronted us when we first arrived in Houston; of course, the accident had ccurred some days prior to that. We know from telemetry that the accident probably centered around the No. 2 oxygen tank in the ervice module. I will tell you a little bit more about that tank in a minute.

Ve did have fairly good although not absolutely complete telemetry. That was a strong plus point to start with. We also had the crew ports. The crew had heard the bang. At least one member felt a shudder. They observed the venting of some form of material into pace, and near the end of the flight when they separated from the service module they observed that the panel covering bay 4 was issing. And they brought back photographs of the service module from which we attempted to glean information about what had appened within that bay.

ow, what then followed I think was aptly described by Dr. Paine as a rather massive detecting job, and to take you through that I would e to begin by describing the system in which the failure occurred and place it in the context of the total Apollo system.

he next slide — of course, you are still familiar with the basic Apollo launch vehicle and where the service module, the Lunar Module d the command module are located. The problem was in the service module. (Slide 2). The next slide (slide 3) gives a little bit more etail on the service module. Bay 4 it is called sector 4 on this slide. This is the Sector in which fuel cells, oxygen tanks, and hydrogen nks are located, and I will show you that a little bit better in another slide.

In addition, the propellants for the service modules are located in adjacent bays and there is a center tunnel here which assume significance in the analysis later on, because as this sector is pressurized by a rupturing tank, the gas flows into this center section an pressurizes it.

If that pressure were to rise enough, it could have separated the command module from the service module — simply blown it off. takes about 10 p.s.i. pressure to do that, whereas it takes something in excess of 20 to 25 p.s.i. to blow the panel off bay 4.

The next slide shows in a little bit more detail the service module with the panel removed from bay No. 4 (slide 4). In the upper she are the three fuel cells that provide power. They in turn are fed by oxygen tanks and hydrogen tanks.

These are two hydrogen tanks with a cylindrical sleeve to fasten them together. The two oxygen tanks sit on this shelf. Oxygen tank is here, and No. 2 is here [indicating]. There is a valve module and servicing panel here. This oxygen tank No. 2 is the one which faile

May we have the next slide, please. (Slide 5.)

This is the lower section of bay 4 showing the hydrogen tanks and the bottom of oxygen tank 2 which failed. All this crinkly materi which looks a little messy, is designed that way. It is a Kapton-coated Mylar insulation.

On the right (slide 6) are shown two of the three fuel cells which are on the shelf above the oxygen tanks. This is oxygen tank No. that failed. This is the dome of the tank which covers the spiraling tubes which penetrate the tank and which carry fluid into and out the tank as well as the electrical wiring.

I might say at this point, Mr. Chairman, that we have brought an oxygen tank of this type with us today, on my left, which is actually or in which we ran a full-scale combustion test. I will describe that later.

Next slide. (Slide 7.)

This is oxygen tank 2. The lines run from the top over to the servicing panel.

The next slide (slide 8) is a cutaway drawing to show you what is inside that tank. Basically, the tank is a double-walled tank. The inn wall carries the loads. There is insulation outside of that between the inner wall and outer wall, and this area is pumped to low vacuu to maintain the proper heat leak rates, which have to be very low.

This tank was a very excellently performing tank thermodynamically. It is a tough problem to just keep the oxygen that long in spac and this was a major development problem in getting this tank built in the first place.

On the right, here, is a heater assembly. There are two heater coils wrapped around the tube here, each of which have a thermostat switch, which you will hear more about later.

The switch was designed to open at 80° F., plus or minus 10°, to protect the heater assembly from overheating. At either end is a electric motor with a fan. The oxygen is drawn into this tube through the small hole shown here, it flows downward through the fa and is blown outward through these little square holes shown here.

The fans were required for several purposes actually, one, when fans are not used the heat from the heater tends to remain in the vicini of the heater and creates a thermal bubble. If this gets too big and then later on is mixed, you get a pressure collapse, which is a sudde drop in pressure which may be undesirable under some conditions. If a thermal bubble exists it interferes with the accuracy of th quantity measurements.

This probe down the middle is a capacitance gage, so-called, which measures the density and hence the amount of supercritical oxyge in that tank. The oxygen is kept in a supercritical state, which is a single phase state. There is never a boundary in it between liquid ar gas. You can think of it as a very heavy gas. Some people like to think of it as a very heavy gas, others as a liquid that never recedes, a sense, it always fills up the volume.

Now, the fans then break up this bubble when the heaters are on and eliminate the stratification, so-called, that takes place there. Th makes possible accurate quantity measurements throughout the entire flight regime. They also make possible the input of larger amoun of heat without the risk of an extra hot region right in here, because they stir the contents up and mix it around.

This tank through the supply line supplies the fuel cells and the environmental control system. The oxygen goes into power generatic and breathing for the crew.

There is a pressure switch on this line, a pressure transducer to measure the pressure, and a relief valve in the event this pressure ge too high. This relief valve actually opened during the course of the accident.

I believe the next slide (slide 9) shows a little more detail on the area where the problem at least got to its worst stage.

There is a bundle of wires that come up from the lower fan motor and joins with the wires from the upper fan motors and the heat wires, and they run over through a loop through some holes in this Teflon and glass collar.

There is a temperature sensor mounted right on the collar to measure the temperature, in the tank. All of this wiring goes up throu the top of this quantity gage and runs out of the tank through an electrical conduit. It spirals around through this conduit and it is broug out here.

I would like to point out that I am going to refer later to a fill tube assembly. This is the fill tube assembly. It consists of three pieces. is possible within the tolerances of manufacture on those three parts to build that so that it fits in there loosely, and in fact it can loosely enough so it can be displaced subsequently by either normal handling or abnormal handling. I will come back to that.

I will show you later that we are convinced that a fire started on wiring either lower down in the heater assembly or in this region. Th fire progressed along these wires through the holes in the Teflon element here, probably igniting that Teflon element, setting up a litt furnace in here which failed at least the conduit, and probably more than the electrical conduit. A temperature rise of not too mu would make it fail at the 1,000 pounds per square inch pressure it was then experiencing. Probably the penetration cap came out at th time.

Now, I would like to go on and show you the next slide. (Slide 10.)

This is the best — I won't say the best as projected because you lose something in the process of making view graphs, but this is an enhanced photograph taken by the crew of the service module showing bay 4. This is the vicinity of the oxygen tank. These are the fuel cells, the hydrogen tanks over here. There is quite a bit of Mylar projecting from this bay which confuses the photography.

The photograph experts, the photo interpreters who have worked days and days on this photography, feel they can find highlights which show that the tank is still there. We all believe that the tank is there. At least half of the people — probably more — who look at it can't see it. It turns out and we have duplicated this in the laboratory — that with the particular geometry of the tanks and with that lighting condition, it is very hard to see in any event. Certainly, we were not able to determine the condition of the tank, so the main value of this photograph, I feel, is the conclusive evidence that the panel is gone. If we hadn't had this observation, I am not sure that we would have concluded that.

I would like to go to the next slide (slide 17) to go back to the telemetry.

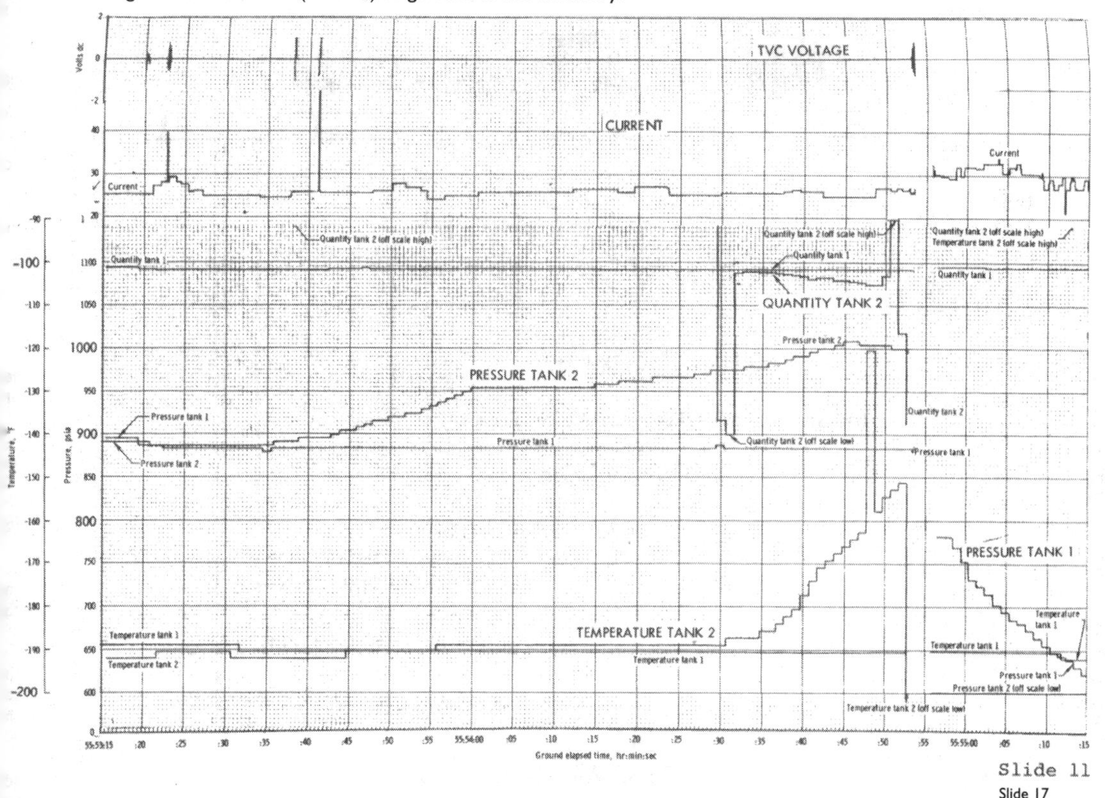

Slide 11

Slide 17

Could I have the next one (slide 11)?

The slide on the right (slide 17) is a portion of the telemetry record and I have extracted events from that record and listed them in an easier manner to follow the slide on the left (slide 11).

Here are some voltage spikes on a system, an attitude control system that was connected to the same electrical bus as the tank on which we had the problem. This was an indication of something happening in the tank. These are referred to as glitches.

These are two large current spikes which occurred here. They both short circuits. The first one is the one which we now feel started the fire in the tank. It occurred immediately after turn on of the fans.

Shortly thereafter, the pressure in the tank began to rise and you can see this pressure rise here up to about a value of 1,008 pounds per square inch. Later on, I will refer to tests that have duplicated in general this condition.

The temperature indication rose much later but, if you recall from the last slide I had on, the temperature sensor is located on the quantity probe.

Here is the quantity probe. Here is the temperature measuring device. You wouldn't expect that to measure a temperature increase until the fire progressed along the wire to that vicinity. That is why the temperature rise is delayed. Everything let go about this point. There was a dropout of telemetry and later on a decrease of the oxygen pressure in tank No. 1, which was also failed by this rupture.

There are the events summarized here. Fans on, short circuit starts the fire, the pressure begins to rise, and this pressure rose, by the way, because as the oxygen is heated by the fire it expands in sort of a bubble around the wire and compresses the rest of the oxygen in the tank.

Temperature begins to rise when the fire gets to the vicinity of the temperature gage. The highest pressure reading occurred here.

We later ran tests which showed that the pressure dropped off at that point because the relief valve opened and it was determined that the relief valve was sufficient to drop the pressure at the observed rate.

The panel separated here. The, telemetry dropped out. The crew reported the problem. And very soon thereafter fuel cell 3 and fuel cell 1 failed, because the supply valves had closed from the shock of the explosion and the blow off of the panel and thus shut off the fuel cells.

The CHAIRMAN — What was about the elapsed time from the time you got the first spike? What was the time elapse between th first and second spikes?

Mr. CORTRIGHT — These were different short circuits. The fire had already begun and in consuming insulation could have contribute to the second one.

The CHAIRMAN — Aren't the lower figures the time element?

Mr. CORTRIGHT — Yes, hours and minutes.

The CHAIRMAN — That I can see, I was looking at the other chart.

Mr. CORTRIGHT — That is hours, minutes, and seconds. The scales are different. This goes from 55 hours 53 minutes and 15 seconc to 55 hours, 55 minutes, over here, if my eyes aren't failing me — so there is 5 seconds between each major block.

The CHAIRMAN — I see.

Mr. FULTON — Is your graph in real time?

Mr. CORTRIGHT — Yes.

Mr. FULTON — That is the real time graph?

Mr. CORTRIGHT — Yes, it is an accurate representation of what happened.

Now, having this telemetry and the crew reports, it didn't take us too long to reach the conclusions I have given you so far. We the decided it was time to prove that this analysis was correct, so we had to ask ourselves a series of questions and then proceed to answe them and prove our answers.

The first question was: Was combustion required to raise the pressure the indicated amount? Analyses showed that you could not ge increases of that amount simply by feeding electrical energy alone into the tank as a heat source. There was no other possible he source, so the conclusion was that yes, combustion was required.

The next question then is: What was there in the tank that could burn? The materials in the tank had passed characteristics of materia of the "COMAT" system which is used to determine acceptability of materials in oxygen environments.

We made an investigation and determined that Teflon can burn under these supercritical oxygen conditions, as can aluminum, solde and other materials present.

Mr. HECHLER — At what temperature does Teflon burn?

Mr. CORTIGHT — The actual temperature of combustion are over 2,000° but ignition energy required is as low as one joule if it done through an electric arc.

Mr. PRICE — Why did not the relief valve take care of that?

Mr. CORTRIGHT — When the pressure built up due to the combustion, the relief valve held the pressure from exceeding about 1,0C pounds per square inch. At the same time the fire progressed up to the metals of the tank wall and its tubing and overheated them, which point they lost their strength and failed. That was the manner of the failure.

Mr. PRICE — It wasn't, pressure in the tank that blew the tank?

Mr. CORTRIGHT — In a way it was. The tank had plenty of strength. It was twice as strong as necessary to hold a thousand poun per square inch at cryogenic temperatures, but when it was heated by the fire it lost its strength and failed locally. So it is a combinatic of effects. The material loses its strength and the pressure blows through the weakened portion.

Mr. DOWNING — At what time on the time scale did the explosion occur?

Mr. CORTRIGHT — Right here where you see pressure drop.

Mr. DOWNING — Transpose that over to the other chart.

Mr. CORTRIGHT — Right here, panel separates, telemetry drops out. All of this occurred in a very brief period of time.

I have not expanded the scale enough to break the details down into milliseconds.

All right, so getting back to the questions, we had to ask ourselves what would burn, and I just told you what would burn.

Then we had to ask ourselves, well, how could you start these materials burning since generally speaking they are considered compatib with oxygen, and I think here we ran into a new phenomenon that was not recognized or widely recognized before, and that is th Teflon can be ignited rather easily if an electric arc is the igniting mechanism, and the combustion will propagate in supercritical oxyge We ran tests to show that these small amounts of electrical energy were sufficient if they were in the form of an electrical arc whi concentrates the heat very locally in the material, and we ran additional tests to show that even through the 1 amp, relatively quick blc fuse that was on that line to protect it, you could get energy 10 to 100 times in excess of what was required to ignite the wire, insulatic

Now, having determined that, we said to ourselves, well, if we had that type of fire would that be consistent with the times of t pressure and temperature rise? So we made measurements of the rate of combustion of Teflon wire insulation in supercritical oxyge and I will show you photographs of this later but, briefly, we found that the rate of burning varies at 1g, because, just like the smoke fro the candle goes up if you are burning it, more fresh oxygen is drawn in by convection and feeds it. If the fire is burning up, it goes fa horizontally, medium, and down slow. When it is burning down, it burns about one quarter of an inch a second. We ran tests at ze gravity and found it burns from one-eight to one-quarter of an inch a second at zero g., so it burns still more slowly.

Going back to the lengths that were in the tank, we were able to confirm that there was correlation here between what we thoug vould happen and what happened, and later on then we took an actual tank, the one sitting over here, and ignited its wiring insulation nd recorded its temperature and pressure history.

1r. FULTON — Mr. Cortright, the question of burning insulation through arcing has come up earlier in our previous Apollo investigation, nd also the question of what kind of insulation.

ome of us went down to Houston and various places and were surprised, amazed and shocked at the way the insulation on wire burned fter there had been an arcing situation. It looked to us as if it were a sparkler on the Fourth of July. You could see it just running along he wire emitting these sparks.

low, when we had the Apollo 204 insulation question had this material been tested for electrical arcing, or was this a new situation hat even the manufacturer had not considered in spite of Apollo 204?

1r. CORTRIGHT — The testing that followed Apollo 204 concentrated on the environment of the crew quarters, which is a much ower pressure. The testing was not extended to include the conditions within the oxygen tanks themselves, Where, supercritical oxygen stored.

lr. FULTON — That is my point. When we had much lower pressure in a crew cabin atmosphere and we could see an arc, and the sulation burning with sparks being emitted, why then, with this oxygen situation and a much higher pressure and possible arcing, wasn't hat gone into?

lr. CORTRIGHT — Mr. Fulton., the wire that caused the problem in the 204 was a polyvinyl chloride, I believe, and we switched to eflon and Teflon was qualified in the pure oxygen and the polyvinyl chloride was totally eliminated.

lr. FULTON — It is no longer qualified?

lr. CORTRIGHT — Teflon is qualified for the cabin, to the best of our knowledge. It is definitely combustible at supercritical conditions high pressure oxygen.

lr. FULTON — Do you need in your estimation, further tests along these lines of combustibility as well as insulation subjected to arcing onditions? Is this just one stage, or should we go much further before we say that the hookup is safe?

ly. CORTRIGHT — We have a recommendation along the lines you are suggesting to conduct considerable additional research on this roblem within NASA and to revise our standards where appropriate. I think more work has to be done, because we have learned so uch in the past 2 months that we tell ourselves we should have known earlier.

lr. FULTON — Was this covered by any technical report of NASA, this particular arcing situation in respect to insulation under high ressure in an oxygen atmosphere?

lr. CORTRIGHT — Not to my knowledge.

lr. FULTON — Why wasn't it?

r. CORTRIGHT — Teflon was felt to be safe under these conditions. In hindsight, if we had conducted research on various manners which Teflon might burn in this very high pressure environment, we would have found out it can be, but this was not done.

r. FULTON — Did NASA accept the statement of industry or the industry catalog or manual on the characteristics of Teflon under cing conditions, or did it do its own separate investigation on it?

r. CORTRIGHT — To answer your question as best I can, I am quite certain I am not familiar with all the governing documentation available literature on the subject. The Teflon was qualified by means of an impact test which is one method widely used for etermining compatibility of materials in all types of oxygen environments.

his test consists of a blow on the material and sensitive materials can be ignited in this manner. If they will survive a blow of a certain tensity, they are considered compatible.

r. FULTON — At what point does that responsibility rest? With NASA? The contractor? The manufacturer? Where does that sponsibility rest? Certainly not with the contractor.

r. CORTRIGHT — I think the responsibility for seeing that our equipment is flight worthy rests with NASA.

ne CHAIRMAN — This might have been unanticipated because in the first place there haven't been a lot of places in NASA or in dustry where you would duplicate such conditions under high oxygen pressure.

r. CORTRIGHT — This is the only case we know of, Mr. Chairman. There may be others. It is the only case we know of where wire sulation of this type, was ignited and burned in an environment like this.

flon is generally considered to be the best of the flexible insulation available.

ne CHAIRMAN — And of course we have had precedence in this at NASA where we found that the generally accepted types of elding, for instance, in the case of the Centaur, proved that they were not fit and they had to go out and do it over again. This is part the progress that we make in the space effort, is it not?

r. CORTRIGHT — That is true.

r. HECHLER — Does your recommended research entail use of different material other than Teflon? Are there other materials?

. CORTRIGHT — Yes sir. The recommendation covers the compatibility of various materials with pure oxygen under other nditions, further research into supercritical oxygen and examination into other modes of propagation.

r. HECHLER — As a layman, I didn't understand your use of the term "one joule."

. CORTRIGHT — That is a measure of electrical energy which would go into a spark. If you heat Teflon, it depends on how you heat it. If you just heat it in an oxygen environment, the first thing that will happen is that the insulation will start to deteriorate.

I am going to give you a roundabout answer to your question and we have run these tests subsequent to the accident and actually, as will show later, that did happen to this wiring before the launch. But when it gets up around 800°F. or 900°F., the insulation will slow oxidize away and disappear off the wires entirely.

If you, say, ignite a local portion by means of a Nichrome wire, which would be a glowing white hot wire, when the Teflon gets to 1,300° in high density oxygen, it will react in a combustible manner, 1,300°F. is the figure you were asking for. It then burns at a high temperature.

Mr. FULTON — Why was this in the oxygen tank when it looks as if there was room outside?

Mr. CORTRIGHT — We need heaters in the tank to keep the pressure up, to keep the oxygen feeding properly into the fuel cells an into the crew compartment. The fans were required to stir the contents.

Now, the particular mechanization that was used was two electric motors in the tank, and we cover that in our report.

Mr. FULTON — Why was everything put inside the tank when there was some danger of arcing?

Mr. CORTRIGHT — I think that is a fair question. This particular design approach chose to do it that way. What was done at th manufacturer's plant is that very meticulous assembly procedures were developed to prevent or minimize the chance of damage to th electrical wiring so that short circuiting could not or probably would not take place later.

We think that the design was deficient. It turns out that the basic design wasn't really at fault in this case, the wire was damaged by a overheating condition that I am going to describe for you.

Mr. FULTON — In summing up the basis of the opinion which you developed, the gentleman from West Virginia and the chairman ar I think you were citing the advance state of art not only on the material but on the circumstances that had developed in the manufactu is that your opinion?

Mr. CORTRIGHT — Mr. Fulton, I believe that the tank when it was built constituted a very advanced tank, and we know much mo today about the sorts of problems you can get into with a tank of this type, and also other ways in which it might be designed and p together that would make a more reliable unit out of it. I think that is what you told me, and I agree, with it.

Mr. FULTON — I say it is not any negligence or any failure either on the person selecting the material, the insulation, the engineerin the design, or the operation of the vehicle it is rather that a combination of circumstances created a requirement for a now advance the art which you are now doing. Is that it?

Mr. CORTRIGHT — Not entirely. I believe there were deficiencies in the tank design and the manner in which the tank was handle and I will go into that in the balance of my statement.

Mr. HECHLER — I also would like to observe, and I will develop it further, I think we need an advance in the art of administration well as technical design, but I will develop that later.

Mr. PRICE — I would like to ask you, do you think that NASA's qualification of material that is assembled into our Apollo equipment sufficient, or do you think it could be improved, not just in this case but in the overall situation?

Mr. CORTRIGHT — Yes, I think we can improve and we are always trying to. In this particular case, or in all cases actually, there a many safeguards to insure that non-flight worthy hardware is weaned out. Sometimes due to an unusual combination of events mistakes, this does not happen. The Agency has a meticulous system to prevent this from happening. No system is perfect. Problems c sneak through.

Mr. PRICE — What progress has been made in this particular area? Has this been altered so we can continue with the next flight, or considerable change necessary?

Mr. CORTRIGHT — The hardware itself is being redesigned for modification at this time, and within the next few weeks I expect th selection of the design will be made.

As far as the procedures are concerned, I think that every element of the organization is taking another look at the procedures it h been using in the. light of what we have learned here, to make sure it doesn't happen again.

Mr. PRICE — When you accept parts like this, do they, for instance, flush the system or try the system before it is put in? Is ea individual part as it comes from the company checked? Do we have controls in accepting the parts?

Mr. CORTRIGHT — Depending upon the particular compound or parts, there are flight acceptance test specifications which norma a subsystem would have to meet. The individual parts may or may not have acceptance testing depending upon what the part is, but th have to meet certain standards.

The specifications for these parts are written at various levels, some are written by the subcontractor to the vendor, others are writt by the prime contractor to the subcontractor, and some top level requirements are written by NASA to the prime contractor, so the are various levels of checks and balances and review at work here.

Mr. PRICE — Do you think we should go further and go into these companies with NASA's own inspectors and where feasible rur test that will meet that standard-in other words, the company will say yes, we have met your requirements, then is it tried out to see it meets requirements? Is this done?

Mr. CORTRIGHT — This was normally done. In this particular case there was a thermal switch which was not tested, and this ca back to bite us. Normally what you asked us for is done. We are all reviewing our procedures.

Mr. PRICE — Whose responsibility was it?

Mr. CORTRIGHT — NASA accepts responsibility for the total system. This cannot be delegated.

Mr. PRICE — When it was built, whose responsibility was it?

Mr. CORTRIGHT — The prime contractor and subcontractor who manufactured this equipment had a certain responsibility to insure that the testing was sufficient. I will come to that in more detail later.

Mr. MOSHER — Mr. Cortright, this thermostatic switch, was that same equipment on Apollo 11 and on Apollo 12?

Mr. CORTRIGHT — Yes; and on the earlier Apollo flights.

Mr. MOSHER — What then was the crucial difference or the crucial event that made the difference between those flights and this flight?

Mr. CORTRIGHT — The crucial difference was a special detanking procedure that took place at the Cape, and I will go into that later.

I am delighted to answer all the questions I can, but it might make a more coherent story if I quickly finish up. It will only take me 10 minutes.

Mr. KARTH (presiding) — With the indulgence of the committee, I think we should wait to ask questions until he finishes.

Mr. CORTRIGHT — Thank you.

There is a coherence to this that hasn't become apparent yet; I hope it will. [Laughter]

I was going through the problems that we faced and the things we have to prove to ourselves. I believe I had gotten down to the point of why the — I had explained that the pressure rise history was consistent with the rates of burning along the wire, and we proved that by test.

We also postulated that the temperature rise delay could reasonably be expected to occur with fire remote from the sensor, and this was later demonstrated in tests.

The maximum pressure correlated with the relief valve operation. The telemetry dropout was the next question. Why did we lose telemetry?

Two things happened. One was that there was a strong shock to the spacecraft at the time the panel blew off, and second, from the photograph, the high-gain antenna on the service module was bent.

It seems reasonable that either parts of the panel coming off, which you will see in a motion picture of a model test, would have done that, or some other part coming out of that bay.

The loss of pressure from oxygen tank No. 1 is surmised to have occurred from one of two causes, either shocking open of the valve or cracking of one of the high pressure lines from the tank.

Now, having put those pieces together and run the tests to validate them, the question was, does it all hang together and make a coherent a story?

As I may have mentioned earlier, and I will repeat it for emphasis, we did bring the total resources of the Agency to bear on the investigation which involved all of our centers, plus the prime contractor, North American Rockwell, the subcontractor who delivered these tanks to North American Rockwell, Beech Aircraft Corp., and a number of other companies helped.

This included about 100 special tests involving several hundred people and I would like to show you a film now which gives you highlights of this testing program.

(Film shown.)

Mr. CORTRIGHT — The first thing we had to do was to demonstrate that we could ignite Teflon with the lower energies that were available. This was a test done in Houston at the Manned Spacecraft Center. An electric arc at the left ignites the Teflon wire, which burns along the wire toward the right. It burned out another wire, and now it will progress across.

This is burning in the very high pressure supercritical oxygen of the type used in tank No. 2. These photographs are taken at normal speed. It burns along like a fuse.

This is combustion taking place in a test rig, at the Lewis Research Center, which later on will be used to demonstrate zero g. combustion. This is one g., to give you a point of comparison. The smoke is going up. At zero g., smoke doesn't know enough to go up because there is no up.

The temperatures of supercritical oxygen for tests like this range from minus 100° to minus 200° F. This is burning in an extremely cold environment.

Here is the same test run at zero g., first to illustrate the type of rig, this container is dropped 500 feet into a silo where it impacts into plastic spheres to absorb the shock. This is combustion at zero g. at a rate about half that that occurs at one g.

These photographs are taken at 400 frames a second because the entire time of the test was 3 to 4 seconds. The apparent out-of-focus nature of it is caused by the refraction of the supercritical oxygen, not by the camera.

We also ran tests to show that wire that had been baked at high temperature burned similarly. This is a bundle that is being ignited in a simulated tank. It will burn through the wires down into the Teflon collar. It ignited the Teflon collar and burned a 2-inch hole at the top.

This is more of a boilerplate tank. The rupture comes right through here, very rapidly. There is the rupture.

Mr. FULTON — Where does the arcing occur on those pictures?

Mr. CORTRIGHT — I will come back to that, Mr. Fulton, if that will satisfy you.

Now, this is moving through a full-scale tank, the one that I have with me this morning. The fire is burning inside — it just blew throug This is an escaping mixture of gaseous and liquid oxygen which took place in that rather confined compartment of bay No. 4 of th spacecraft.

Mr. Fulton, to come back to your question, the last test you saw, ignition was achieved either with a Nichrome wire or a squib rathe than an electric arc. The electric arc tests were a separate, series of tests that we ran.

Depending upon the size of the hole, it may be necessary to get additional pressure in the bay to get it off. These tests were run Langley to demonstrate that the oxygen products accompanied by sparks and burning material from the tank are sufficient to ignite th Mylar insulation that you saw earlier fill the bay.

We have measured augmented pressure rises as much as a factor of 6, it probably wouldn't be quite that in high flight. A combinatio of a hole in the tank between 1 and 2 inches in diameter supplemented by this combustion which has been demonstrated, would b sufficient to take the panel off the tank.

This particular test is a slow motion film of oxygen combustion with the Mylar. At the same time, we ran analyses to determine wha type of pressure pulse would be required to take a panel off, and then ran tests which you are about to see to measure, in fact, whethe our calculated pulses would in fact take the panel off.

Here is a film taken at 2,000 frames a second, and this panel is a half scale honeycomb panel blowing off with a simulated tank ruptur the tank being in this location. That all takes place in a few milliseconds.

It can occur in such a way that the pressure builds up highest right in this vicinity and to a lesser extent in the rest of the bay and to still lesser extent in the rest of the service model, so that it would not have blown off the command module.

The combination of analyses and tests in this point of time provide a fairly good reproduction of what probably happened.

That concludes the film.

At this point, Mr. Chairman, I would like to wrap up what I have been telling you for the past 45 minutes or so, by reading a portion the Apollo 13 Review Board report. I am turning to the report here because these words are carefully chosen and I think words shoul be carefully chosen when we come right down to the point of what most likely happened and what roles various organizations playe in the problem.

In reaching its findings, determinations, and recommendations, it was necessary for the Board to review critically the equipment and th organizational elements responsible for it. It was found that the accident was not the result of a chance malfunction in a statistical sens but rather resulted from an unusual combination of mistakes, coupled with a somewhat deficient and unforgiving design. In brief, this what happened:

a). After assembly and acceptance testing, the oxygen tank No. 2 which flew on Apollo 13 was shipped from Beech Aircraft Corporatio to North American Rockwell (NR) in apparently satisfactory condition.

b). It is now known, however, that the tank contained two protective thermostatic switches on the heater assembly, which wer inadequate and would subsequently fail during ground test operations at Kennedy Space Center (KSC).

c). In addition, it is probable that the tank contained a loosely fitting fill tube assembly. This assembly was probably displaced durin subsequent handling, which included an incident at the prime contractor's plant in which the tank was jarred.

d). In itself, the displaced fill tube assembly was not particularly serious, but it led to the use of improvised detanking procedures at KS which almost certainly set the stage for the accident.

e). Although Beech did not encounter any problem in detanking during acceptance tests, it was not possible to detank oxygen tank No 2 using normal procedures at KSC. Tests and analyses indicate that this was due to gas leakage through the displaced fill tube assembl

f). The special detanking procedures at KSC subjected the tank to an extended period of — actually about 8 hours — heater operatio and pressure cycling for about 2 hours. These procedures had not been used before, and the tank had not been qualified by test for th conditions experienced. However, the procedures did not violate the specifications which governed the operation of the heaters at KSC

g). In reviewing these procedures before the flight, officials of NASA, NR, and Beech did not recognize the possibility of damage due t overheating. Many of these officials were not aware of the extended heater operation. In any event, adequate thermostatic switche might have been expected to protect the tank.

h). A number of factors contributed to the presence of inadequate thermostatic switches in the heater assembly. The original 196 specifications from NR to Beech Aircraft Corporation for the tank and heater assembly specified the use of 28 V dc power, which used in the spacecraft In 1965, NR issued a revised specification which stated that the heater should use a 65 V dc power supply fo tank pressurization; this was the power supply used at KSC to reduce pressurization time. Beech ordered switches for the Block II tank but did not change the switch specifications to be compatible with 65 V dc.

Mr. FULTON — Would you say that again?

Mr. CORTRIGHT (continuing) — Beech ordered switches for the Block II tanks but did not change the switch specifications to b compatible with 65 V dc.

i). The thermostatic switch discrepancy was not detected by NASA, NR, or Beech in their review of documentation, nor did tests identif the incompatibility of the switches with the ground support equipment (GSE) at KSC, since neither qualification nor acceptance testin required switch cycling under load as should have been done. It was a serious oversight in which all parties shared.

j). The thermostatic switches could accommodate the 65 V dc during tank pressurization because they normally remained cool an closed. However, they could not open without damage with 65 V dc power applied. They were never required to do so until the specia detanking. During this procedure, as the switches started to open when they reached their upper temperature limit, they were welde permanently closed by the resulting arc and were rendered inoperative as protective thermostats.

By the way, I do have a failed switch here, one from the test, which I can show you later.

). Failure of the thermostatic switches to open could have been detected at KSC if switch operation had been checked by observing heater current readings on the oxygen tank heater control panel. Although it was not recognized at that time, the tank temperature readings indicated that the heaters had reached their temperature limit and switch opening should have been expected.

). As shown by subsequent tests, failure of the thermostatic switches probably permitted the temperature of the heater tube assembly to reach about 1000° F in spots during the continuous 8-hour period of heater operation. Such heating has been shown by tests to severely damage the Teflon insulation on the fan motor wires in the vicinity of the heater assembly. From that time on, including pad occupancy, the oxygen tank No. 2 was in a hazardous condition when filled with oxygen and electrically powered.

Just to digress for a moment, I would like to show my last three view-graphs (slides 14, 15, and 16) which show the manner in which the contacts of the thermostatic switch can weld together.

Here are the welded together electrical contacts of the thermostatic switch when subjected to about 1½ amperes, all that it will carry at the 65 volt d.c. This is the condition of the wire when removed from a test heater assembly. These wires run up through a conduit inside the heater assembly, which runs from the other side of the tube from the electrical heater. This was liquid nitrogen. All you see is thermal damage. In oxygen, one would have expected some of that to slowly oxidize away. In some cases it can totally disappear.

n). It was not until nearly 56 hours into the mission, however, that the fan motor wiring, possibly moved by the fan stirring, short circuited and ignited its insulation by means of an electric arc. The resulting combustion in the oxygen tank probably overheated and failed the wiring conduit where it enters the tank, and possibly a portion of the tank itself.

). The rapid expulsion of high-pressure oxygen which followed, possibly augmented by combustion of insulation in the space surrounding the tank, blew off the outer Panel to bay 4 of the SM, caused a leak in the high-pressure system of oxygen tank No. 1, damaged the high-gain antenna, caused other miscellaneous damage, and aborted the mission.

The accident is judged to have been nearly catastrophic. Only outstanding performance on the part of the crew, Mission Control, and other members of the team which supported the operations successfully returned the crew to Earth.

A large amount of material is included in our report and in Appendix B (Report of the Apollo 13 Review Board) to show the manner in which the mission control and the crew coped with this in-flight emergency. I think it was truly admirable. I would commend it to your reading. I have not taken time to go through that this morning. I would say this also: In investigating the accident to Apollo 13, the Board has also attempted to identify those additional technical and management lessons which can be applied to help assure the success of future space flight missions; several recommendations of this nature are included.

I will now, on behalf of the Board, state that we recognize our report as being preoccupied with deficiencies, that is the nature of a review board. We feel that the deficiencies we have uncovered will help the program to do a better job in the future, and that they should be viewed in the light of the considerable successes that this equipment and the people who build and operate it have achieved today.

Mr. Chairman, I now would like to read the recommendations of the Board, and this will conclude my statement. This is on page 540 of the summary report of the Apollo 13 Review Board, this particular volume, if you want to read along with me:

1. The cryogenic oxygen storage system in the service module should be modified to:

(a) Remove from contact with the oxygen all wiring, and the unsealed motors, which can potentially short circuit and ignite adjacent materials

Incidentally, page 540 is the very last thing in the volume.

Mr. KARTH — Page 42 in the copies of your statement the members have?

Mr. CORTRIGHT — It is the statement on 42, it is in the report at page 540. They both say the same thing.

(a) Remove from contact with the oxygen all wiring, and the unsealed motors, which can potentially short circuit and ignite adjacent materials; or otherwise insure against a catastrophic electrically induced fire in the tank.

(b) Minimize the use of aluminum, and other relatively combustible materials in the presence of the oxygen and potential ignition sources.

2. The modified cryogenic oxygen storage system should be subjected to a rigorous requalification program, including careful attention to potential operational problems.

3. The warning systems onboard the Apollo spacecraft and in the Mission Control Center should be carefully reviewed and modified where appropriate, with specific attention to the following:

(a) Increasing the differential between master alarm trip levels and expected normal operating ranges to avoid unnecessary alarms.

(b) Changing the caution and warning system logic to prevent an out-of-limits alarm from blocking another alarm when a second quantity in the same subsystem goes out of limits.

(d) Establishing a second level of limit sensing in Mission Control on critical quantities with a visual or audible alarm which cannot be easily overlooked.

(e) Providing independent talkback indicators for each of the six fuel cell reactant valves plus a master alarm when any valve closes.

4. Consumables and emergency equipment in the LM and the CM should be reviewed to determine whether steps should be taken to enhance their potential for use in a "lifeboat" mode.

5. The Manned Spacecraft Center should complete the special tests and analyses now underway in order to understand more completely the details of the Apollo 13 accident. In addition, the lunar module power system anomalies should receive careful attention. Other NASA Centers should continue their support to MSC in the areas of analysis and test.

6. Whenever significant anomalies occur in critical subsystems during final preparation for launch, standard procedures should require

a presentation of all prior anomalies on that particular piece of equipment, including those which have previously been corrected c explained. Furthermore, critical decisions involving the flight worthiness of subsystems should require the presence and f participation of an expert who is intimately familiar with the details of that subsystem.

7. NASA should conduct a thorough reexamination of all its spacecraft, launch vehicle, and ground systems which contain high-densi oxygen, or other strong oxidizers, to identify and evaluate potential combustion hazards in the light of information developed in th investigation.

8. NASA should conduct additional research on materials compatibility, ignition, and combustion in strong oxidizers at various g level and on the characteristics of supercritical fluids. Where appropriate, new NASA design standards should be developed.

9. The Manned Spacecraft Center should reassess all Apollo spacecraft subsystems, and the engineering organizations responsible fc them at MSC and at its prime contractors, to insure adequate understanding and control of the engineering and manufacturing detai of these subsystems at the subcontractor and vendor level. Where necessary, organizational elements should be strengthened an in-depth reviews conducted on selected subsystems with emphasis on soundness of design, quality of manufacturing, adequacy of tes and operational experience.

Mr. Chairman, that completes my presentation.

Mr. KARTH — Thank you very much, Mr. Cortright, for your summary report. Congratulations to you and the Review Board are i order for having made what I consider to be a very positive, definitive and candid analysis of the accident.

Certainly it gives me confidence that an in-house investigation can be made, which in the final analysis can result in criticism if th investigation merits criticism, so I want to congratulate you and the Board.

I am going to ask the members to adhere to the 5-minute rule to give everyone on the committee an opportunity to ask questions.

The Chair recognizes Mr. Fulton.

Mr. FULTON — I believe you have done a careful and excellent job on the Review Board.

I would like, with the chairman's permission, to have the recommendations of the Review Board of Apollo 204 put immediately after th recommendations that have just been made by this Apollo 13 Review Board.

Mr. KARTH — Without objection, it is so ordered.

(The recommendations of the Review Board of Apollo 204 are as follows:)

BOARD FINDINGS, DETERMINATIONS AND RECOMMENDATIONS

In this Review, the Board adhered to the principle that reliability of the Command Module and the entire system involved in its operatio is a requirements common to both safety and mission success, Once the Command Module has left the earth's environment th occupants are totally dependent upon it for their safety. It follows that protection from fire as a hazard involves much more than quic egress. The latter has merit only during test periods on earth when the Command Module is being readied for its mission and not durin the mission itself. The risk of fire must be faced; however, that risk is only one factor pertaining to the reliability of the Command Modul that must receive adequate consideration. Design features and operating procedures that are intended to reduce the fire risk must no introduce other serious risks to mission success and safety.

1. FINDING

(a) There was a momentary power failure at 23:30:55 GMT.

(b) A detailed design review be conducted on the entire spacecraft communications

(c) No single ignition source of the fire was conclusively identified.

Determination

The most probable initiator was an electrical arc in the sector between the -Y and +Z spacecraft axes. The exact location best fittin the total available information is near the floor in the lower forward section of the left-hand equipment bay where Environmenta Control System (ECS) instrumentation power wiring leads into the area between the Environmental Control Unit (ECU) and th oxygen panel. No evidence was discovered that suggested sabotage.

2. FINDING

(a) The Command Module contained many types and classes of combustible material in areas contiguous to possible ignition sources.

(b) The test was conducted with a 16.7 pounds per square inch absolute, 100 percent oxygen atmosphere.

Determination

The test conditions were extremely hazardous.

Recommendation

The amount and location of combustible materials in the Command Module must be severely restricted and controlled.

3. FINDING

(a) The rapid spread of fire caused an increase in pressure and temperature which resulted in rupture of the Command Modul and creation of a toxic atmosphere. Death of the crew was from asphyxia due to inhalation of toxic gases due to fire. A contributory cause of death was thermal burns.

(b) Non-uniform distribution of carboxyhemoglobin was found by autopsy.

Determination

Autopsy data leads to the medical opinion that unconsciousness occurred rapidly and that death followed soon thereafter.

4. FINDING

Due to internal pressure, the Command Module inner hatch could not be opened prior to rupture of the Command Module.

Determination

The crew was never capable of effecting emergency egress because of the pressurization before rupture and their loss of consciousness soon after rupture.

Recommendation

The time required for egress of the crew be reduced and the operations necessary for egress be simplified.

5. FINDING

Those organizations responsible for the planning, conduct and safety of this test failed to identify it as being hazardous. Contingency preparations to permit escape or rescue of the crew from an internal Command Module fire were not made.

(a) No procedures for this type of emergency had been established either for the crew or for the spacecraft pad work team.

(b) The emergency equipment located in the White Room and on the spacecraft work levels was not designed for the smoke condition resulting from a fire of this nature.

(c) Emergency fire, rescue and medical teams were not in attendance.

(d) Both the spacecraft work levels and the umbilical tower access arm contain features such as steps, sliding doors and sharp turns in the egress paths which hinder emergency operations.

Determination

Adequate safety precautions were neither established nor observed for this test.

Recommendations

(a) Management continually monitor the safety of all test operations and assure the adequacy of emergency procedures.

(b) All emergency equipment (breathing apparatus, protective clothing, deluge systems, access arm, etc.) be reviewed for adequacy.

(c) Personnel training and practice for emergency procedures be given on a regular basis and reviewed prior to the conduct of a hazardous operation.

(d) Service structures and umbilical towers be modified to facilitate emergency operations.

6. FINDING

Frequent interruptions and failures had been experienced in the overall communication system during the operations preceding the accident.

Determination

The overall communication system was unsatisfactory.

Recommendations

(a) The Ground Communication System be improved to assure reliable communications between all tests elements as soon as possible and before the next manned flight.

(b) A detailed design review be conducted on the entire spacecraft communication system.

7. FINDING

(a) Revisions to the Operational Checkout Procedure for the test were issued at 5:30 p.m. EST January 26, 1967 (209 pages) and 10:00 am EST January 27, 1967 (4 pages).

(b) Differences existed between the Ground Test Procedures and the In-Flight Check Lists.

Determination

Neither the revision nor the differences contributed to the accident. The late issuance of the revision, however, prevented test personnel from becoming adequately familiar with the test procedure prior to its use.

Recommendations

(a) Test Procedures and Pilot's Checklists that represent the actual Command Module configuration be published in final form and reviewed early enough to permit adequate preparation and participation of all test organization.

(b) Timely distribution of test procedures and major changes he made a constraint to the beginning of any test

8. FINDING

The fire in Command Module 012 was subsequently simulated closely by a test fire in a full-scale mockup.

Determination

Full-scale mock-up fire tests can be used to give a realistic appraisal of fire risks in flight-configured spacecraft.

Recommendation

Full-scale mock-ups in flight configuration he tested to determine the risk of fire

9. FINDING

The Command Module Environmental Control System design provides a pure oxygen atmosphere.

Determination

This atmosphere presents severe fire hazards if the amount and location of combustibles in the Command Module are not restricted and controlled.

Recommendations

(a) The fire safety of the reconfigured Command Module be established by full-scale mockup tests.

(b) Studies of the use of a diluent gas be continued with particular reference to assessing of problems of gas detection and control and the risk of additional operations that would be required in the use of a two gas atmosphere.

10. FINDING

Deficiencies existed in Command Module design, workmanship and quality control such as:

(a) Components of the Environmental Control System installed in Command Module 012 had a history of many removals and of technical difficulties including regulator failures, line failures and Environmental Control Unit failures. The design and installation features of the Environmental Control Unit makes removal or repair difficult

(b) Coolant leakage at solder joints has been a chronic problem.

(c) The coolant is both corrosive and combustible.

(d) Deficiencies in design, manufacture, installation, rework and quality control existed in the electrical wiring.

(e) No vibration test was made of a complete flight-configured spacecraft.

(f) Spacecraft design and operating procedures currently require the disconnecting of electrical connections while powered.

(g) No design features for fire protection were incorporated.

Determination

These deficiencies created an unnecessarily hazardous condition and their continuation would imperil any future Apollo operations.

Recommendations

(a) An in-depth review of all elements, components and assemblies of the Environmental Control System be conducted to assure its functional and structural integrity and to minimize its contribution to fire risk.

(b) Present design of soldered joints in plumbing be modified to increase integrity or the joints be replaced with a more structurally reliable configuration.

(c) The coolant is both corrosive and combustible.

(d) Review of specifications be conducted, 3-dimensional jigs be used in manufacture of wire bundles and rigid inspection at all stages of wiring design, manufacture and installation be enforced.

(e) Vibration tests be conducted of a flight-configured spacecraft.

(f) The necessity for electrical connections or disconnection with power on within the crew compartment be eliminated.

(g) Investigation be made of the most effective means of controlling and extinguishing a spacecraft fire. Auxiliary breathing oxygen and crew protection from smoke and toxic fumes be provided.

11. FINDING

An examination of operating practices showed the following examples of problem areas:

(a) The number of the open items at the time of shipment of the Command Module 012 was not known. There were 113 significant Engineering Orders not accomplished at the time Command Module 012 was delivered to NASA; 623 Engineering Orders were released subsequent to delivery. Of these, 22 were recent releases which were not recorded in configuration records at the time of the accident.

(b) Established requirements were not followed with regard to the pre-test constraints list. The list was not completed and signed by designated contractors and NASA personnel prior to the test, even though oral agreement to proceed was reached.

(c) Formulation of and changes to prelaunch test requirements for the Apollo spacecraft program were unresponsive to changing conditions.

(d) Non-certified equipment items were installed in the Command Module at time of test.

(e) Discrepancies existed between NAA and NASA MSC specifications regarding inclusion and positioning of flammable materials.

(f) The test specification was released in August 1966 and was not updated to include accumulated changes from release date to date of the test.

Determination

Problems of program management and relationships between Centers and with the contractor have led in some cases to insufficient response to changing program requirements.

Recommendation

Every effort must be made to insure the maximum clarification and understanding of the responsibilities of all the organizations involved, the objective being a fully coordinated and efficient program.

Mr. FULTON — I would like to ask General Hedrick, since you are Director of Space Headquarters at USAF, do you have there an Inspector General under the U.S. Air Force who is independent and makes independent inspections?

General HEDRICK — Yes; we do.

Mr. FULTON — Do you need him in Space in the U.S. Air Force? Is he valuable?

General HEDRICK — Yes.

Mr. FULTON — I thank you for trying to get an Inspector General set up for NASA. Either he is not needed in space in the U.S. Air Force or else he is badly needed in NASA. I do feel we need an outside independent inspection system that is reportable to the top management of NASA.

As of now, anyone who hangs a lemon on a capsule can only complain to a contractor, a subcontractor, a man working for the Manned Space Flight Center, or the particular Center where this project is being developed, or he is required to report to a program director, of course, who wants to get along with the job.

If he is down at the launch site, he will be holding up the launching if he thinks there is something which might slightly go wrong and probably won't. So again I recommend to NASA a strongly independent Inspector General setup so that we can do especially the No. [illegible] recommendation to insure adequate understanding and control of the engineering and manufacturing details of these subsystems at the sub-contractor and vendor level.

On recommendation No. 5, it would help.

On recommendation No. 6, it would help.

On recommendation No. 7, it would help.

I am, of course, interested in the use of Mylar insulation as a blanket, and also interested in the insulation on wires carrying electrical currents under oxygen conditions.

Has the manufacturer taken off his list or his catalog or limited for these purposes these two materials? What has been the result for the general public and general business on the investigation? Are we going to limit Teflon and Mylar insulation blanketing?

Mr. CORTRIGHT — If I understand your question right, the combustion of the Mylar insulation occurred in a very unusual circumstance, namely —

Mr. FULTON — I agree with that. What protection is there for the general public and general business with the new information we have? I believe I will answer it: The manufacturer, I understand, has taken off its catalog lists for these purposes at least the Teflon.

Mr. CORTRIGHT — I am sorry, Mr. Fulton, I am not aware of that.

Mr. FULTON — One other point I would like to ask about is this: When there is a combination of circumstances resulting in one warning, it seems to me incredible that there is not an alternative system that might turn up a second warning so that the first warning system doesn't smother the second.

Mr. CORTRIGHT — That is a reasonable observation. The compromise always is: how complex can you make the system? The more modes of failure the system can handle, the more complex the system gets. This particular system has certain situations of the type you described. We have asked that they be reexamined to see if anything can be done about it. It may not be practical to do so. By the way, Mr. Chairman, I would like to welcome the last missing member of our Board, Mr. Neil Armstrong, who has now arrived. He is familiar with one alarm overriding another. He may comment on that question.

Mr. FULTON — We think that astronauts had better not be cross-examined too closely. We would rather have you fellows respond, and while they are orbiting the White House and the Capital now more than the moon, we nevertheless give them a little immunity, which I think we owe them. The point I have always made, and made especially on Apollo 204, there is no failure on the part of the astronauts in handling the equipment, on running the mission, on the decision to take certain rescue operations and the return. The astronauts all performed well without any negligence or failure whatever. Is that correct?

Mr. CORTRIGHT — I guess I don't know of any.

Mr. FULTON — How about the Administrator?

Dr. PAINE — I certainly concur in that statement.

Mr. FULTON — How about Mr. Armstrong?

Mr. ARMSTRONG — I would have to say that there were a number of options available to the crew and they didn't investigate ever option, which in hindsight could have been investigated, but there isn't any reason to believe they should have with the information the had available to them either.

Mr. FULTON — How about General Hedrick?

General HEDRICK — I think they performed admirably.

Mr. FULTON — Thank you.

Mr. HECHLER — Technically, I think this is an outstanding report and I like its forthrightness. We can call it the Forthright Cortrigh Report. Seriously, the recommendations are almost entirely technical in nature with the possible exception of parts of No. 6. In an organization like NASA where you have individuals of high technical competence planning for a very hazardous mission, there has to b mutual respect and confidence on the part of those that are using the equipment that everything will go right, whereas Murphy's lav occasionally crops up. So what really concerns me about both the 1967 fire and this accident is that although we have devise recommendations which take care of correcting the technical aspects, we have done little to correct administrative deficiencies.

You really need some critical people who may not be very popular in NASA, they may not get many invitations to social events, bu there are people who have a critical, skeptical bent in their questions about whether or not the contractor has produced safe equipmen They must ask the kind of questions like, what about the hazards of all-oxygen environment, questions like we by hindsight asked i 1967, why couldn't you open a hatch from the inside a little quicker during the test.

I think you need a group of people with this type of inquiring, critical mind, that can ask these questions consistently and continuousl as the equipment comes from the contractor, to not only watch the development and review the procedures according to the manua but to find out if there were any unusual events like the dropping of 2 inches onto the cement floor and what effect this had.

I would like to ask Mr. Armstrong if one of the astronauts who has made a flight could be placed at the head of the team who coul independently ask the kind of questions that the ordinary experts within NASA do not ask because they have confidence that everythin will go right?

Mr. CORTRIGHT — If I can interrupt before Mr. Armstrong can answer that question, I would like to correct the error of droppin onto the cement floor which appeared in the newspaper. This so-called drop incident was not like that at all. The tank was assemble in a shelf, the shelf was being lifted out of the bay No. 4 and one bolt had not been removed and as a result, the lifting device broke an the shelf containing the two oxygen tanks dropped 2 inches back onto its mounting brackets.

Mr. HECHLER — I am glad we got that correction.

Mr. ARMSTRONG — I am sure that astronauts who really spend very little time in space compared to the amount of time they spen asking questions in the course of their job could do such a job as Mr. Hechler suggested, and we find many other individuals within ou Agency and without, who are also very penetrating in their inspections and could also do such jobs.

Mr. HECHLER — Would it take someone outside of the Agency coming in or could it be done by someone who is necessarily an exper and maybe going around with the wheels and have the confidence in the equipment which results from just being an expert?

Mr. ARMSTRONG — I should think there is always some advantage to people who are put in this position of having some independen authority.

Mr. HECHLER — I want to ask Dr. Paine if he had any further comment on this.

Dr. PAINE — Of course, Mr. Hechler, we have the Aerospace Safety Advisory Panel which is specifically designed to report directly t me outside of any other channel. It includes people outside of NASA who sit in and review the procedures we are using. They ar penetrating. We need — not any one magic solution — but we need to take a number of different approaches.

In order to penetrate a system as complex as Apollo to the tremendous depths in which it must be penetrated and I think we have beautiful example in this very small thermal switch which was certainly one of the major contributors, it is necessary to have a ver large organization working on a full-time basis with no other responsibilities such as our Apollo management system.

In addition to that, we do need outside people to come in and ask the very different overall kinds of question, whether or not we indee have got this set up properly, whether or not the channels of reporting are correct, whether or not we are indeed using the best an most modern techniques to attempt to have the entire system ferret out such questions.

What we have here before us today is an example of a breakdown of a system in which we failed with the kinds of gates that we hav assembled to prevent these things from going through. We failed to detect the fact that in the change from a 28-volt to a 65-volt GS power supply, we failed to test the switch specification. We then have failures on the part of additional people later to see to it that thi system got an adequate test which would expose the fact that this switch, which is never called upon to operate in flight, under th ground conditions we encountered would fail to operate successfully.

We have a number of such failures. I haven't had an opportunity to go through the report in detail — which I will do — which calls fo us to reexamine the systems we have in place and ask ourselves in detail what must be done to make it such that in the future we catc things of this nature.

The fact that we did this special detanking procedure on this tank which had never been done on a previous Apollo mission indicate why it was that on Apollo 13 we encountered this difficulty when we had successfully flown all the previous Apollo missions. In no cas had this switch ever had the opportunity to operate. It was the special detanking proceeding.

The lesson that we have got to examine here is how it could be that we would indeed carry out this special detanking procedure i Cape Kennedy — when we ran into difficulties in detanking this tank during the test period, how it would be that we would carry ou the procedure and not fully examine all the consequences of this.

There are many questions in the administration end which we must reexamine as a follow-up to the job that Mr. Cortright and his team have so ably done.

Mr. HECHLER — Thank you, Dr. Paine.

Mr. KARTH — Thank you, doctor.

The Chair recognizes Mr. Mosher.

Mr. MOSHER — Are you saying that NASA as yet has not precisely identified the point of procedure or the persons in the procedures who should have asked the right questions about the effect of the special testing on the pad which fused the thermal switch? You haven't yet precisely identified the point or the person where the crucial question should have been raised?

Dr. PAINE — Mr. Mosher, it is my guess that we will never identify one particular person that might be called the villain of Apollo 13. There are many different failures that have come to light. There was the failure in the switch area. There was the failure in assembling the fill system, which then, in turn, led to the necessity at the Cape for the special detanking procedures. There were a number of different events which happened along the line. Each one of these was necessary.

Mr. MOSHER — I wasn't looking for the villain of the piece. I was raising essentially the same question that Mr. Hechler and Mr. Fulton raised — you haven't precisely identified the person in the future who is going to ask these embarrassing or these crucial questions?

Dr. PAINE — That is right. We have not yet made our decision as to what changes are necessary in order to preclude such a thing in the future.

Mr. MOSHER — Are there any aspects of this, or any event, not yet identified? Is there any remaining mystery as to what happened still unexplained?

Mr. CORTRIGHT — I guess it is pretty dangerous to say "No" to that question. But, at the moment, we don't know of any remaining mysteries. There was one test which didn't turn out quite the way we thought it would turn out. It is being rerun at Beech. That was a full-scale duplication of the detanking that took place at the Cape in all respects. When that took place, the switches failed in a different manner than they did in our test setup and, as a result, one switch remained closed and one open by virtue of the fact that the terminals melted and fell out.

Mr. MOSHER — So you will still be doing some work?

Mr. CORTRIGHT — That is right. One heater stayed on and one did not; as a result, the temperatures didn't get as high as it did in the Apollo 13, and so the insulation wasn't damaged, although we had done other tests — there are other details that need to be cleaned out. We pointed out in our letter of transmittal that we plan to reconvene a little later in the year to look over any additional tests and analyses to see if what we have said here still stands up.

Mr. MOSHER — You have made several recommendations that will take time. What about the impact of this on Apollo 14? How much postponement is there going to be?

Mr. CORTRIGHT — I don't know that there will be any. The recommendations we have made are generally cast in a two-level type review, for example, where we ask the subsystems to be reviewed, we are first essentially asking for a screening to identify those that we are not so much on top of. It is our feel it can be concluded before Apollo 14.

Mr. MOSHER — So December is still a good time?

Mr. CORTRIGHT — Yes. I think it will be a hard point to meet from the changes in hardware that will be selected. Whether it is possible or not, I am not qualified to say.

Mr. MOSHER — Thank you.

Mr. DOWNING — I would like to congratulate Mr. Cortright and the board for what I think is an excellent report and a practical one. It reminds me of the one we had several years ago. We have complete confidence in it. Was this the first time that the fan in oxygen tank No. 2 was turned on?

Mr. CORTRIGHT — No, sir; the fans and heaters are used whenever the tank is filled with cryogenic oxygen. They are not used continuously.

Mr. DOWNING — During the flight?

Mr. CORTRIGHT — No. Pardon me. The fans had been turned on several times before during the flight.

Mr. DOWNING — Were there other tanks on board which had the same switches and thermostats and which did operate properly during the flight?

Mr. CORTRIGHT — Oxygen tank No. 1 is essentially identical and it operated properly. The hydrogen tanks are similar and they operated properly. Actually, as I point out in the board report, these particular tanks accumulated nearly 3,000 hours of space flight without significant problems.

Mr. DOWNING — They had not been redesigned with the 65-volt switch?

Mr. CORTRIGHT — No. They have used the 65 volt at the Cape for checkout of all of these tanks for pressurization, but not under the circumstances of this detanking procedure. Let me make sure that is clear. I am not sure that I did this. The difference is that when the heaters are left on during detanking, they are running when the tank is almost empty and you don't have that large quantity of very cold oxygen to keep things cool, so at this point they get very hot. That had never happened before.

Mr. DOWNING — If I read the time chart correctly, there was something more than a minute from the time the fan turned on until the explosion occurred. Is there anything that the crew could have done, in hindsight, or that the ground crew could have done?

Mr. CORTRIGHT — No, sir; the only thing that could have been done was to observe the increase in pressure and reduce the troubleshooting time afterward to identify why it happened, but there was nothing that could have been done to save the mission.

Mr. DOWNING — You termed this a near disaster, which it was. What could have happened? What did you fear the most?

Mr. CORTRIGHT — I think that in space it might have been possible to rupture a propellant tank in an adjacent bay. It might have failed oxygen tank No. 1 more rapidly, not giving the crew adequate time to make the transition that they did to the LEM lifeboat mode. It might have occurred at a different point in the mission when recovery would not have been possible.

Mr. DOWNING — Was it more of an explosion than an implosion?

Mr. CORTRIGHT — Yes, sir; I think it is most easily understood as a failure in the pressure vessel or its high-pressure tubing due to overheating, a rupture, if you will, through which high-pressure oxygen bursts or streams very rapidly.

Mr. DOWNING — Thank you very much.

Mr. KARTH — Mr. Winn.

Mr. WINN — Thank you. The review board has done an excellent job in which I concur with the remarks of the other members of the committee. I would like to follow up the thought that Mr. Mosher pursued. Did anything else in your various tests that you ran give you great concern, other than the additional switch, when you were really putting some of these pieces to extreme tests which were shown in the movie? Did anything else show up that really bothered you?

Mr. CORTRIGHT — Yes, sir; these are all spelled out in the board report We were concerned with certain aspects of the basic tank design which indicated to us that this ignition might have taken place with a tank with good switches in it, in the event the insulation would be damaged in the assembling. Until we found the switches, we concentrated very hard on the manner in which wiring insulation could be damaged and convinced ourselves to the point it is still in the report that yes, this could happen with this type of tank design.

We also were concerned with the amount of potentially combustible material in close proximity to electrical sources which could become ignition sources. We recommended that something be changed there. There was a battery problem on the lunar module which was not related to this accident, but it occurred on the way back, and that has to be run down. So I think it is not just as simple as this thermostatic switch.

Mr. WINN — That is what I gathered that you were saying in your recommendations which looked to me as if they were very thorough and you made a statement on page 8-44, "Where appropriate, NASA designed standards should be developed." In part of your recommendations you say that the review board will be called together again shortly. Did I understand you to say that?

Mr. CORTRIGHT — We plan one more session ourselves. We are at the disposal of the Administrator to reconvene any time he thinks he needs us.

Mr. WINN — If you haven't developed a program yet, Dr. Paine, who in NASA, is going to follow through on these recommendations and if additional recommendations are to be made, I don't see how you are going to be able to keep the time schedule for Apollo 14 when everything is still up in the air.

Dr. PAINE — This will be examined. After every Apollo mission, a great deal of attention is given to going back over all the anomalies that have happened. In each mission there have been certain things that were unexplained, which had to be dug into, and Mr. Cortright has mentioned several additional ones in Apollo 13. In no case do we ever fly a mission until we have cleaned up all the things to our satisfaction which we have been shown in previous flights.

Mr. WINN — If new parts are needed and new parts have to be designed, built, and tested, I suppose in that case it would depend on what it is and how important a part it plays in the overall production. When we get down to little wires and switches, it looks as if everything is just as important as the things we hear about.

Dr. PAINE — The smallest component is just as important as the largest, and we have just had a very dramatic demonstration of that. I can assure you we will not fly Apollo 14 until we are satisfied that we have fixed up everything that has come to light.

Mr. WINN — Thank you very much.

Mr. KARTH — Mr. Goldwater?

Mr. GOLDWATER — Why was it necessary to detank this particular vehicle?

Mr. CORTRIGHT — The procedures at the Cape require that when the countdown demonstration test is complete, that the tanks be emptied and then filled again prior to launch at a later date. I think this is partly for safety reasons as a matter of fact.

Mr. GOLDWATER — This was done on 11 and 12?

Mr. CORTRIGHT — It is done always. In this particular case, the tank would not expel its oxygen in a normal fashion. The way that is done is to take the vent line and pressurize the inside of the tank through that vent line and that pushes down on the oxygen which pushes it up through the fill line and out the fill tube. When there is a loose connection at the top, the gases you are using to pressurize the tank would go in one line and out the other and don't pump fluid out with them. That is the problem that was run into.

Mr. GOLDWATER — This happened during detanking?,

Mr. CORTRIGHT — Yes.

Mr. GOLDWATER — Nothing was done about it?

Mr. CORTRIGHT — It was not recognized that the heater operation was a problem.

Mr. GOLDWATER — I see.

Mr. CORTRIGHT — It was not known that the wires had been damaged and that the heaters stayed on continuously, as I told you in the outline of what happened. That was not recognized before launch.

Mr. GOLDWATER — Did you feel that during this detanking period that when the temperatures built up, they burned the wires?

Mr. CORTRIGHT — Yes; the heating damaged wire insulation.

Mr. GOLDWATER — Could you clarify this change in the provision from the 28 to the 65 volts power switch specification — why this was important?

Mr. CORTRIGHT — Yes. The spacecraft flies on 28 volts and North American Rockwell uses 28 volts. The Kennedy Cape Center uses 65 volts d.c., Beech uses 65 volts a.c. At the Cape they have a 65 d.c., volt system. The higher voltages or currents are used to accelerate the tank pressurization. When you first fill the tank at low temperature, then to build up the pressure at the operating range, you have to put heat in and you can save several hours by accelerating this, and it seems to be an acceptable and desirable procedure from my point of view, provided everything is protected from the higher voltage power supplies.

In this case, that was a change, back in 1965, but the subcontractor, Beech, did not change the switches at that time. They left the switches in, or essentially the same switches that were in and these were not capable of protecting against an overheat condition, which they never should have encountered in this detanking procedure.

Mr. GOLDWATER — You are running a 28 volt switch on 65 volts?

Mr. CORTRIGHT — Yes.

Mr. GOLDWATER — You feel that is what melted the contact?

Mr. CORTRIGHT — While the switch was closed the 28-volt switch will take it. If you attempt to break a d.c. current, it is difficult to do. That arc starts and it wants to hang on, stay there, persist. In the process, it erodes, melts, and displaces and, in this case, welds across the two contact points.

Mr. GOLDWATER — Even before the liftoff?

Mr. CORTRIGHT — Yes.

Mr. GOLDWATER — Why did it take so long, 56 hours before the explosion, took place?

Mr. CORTRIGHT — We will probably never know.

Mr. GOLDWATER — You said you were going to elaborate on, which I don't think you did, the tremendous pressures that were built up during this explosion, it took some 20-some p.s.i., yet 10 p.s.i. through the center section could blow the command module off the top. Could this happen again with some other system failure?

Mr. CORTRIGHT — Any time the entire face of the command module is subjected to about 10 pounds per square inch, it will tear loose from the service module. This was one of the problems we faced in trying to rationalize or understand what happened, and current views based on the Langley tests and analyses are that the pressure buildup took place rapidly and did not have time to build up against the face of the command module, so actually you had high pressures in one part of the structure and lesser pressures in the other.

Mr. GOLDWATER — If the pressures didn't release out the side, it could have gone to the top?

Mr. CORTRIGHT — It could have.

Mr. KARTH — Mr. Price?

Mr. PRICE — Thank you, Mr. Chairman. I want to commend you on your effort. You have certainly pointed up a lot of things that needed to be pointed up. Perhaps such a board should look into the operations of every flight as a means of bettering our operation. Dr. Paine, doesn't this point up the need for a rescue system, or the thing we have been talking about, the following-on of the shuttle and a space station? Had we had such a system in space, there was a possibility with the correct modifications that they could have attached to a space station and saved their lives?

Dr. PAINE — This particular accident, and the manner of its occurrence 205,000 miles out on the mission, probably would not have been affected by the capability to launch a rescue mission as we look at it. On the other hand, had the accident occurred at another part of the mission or in another manner, it is certainly possible that the existence of a space shuttle system or a space station system might have been able to provide some assistance.

I think it is correct to say that when such systems are available, we will all feel a good deal easier about flying men in space.

You have to recognize we are still in the early days of the Space Age, and at the present time we are flying missions with pioneers out to explore these new areas, and we do not have a rescue capability for most parts of the mission, particularly, of course, including the lunar surface activities.

Mr. PRICE — Mr. Cortright, in your first paragraph of your closing remarks, you said something about an unforgiving design. Could you elaborate on that a little bit? Who is responsible for an unforgiving design?

Mr. CORTRIGHT — The prime thought we had in mind, in using that word, was the presence of sufficient combustibles in the tank to support a rather strenuous fire in there and the combustion paths which permitted this burning to get to the vicinity of thin walled, high-pressure metal.

Mr. PRICE — Is NASA responsible for the design as they pass it on for bidding? Is it a factor in the specifications?

Mr. CORTRIGHT — The process was to have a competition in which a number of contractors bid and proposed their design. A particular subcontractor won. The competition was conducted by the prime contractor and NASA had an overview responsibility on all of it. The ultimate responsibility for accepting the design approach is NASA's.

Dr. PAINE — I certainly would like to emphasize that, the ultimate responsibility for the safety of all our missions is NASA's and we fully accept it.

Mr. PRICE — Neil, I notice on page 5-39 of the report, the finding:

"The crew maneuvered the spacecraft to the wrong LM roll attitude in preparation for LM jettison. This attitude put the CM very close to gimbal lock which, had it occurred, would have lost the inertial attitude reference essential for an automatic guidance system control of reentry." Was this sent up from ground control or could you explain it?

Mr. ARMSTRONG — No, Sir, Mr. Price. It was not bad information on the part of the ground. It, in this case, was an error on the part of the crew. However, I suspect I might have been guilty of making that same type of error, since it was an attitude control situation with which they were not familiar as a crew. It was one that was improvised during flight, and there is a certain amount of learning involved in this particular control method and the interpretation of the displays, and they just made a mistake.

Mr. PRICE — Also, in the testimony here, it speaks of Manned Spacecraft Center engineers "devised and checked out a procedure for using the CM LiOH canisters to achieve carbon dioxide removal." Mr. Low and I have been doing some deep sea diving and we began finding out, about carbon dioxide. This was a critical area in not having enough air and re-breathing carbon dioxide, was it not? And why cannot provisions be made in the future, subject to such an eventuality, and make it a part of the equipment? It might just mean their survival.

Mr. CORTRIGHT — Both systems, command module and the LM, were designed with sufficient carbon dioxide removal for their own purposes. The particular failure with the LM lifeboat did not receive much attention. One of our recommendations is that this be examined to see if the consumables should be handled or planned in a little different manner to enhance this lifeboat capability. That is what you are suggesting, and we agree.

Mr. PRICE — So, if something should happen in the future, we should have longer life capability in the LM, even though we don't now have the capability — in the future we should develop this so that we can well give them a chance of possible rescue. Neil, would you have any comments on that?

Mr. ARMSTRONG — I agree with our board chairman that such a thing is desirable. This situation was a product of the timing. This particular configuration, the so-called LM lifeboat, was not included as a design specification. It was not an intent in the original design. It was something developed after we had the vehicle and said now, if we really get into a problem what we actually could do is use the LM as an aid to help us in an emergency situation. That being the case, it is understandable that the particular fittings and so on were not compatible, and we recognize now that it would certainly be an aid to have them so.

Mr. PRICE — Recommendation 1 (a) states:

Remove from contact with oxygen all wiring and the unsealed motors which can potentially short circuit and ignite adjacent materials. What are the potentials of these unsealed motors? It would seem to me they should be developed to get away from any potential short circuit. Shouldn't you really bear down on this area?

Mr. CORTRIGHT — Yes, Sir; we should. Those motors and the wiring are being looked at very hard.

Mr. PRICE — No. 2 recommendation states: "The modified cryogenic storage system should be subjected to rigorous requalification program," and so forth.

Shouldn't all these systems be subjected to a rigorous requalification program throughout the Apollo?

Mr. CORTRIGHT — Any system that is changed has to be properly re-qualified, and I suppose, in a sense that recommendation was unnecessary because all systems go through a qualification program, but we put it in for emphasis.

Mr. PRICE — In closing, also on recommendation No. 6, down in the middle of the paragraph you state.
Furthermore, critical decisions involving the flightworthiness of subsystems should require the presence and full participation of an expert. I am amazed that we don't have that at present.

Mr. CORTRIGHT — In the present case, experts were contacted by phone, which is done sometimes, and, in this case, it resulted in some confusion and misinformation so that people overlooked the potential of an overheating damage, and the board is speculating that this might not have happened if someone who really knew the inside of that tank and all its idiosyncrasies had been down there in the conference on detanking.

Mr. PRICE — It would seem to me it would be advisable for this type of man to be there — that knew the interior workings of every joint — if I were flying I would want that. That is all, Mr. Chairman.

Mr. KARTH — Mr. Cortright, in addition to your objectivity, I am sure that Dr. Paine chose you to be head of the review board because of your competence, and, retrospectively, I would say probably it was the best choice that could be made. In your opinion, in view of your competence, what was most responsible for the accident — design, manufacturing tests, or management?

Mr. CORTRIGHT — I don't think I can answer that by selecting one. I think it was an unusual combination of things that made this accident happen.

Mr. KARTH — Could you grade those 1, 2,3,4?

Mr. CORTRIGHT — I am afraid I could not.

Mr. KARTH — In your list of recommendations, recommendation No. 6, let me just reread that first sentence of the paragraph:

Whenever significant anomalies occur in critical subsystems during final preparation for launch, standard procedures should require a presentation of all prior anomalies on that particular piece of equipment, including those which have previously been corrected or explained. Isn't that standard operating procedure?

MR.CORTRIGHT — No, sir; presentation is not necessarily required.

Mr. KARTH — Don't you think it ought to be?

Mr. CORTRIGHT — That is what we are suggesting here; yes, we do.

Mr. KARTH — It is rather amazing to me that up to this point in time that hasn't been standard operating procedure.

No. 7: "NASA should conduct a thorough reexamination of all of its spacecraft, launch vehicle, and ground systems which contain high density oxygen," et cetera, et cetera.

Does my memory serve me properly, after the Apollo 204 fire, this was essentially a recommendation which had been made at that time. Because the record doesn't show the shaking of a witness' head one way or another, let me point out that one of the witnesses indicates the answer to that question is yes; is that right?

Mr. CORTRIGHT — This is our general counsel, George Malley, from the Langley Research Center, who was also counsel to 204. He seems to concur that that was the case.

Mr. KARTH — As a result of this recommendation, it is obvious that that procedure was not previously followed. Is that correct?

Mr. CORTRIGHT — It is correct to say, here we are with another oxygen fire on our hands after having gone back to look the system over — yes, that is correct.

Mr. KARTH — Was a thorough reexamination of this particular piece of equipment made after the Apollo 204 accident?

Mr. CORTRIGHT — In view of myself and most members of the board, it was not a thorough review. There was some review made of this tank and the materials were once again checked against the so-called COMAT standard, but I don't believe it was as penetrating as it should have been.

Mr. KARTH — In your judgment are present management procedures entirely adequate to preclude similar future occurrences?

Mr. CORTRIGHT — I wouldn't say that with 100-percent confidence. We found the procedures themselves, in general, good, but it was possible to got a non-flight worthy piece of equipment through, even with those procedures, so, until we complete our reexamination of how we are doing our business on the subsystems, I would not say that with confidence. On the other hand, I think the procedures are good and the management panel was quite complimentary in its review of both the procedures and the rigor with which people stick to them and sign off all the proper forms and do all the proper things that are supposed to prevent this.

Mr. KARTH — The hour is late and we have already started a quorum call. I had a list of questions I wanted to ask you. Because of the press of time, we will not have an opportunity to do so. Would you prepare answers to them?

Mr. CORTRIGHT — Yes, Sir.

Mr. KARTH — And submit them for inclusion in the record

Mr. CORTRIGHT — Yes, Mr. Karth.

(The following information is provided for the record:)

NATIONAL AERONAUTICS AND SPACE ADMINISTRATION, OFFICE OF THE ADMINISTRATOR, Washington, D.C., June 30, 1970.

Hon. Joseph E. Karth, House of Representatives, Washington, D.C.

DEAR MR. KARTH: This is in response to the questions you submitted to me during the hearing held before the Committee on Science and Astronautics on June 17, 1970.

As Dr. Low and I requested, Dr. Charles D. Harrington, Chairman of the Aerospace Safety Advisory Panel, submitted the report of the Panel to us on June 25, 1970, in the form of a letter, a copy of which is attached (TAB A), on the procedures and findings of the Review Board. Based on these reports and on extensive discussions at reviews and meetings held since June 25, Dr. Dale Myers, Associate Administrator for Manned Space Flight, has formally submitted to me with his endorsement the final recommendations of Dr. Petrone, the Apollo Program Director, to prepare for the Apollo 14 mission. These recommendations are embodied in Dr. Petrone's memorandum to me of June 27, 1970, a copy of which is also enclosed (TAB B).

On the basis of these reports and recommendations, Dr. Low and I have approved the following actions to implement the recommendations of the Apollo 13 Review Board and to carry out the steps recommended by Dr. Petrone and Dr. Myers.

First, the recommendations of the Apollo 13 Review Board will be implemented before the Apollo 14 mission is approved for launch. This will require postponing the launch date to no earlier than January 31, 1971.

Secondly, the Associate Administrators in charge of the Offices of Space Science and Applications, Manned Space Flight, and Advanced Research and Technology, have been directed to review the Apollo 13 Review Board Report to apply throughout NASA the lessons learned in their areas of responsibility. In addition, we will take steps to disseminate widely throughout Industry and the technical community the lessons of Apollo 13 to prevent recurrences in other areas.

Third, the Aerospace Safety Research and Data Institute (ASRDI) at the NASA Lewis Research Center has been directed to conduct additional research on materials compatibility, ignition, and combustion at various G levels, and on the characteristics of supercritical fluids, as recommended by the Apollo 13 Review Board.

Fourth, I have requested that the Aerospace Safety Advisory Panel conduct a review of the management processes utilized by NASA in implementing the recommendations of the Apollo 13 Review Board and report to me their views no later than the Apollo 14 Flight Readiness Review. I have also asked Mr. Cortright to reconvene the Apollo 13 Review Board later this year, as he suggested, to review the results of continuing tests to determine whether any modifications to the Board's findings, determinations, or recommendations are necessary in light of additional evidence which may become available.

The assessment of the Office of Manned Space Flight, in which Dr. Low and I concur, is that the reasonable time required for the design, fabrication, and qualification testing of the modifications to the Apollo system we have determined to be necessary, and for the other actions outlined above which must be taken before the next Apollo mission, will permit us to launch Apollo 14 to the Fra Mauro region of the moon at the January 31, 1971 launch opportunity. This will also move the planned launch date for Apollo 15 several months to July or August 1971, maintaining the six month interval between launches on which our operations in the Apollo program are now based. However, we will not launch Apollo 14 or any other flight unless and until we are confident that we have done everything necessary to eliminate the conditions that caused or contributed to the problems we encountered on Apollo 13 and are ready in all other respects.

QUESTION — Are the circumstances of the accident sufficiently well understood at this time to proceed on a firm basis with the Apollo 14 flight?

ANSWER — Yes. Dr. Low and I have now had an opportunity to study the report in detail and to review carefully its recommendations. In our view it is an excellent report based on a thorough and objective investigation and highly competent analysis. It clearly pinpoints the causes of the Apollo 13 accident and sets forth a comprehensive set of recommendations to guide our efforts to prevent the occurrence of similar accidents in the future.

QUESTION — What is your best estimate of the time and cost to recover from the Apollo 13 accident?

ANSWER — The assessment of the Office of Manned Space Flight, in which Dr. Low and I concur, is that the reasonable time required for the design, fabrication, and qualification testing of the modifications to the Apollo system we have determined to be necessary, and for the other actions outlined above which must be taken before the next Apollo mission, will permit us to launch Apollo 14 to the Fra Mauro region of the moon at the January 31, 1971 launch opportunity.

It is too early to present to you our detailed estimates of the costs and budgetary impact of the spacecraft modifications and program changes that we are making. Our best current estimate is that the modifications and changes related to the actions resulting from the Apollo 13 accident will be in the range of $10 to $15 million of increased costs, which we plan to handle within our total Apollo budget.

QUESTION — Do you see the need for any major changes in your method of operation or procedures based on the Apollo 13 accident experience?

ANSWER — NASA's actions in response to the Board's recommendations will avoid those specific things which led or contributed to the Apollo 13 accident; and the reviews and research we have undertaken will help us avoid future potential hazards throughout our programs. The reviews now underway throughout NASA in response to the Board's recommendations will, in my view, help us to further strengthen the management of Apollo and other NASA programs.

QUESTION — To what extent would you expect the results of the Apollo 13 accident to affect other NASA programs such Skylab?

ANSWER — The broad effects of the Apollo 13 accident on programs such as Skylab have not been determined. Time and cost impact on Skylab, for example, will depend on results of decisions and actions taken in the Apollo program and the reviews now underway. We do not anticipate any serious implications an Skylab at this time, but we will be continually assessing the situation as these actions are taken.

Certain specific effects have already been evaluated and actions taken relative to the Skylab Program. These include: assuring that the modifications made to the Apollo Service Modules to eliminate the Apollo 13 failure mode will be incorporated on Skylab to the extent that the designs are similar; and applying the experience, insight and data gained from Apollo 13 to the Failure Mode and Effects Analyses and Single Failure Point Analyses being performed on all Skylab flight hardware.

QUESTION — Do you believe that NASA can carry out its currently planned fiscal year 1971 programs including costs of the Apollo 13 accident within your original budget request to the Congress?

ANSWER — As noted above, we now plan to handle the estimated $10 to $15 million of increased costs within our total Apollo budget.

QUESTION — To what extent are other systems in the Apollo vehicle and spacecraft liable to a similar sequence of events leading to the Apollo 13 accident?

ANSWER — We have now instituted a review of all oxidizer systems in all elements of the Apollo system to be sure, in the light of what we have learned in Apollo 13, that materials and energy sources are compatible in these systems, and modifications will be made where appropriate. For example, the fuel cell oxygen supply valve which now has Teflon-insulated wires in high pressure oxygen will be redesigned to eliminate this hazard.

I am enclosing (TAB C) for your information a statement which I am presenting to the Senate Committee on Aeronautical and Space Sciences at a hearing this morning which discusses these actions in greater detail the actions we plan to take in response to the Board's recommendations. I have the utmost confidence that the NASA team can fix the Apollo 13 problem and strengthen its operations to minimize the chances of future problems. We will keep you and the Committee informed of developments.

MANUFACTURING AND TEST

QUESTION — Did the manufacture, qualification and testing of the Service Module oxygen system conform to best practices at the time of its development?

ANSWER — The design was difficult to manufacture, but good practices were followed to help insure against manufacturing defects. Good testing procedures were followed, but the tests did not include a test of the thermostatic switches functioning under load.

QUESTION — Were the latest improvements in manufacture and test incorporated in the manufacture of the Service Module Oxygen system during the progress of the program?

ANSWER — Many improvements were incorporated in the manufacture of the oxygen tanks during the progress of the program. These included the use of special tools, jigs, and fixtures; improved assembly, cleaning and inspection procedures, and more thorough and improved testing and checkout operations.

QUESTION — Could the problem of the thermal switches have been anticipated and corrected in the original testing and manufacture of the oxygen tanks?

ANSWER — If the design qualification or the flight unit acceptance testing of the oxygen tanks had included a functional test of the thermostatic switch interrupting the 65 volt DC, 6 amp ground power load, the potential problem could have been uncovered and corrected.

QUESTION — The launch crews handling the oxygen system tests prior to launch of Apollo 13 were unaware of the potential problem of the thermal switches in the oxygen tank. Was documentation and expert support personnel from industry and NASA available to diagnose this problem?

ANSWER — Although adequate documentation and expert personnel necessary to uncover the potential thermostatic switch problem were not available at KSC, they did not exist among MSC, North American Rockwell and Beech Aircraft. However, the switch problem probably could not have been readily uncovered as demonstrated by the fact that it took considerable time and effort of many people to uncover after the flight, when it was not just a potential problem.

MISSION ANALYSIS

QUESTION — Had the Apollo 13 accident occurred in other portions of the flight do you believe it would have been possible to have recovered the astronauts?

ANSWER — The Board did not review in detail the possible consequences of SM oxygen system failure at other times during the mission. Launch pad abort procedures and the launch escape system are designed to cope with emergencies on the pad or during the early portion of boost. Obviously, recovery would have been earlier and more simple had the accident occurred in earth orbit. Once the LM, separated from the CSM in lunar orbit, recovery would have been more difficult, and, in some cases, perhaps impossible. However, as pointed out in our testimony, the possibilities of recovery would have depended on the actions which could be taken under the precise circumstances involved.

QUESTION — Based on the outstanding performance of the astronauts, ground controllers and supporting personnel do you believe that new or changed procedures equipment or techniques should be provided to improve the probability of recovery in the event of an accident?

ANSWER — Recommendations 3 and 4 of the Board's report recommend that the Manned Spacecraft Center (MSC) consider several specific changes in equipment and operating procedures to improve the possibility of recovery in the event of an accident, and that consumables and equipment in the LM and CM be reviewed to determine if their potential utility in the "LM-lifeboat" mode should be enhanced. Certain tradeoffs must be considered with regard to these recommendations since the addition of further redundancy or complexity might reduce the probability of mission success and crew safety.

QUESTION — Are we taking advantage of our new extensive operational experience to assure maximum safety for the astronauts both in terms of survival equipment and procedures?

ANSWER — We learned a great deal from the Apollo 13 accident regarding the ability of the spacecraft, Mission Control and the crew to function under extremely adverse conditions. The knowledge gained from this experience is being used to enhance and improve simulation and training methods to better prepare future crews for dealing with emergencies, and the Board has recommended review of equipment and procedures in light of this experience.

MANAGEMENT

QUESTION — In the sequence of design, manufacture and test were procedures for quality assurance and reliability fully complied with by all levels of contractor and NASA management?

ANSWER — The review of the Board and its Panels of the oxygen tank system indicated that the procedures for quality assurance and reliability were fully complied with.

QUESTION — Where modifications were required to the Service Module oxygen system, was management visibility within NASA and the contractors sufficient to understand potential problems areas?

ANSWER — Change control procedures were in effect and followed in the course of design, manufacture and test of the oxygen tank system. Visibility was afforded to appropriate levels of management during the course of the work. As the Report of the Review Board states, less detailed procedures were in effect in the early history of the oxygen tank system than are now in effect. The Board further concluded that attention in the design of the system was primarily devoted to its thermodynamic performance, with relatively less attention given to other design details.

QUESTION — Are the management procedures currently in effect sufficient to provide NASA and contractor management adequate information to preclude similar occurrences on future flights?

ANSWER — The management procedures in effect provide a great deal of information and our review indicated that the procedures were followed. Essentially all the information which the Board used in tracing the history of the oxygen tank system was available in the records of NASA or its contractors. We found that there are extensive documentation and procedural controls in effect and it was not obvious to us that major additional procedures are necessary to add to the information that is available.

It should be noted, however, that the Board recommended a reassessment of subsystems to insure adequate understanding and control of the details of the subsystems at the subcontractor and vendor level. The Board also believes that some specific procedural improvements are warranted and made recommendations on those points.

DESIGN

QUESTION — Was the basic design of the oxygen system of the Service Module sound in concept?

ANSWER — The basic design of the oxygen system of the Service Module is considered sound in concept, and no changes in basic system design have been recommended. The detailed design of the interior components of the oxygen tank included a number of deficiencies which are identified in the Board Report and Appendices. The design of these components should be modified, and this redesign is underway.

QUESTION — At the time of design of the oxygen system in the Service Module in 1965-1966 were all of the relevant factors of design known at the time taken into consideration?

ANSWER — The oxygen tank was originally designed in the 1962 to 1963 time period. This was designated the Block I system. In 1965-1966, slight modifications were made — primarily to enhance reliability. This modified system was designated Block 2. The principal change from Block I was the provision of independent circuits for each of the fan motors and heater elements, thus providing functional redundancy for each of these motors.

In general, the relevant factors of design representing the state-of-the-art at the time were incorporated in the design. To cite a few examples:

(1) The material of the pressure vessel is most suitable for this service.

(2) Storing the oxygen in the supercritical state was appropriate. By maintaining the oxygen in this single phase high density state, withdrawing the oxygen for use in simplified, high storage efficiency is obtained, and slosh during acceleration is avoided.

(3) Providing a means for mixing or stirring the fluid was required to assure a homogenous fluid. This avoided the uncertainties associated with the then imperfectly understood behavior of fluids under zero-G conditions.

On the other hand, the factor variously termed manufacture-ability or producability was not taken into account appropriately. This factor includes such considerations as inspectability and testability. It is difficult to install the internal components of the tank system, part of the procedure being "blind." 'This process is conducive to wire damage that can go undetected without visual inspection. Such inspection is not possible with this configuration.

Thus, in this respect it may be said that all of the factors of the design were not taken into account appropriately.

QUESTION — Did NASA at the time of design of the oxygen tank system have a definite procedure for updating the equipment as new knowledge was gained through operation and tests?

ANSWER — Yes, the management procedures in use in Apollo did provide for updating designs as required.

QUESTION — Is it necessary to completely redesign the Service Module oxygen system or can changes be made which will eliminate potential causes of the Apollo 13 accident?

ANSWER — No, a complete redesign is not necessary. Changes to the internal components of the oxygen tank and the fuel cell shut-off valves have been recommended and work is proceeding on these changes.

QUESTION — Are other oxygen tanks within the Apollo vehicle and spacecraft subject to the same problem?

ANSWER — Each remaining Service Module presently includes tanks identical to oxygen tank number 2 in Apollo 13. These will be modified. No other oxygen tanks in the Apollo spacecraft are closely similar to these tanks. The Board recommended that all high pressure oxygen systems in the spacecraft be reexamined.

Mr. KARTH — Had the accident occurred at any other time during the mission, when would it have been unrecoverable?

Mr. CORTRIGHT — After separation from the lunar module for one I will ask Mr. Armstrong to answer that question.

Mr. ARMSTRONG — I think in general that answer is probably sufficient as it stands. I have found in these kinds of situations that people when pressed can usually come up with some effective survival procedures which are completely nonstandard and would be unacceptable before the fact, but when they are the only last ditch effort, that you find, in fact, they will work, and we really don't know how long people will live. We are talking about running out of consumable oxygen, coolants, and, in order to say how long one might live in those conditions, you have to predict physiological factors of individuals and when and how long in a high CO_2 atmosphere they might exist, we don't have good data. There might be some cases where they might, survive, but to predict their survival would be difficult. I would say his answer as it stood from the point of view of rigor is correct in itself.

Mr. KARTH — Are there any further questions?

Mr. Fulton.

Mr. FULTON — This brings up the question that this was actually the same equipment that was operative in both Apollo 11 and 12, was it not ?

Mr. CORTRIGHT — Yes, Sir.

Mr. FULTON — But if something else had happened, it would seem to me that the equipment would have operated all right. That would pretty well eliminate the equipment as an individual inducing cause on present without something else having occurred. Could I ask Mr. Armstrong to comment?

Mr. ARMSTRONG — Yes, Sir; Mr. Fulton. As you know, we spend a great deal of our time in the preparation for emergencies in our training and in our thought processes, planning for these flights.
In the case of crew members, certainly about 75 percent of their time is involved in planning for these emergency situations, so we are not at all surprised when they occur; as a matter of fact, we are probably surprised that so few of them occur in our real flights.

Mr. FULTON — I am ready to go on any trip. Please note.

Mr. KARTH — We are ready to send him too.

Mr. FULTON — I have had one person recommend a one-way trip.

Mr. ARMSTRONG — The problem occurs when you have a combination of circumstances, and that is the situation which existed here. This supersedes our ability to actually, substantially and correctly react and predict those kinds of combinations of failure circumstances.

Mr. FULTON — May I commend Mr. Paine, the Administrator, on his good comments on the safety panel, and may I ask that the accomplishments of the safety panel be put in the record at this point?

Mr. KARTH — No objection.

(Information requested for the record follows:)

The Panel reinforces the continuing attention of NASA and its contractors to risk assessment and the formalization of the hazard identification and control process. Given the dynamics of the development process, the multitude of design and operational decisions and the broad span of technology inherent in NASA's programs it was recognized that the Panel could assess at most a very limited number of these decisions. Therefore it was mutually agreed that the Panel's effectiveness would lie in focusing on the evolution of the risk management systems and policies.

The Panel's first year was spent in a survey of the Apollo program management system and the system for hazard identification and risk assessment. This also enabled the Panel to assess the impact of agency staff activities. The Panel reviewed technical management policies and controls at the system level. Attention was focused on configuration management because of the importance of a system to define the configuration "as designed" and "as built," its test history and the waivers and deviations accepted as risks. The Panel was also particularly interested in the institutionalization of system safety given the structure of the fundamental risk management system. Because the Apollo program was in an advanced stage when the Panel was established it was difficult to evaluate the historical adequacy of the Apollo risk management system. Therefore, the Panel monitored the system as it provided an assessment of mission risks. The Panel gave specific attention to the processes for reevaluation of possible worst case failure modes and definition of the safety factors in life support systems and consumables.

The Panel's review of the Apollo program involved staff and program elements at NASA Headquarters, the manned space flight centers, and the majority of principal contractors for the spacecraft, launch vehicles and Apollo mission support. The Panel met in session twenty-two days. While the Panel had not studied any area sufficiently to evaluate it in depth, the technical management background of the members permitted them to comment selectively on the described systems.

The Panel has recently completed an assessment of the management process for the evaluation of risks inherent in reducing Saturn static testing and launch operations, as well as a review of the investigation process involved in the LLTV/LLRV accidents.

The Panel has also been asked by the Administrator to review the hazard identification and risk assessment system on the NERVA/nuclear stage, the space station and space shuttle. Involvement in the definition phases of program development promises increasing effectiveness for the Panel as the programs mature.

Currently the Panel is involved in an assessment of the procedures, and the findings, determinations and recommendations of the Apollo 13 Review Board.

Mr. FULTON — I would like to commend Mr. Rumsfeld and those of us who put in the bill the recommendations of the Safety Panel. I would like, along the lines that have been discussed here, some further management inquiry and a report to be made on how we can get better inspection procedures so that it is independent, so that somebody outside the line of either production, or the management, or the program director, the Center or the launch area, can be appealed to by anyone who feels that something should be looked into further. At the present time, I feel that there is the pressure to get the job done, and it is too much to expect any particular individual to step clear out of line if he has some ideas.

If I could make that suggestion, I would hope that further safety procedures will be looked into on the management level. I think NASA is doing a fine job, and I am pleased this committee has met this morning, looking into this excellent report.

Mr. KARTH — On that note, I think, Dr. Paine, it will be necessary for us to conclude the hearings today, and I want to thank you again, Mr. Cortright, and members of the review board for preparing this report. If the chairman feels it is necessary to go further, I am sure they will be in touch. Thank you.

Whereupon, at 12:20 p.m., the committee was adjourned.

APPENDIX

REPORT OF APOLLO 13 REVIEW BOARD

NATIONAL AERONAUTICS AND SPACE ADMINISTRATION

June 15, 1970

The Honorable Thomas O. Paine. Administrator
National Aeronautics and Space Administration Washington, D.C. 20546

Dear Dr. Paine:

Pursuant to your directives of April 17 and April 21, 1970, I am transmitting the final Report of the Apollo 13 Review Board. Concurrent with this transmittal, I have recessed the Board, subject to call.

We plan to reconvene later this year when most of the remaining special tests have been completed, in order to review the results of these tests to determine whether any modifications to our findings, determinations or recommendations are necessary. In addition, we will stand ready to reconvene at your request.

Sincerely yours,

Edgar M. Cortright
Chairman

PREFACE

The Apollo 13 accident, which aborted man's third mission to explore the surface of the Moon, is a harsh reminder of the immens difficulty of this undertaking.

The total Apollo system of ground complexes, launch vehicle, and spacecraft constitutes the most ambitious and demanding engineerin development ever undertaken by man. For these missions to succeed, both men and equipment must perform to near perfection. Tha this system has already resulted in two successful lunar surface explorations is a tribute to those men and women who conceive designed, built, and flew it.

Perfection is not only difficult to achieve, but difficult to maintain. The imperfection in Apollo 13 constituted a near disaster, averted on by outstanding performance on the part of the crew and the ground control team which supported them.

The Apollo 13 Review Board as charged with the responsibilities of reviewing the circumstances surrounding the accident, of establishin the probable causes of the accident, of assessing the effectiveness of flight recovery actions, of reporting these findings, and of developir recommendations for corrective or other actions. The Board has made every effort to carry out its assignment in a thorough, objectiv and impartial manner. In doing so, the Board made effective use of the failure analyses and corrective action studies carried out by th Manned Spacecraft Center and was very impressed with the dedication and objectivity of this effort.

The Board feels that the nature of the Apollo 13 equipment failure holds important lessons which, when applied to future missions, w contribute to the safety and effectiveness of manned space flight.

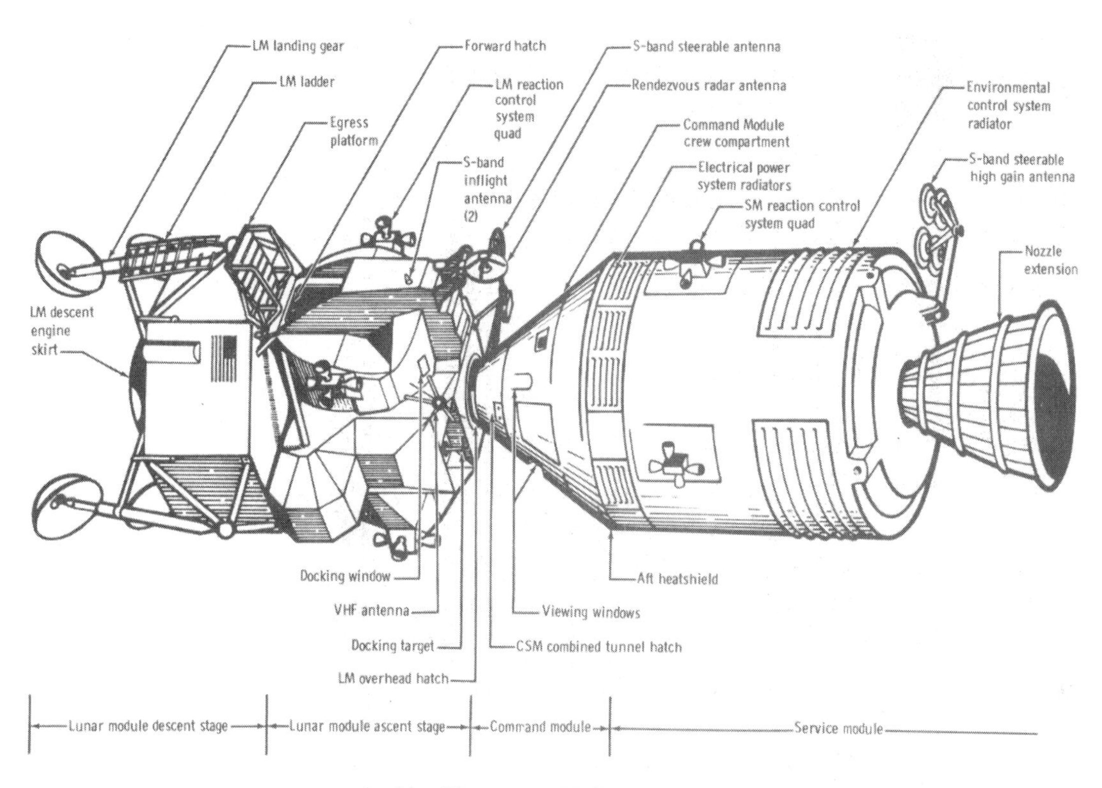

Apollo 13 space vehicle configuration.

CSM in ground test with bay 4 panel removed.

Inflight photograph of service module showing damage to bay 4.

CHAPTER I
AUTHORITIES

NATIONAL AERONAUTICS AND SPACE ADMINISTRATION, WASHINGTON D.C. 20546
OFFICE OF THE ADMINISTRATOR April 17, 1970

To: Mr. Edgar M. Cortright
SUBJECT Establishment of Apollo 13 Review Board

REFERENCES: (a) NMI 8621.1 - Mission Failure Investigation Policy and Procedures
(b) NMI 1156.14 - Aerospace Safety Advisory Panel

1. It is NASA policy " stated in Reference (a) "to investigate and document the causes of all major mission failures which occur in the conduct of its space and aeronautical activities and to take appropriate corrective actions as a result of the findings and recommendations."

2. Because of the serious nature of the accident of the Apollo 13 spacecraft which jeopardized human life and caused failure of the Apollo 13 lunar mission, we hereby establish the Apollo 13 Review Board (hereinafter referred to as the Board) and appoint you Chairman The members of the Board will be qualified senior individuals from NASA and other Government agencies. After consultation with you we will:

(a) Appoint the members of the Board and make any subsequent changes necessary for the effective operation of the Board and

(b) Arrange for timely release of information on the operations, findings, and recommendations of the Board to the Congress and, through the NASA Office of Public Affairs, to the public. The Board will report its findings and recommendations directly to us.

3. The Board will:

(a) Review the circumstances surrounding the accident to the spacecraft which occurred during the flight of Apollo 13 and the subsequent flight and ground actions taken to recover, in order to establish the probable cause or causes of the accident and assess the effectiveness of the recovery actions.

(b) Review all factors relating to the accident and recovery actions the Board determines to be significant and relevant including studies, findings, recommendations, and other actions that have been or may be undertaken by the program offices field centers, and contractors involved.

(c) Direct such further specific investigations as may be necessary.

(d) Report as soon as possible its findings relating to the cause or causes of the accident and the effectiveness of the flight and ground recovery actions.

(e) Develop recommendations for corrective or other actions, based upon its findings and determinations or conclusions derived therefrom.

(f) Document its findings, determinations, and recommendations and submit a final report.

4. As Chairman of the Board you are delegated the following powers:

(a) To establish such procedures for the organization and operation of the Board as you find most effective; such procedures shall be part of the Board's records. The procedures shall be furnished the Aerospace Safety Advisory Panel for its review and comment.

(b) To establish procedures to assure the execution of your responsibilities in your absence.

(c) To designate such representatives, consultants, experts, liaison officers, observers, or other individuals as required to support the activities of the Board. You shall define their duties and responsibilities as part of the Board's records.

(d) To keep us advised periodically concerning the organization, procedures, operations of the Board and its associated activities.

5. By separate action we are requesting the Aerospace Safety Advisory Panel established by Reference (b) to review both the procedures and findings of the Board and submit its independent report to us.

6. By separate action we are directing the Associate Administrator for Manned Space Flight to:

(a) Assure that all elements of the Office of Manned Space Flight cooperate fully with the Board and provide records, data and technical support as requested.

(b) Undertake through the regular OMSF organization such reviews, studies, and supporting actions as are required to develop recommendations to us on corrective measures to be taken prior to the Apollo 14 mission with respect to hardware, operational procedures, and other aspects of the Apollo program.

7. All elements of NASA will cooperate with the Board and provide full support within their areas of responsibility.

George M. Low *T. O. Paine*
Deputy Administrator Administrator

NATIONAL AERONAUTICS AND SPACE ADMINISTRATION, WASHINGTON D.C. 20546
OFFICE OF THE ADMINISTRATOR April 21, 1970

TO: Mr. Edgar M. Cortright

SUBJECT: Membership of Apollo 13 Review Board

Reference: Memorandum to you of April 17, subject: Establishment of Apollo 13 Review Board

n accordance with paragraph 2(a) of Reference (a), the membership of the Apollo 13 Review Board is established as follows:

lembers:
Ir. Edgar M. Cortright, Chairman (Director, Langley Research Center) Mr. Robert P. Allnutt (Assistant to the Administrator NASA Hqs.) Ir. Neil Armstrong (Astronaut, Manned Spacecraft Center) Dr. John F. Clark (Director, Goddard Space Flight Center) Brig. General Valter R. Hedrick, Jr. (Director of Space, DCS/R&D, Hqs., USAF) Mr. Vincent L. Johnson (Deputy Associate Administrator-Engineering, Office of Space Science and Applications)

Ir. Milton Klein (Manager, AEC-NASA Space Nuclear Propulsion Office) Dr. Hans M. Mark (Director, Ames Research Center)

Counsel:
Mr. George Malley (Chief Counsel, Langley Research Center)

OMSF Technical Support:
Mr. Charles W. Mathews (Deputy Associate Administrator, Office of Manned Space Flight)

Observers:
Mr. William A. Anders (Executive Secretary, National Aeronautics and Space Council)
Dr. Charles D. Harrington (Chairman, NASA Aerospace Safety Advisory Panel)
Mr. I. I. Pinkel (Director, Aerospace Safety Research and Data Institute, Lewis Research Center)

Congressional Liaison:
Mr. Gerald J. Mossinghoff (Office of Legislative Affairs, NASA Hqs.)

ublic Affairs Liaison;
Mr. Brian Doff (Public Affairs Officer, Manned Spacecraft Center)

n accordance with applicable NASA instruction, you are authorized to appoint such experts and additional consultants as are required or the effective operations of the Board.

George M. Low — Deputy Administrator
T. O. Paine — Administrator

NATIONAL AERONAUTICS AND SPACE ADMINISTRATION, WASHINGTON, D.C. 20546
OFFICE OF THE ADMINISTRATOR April 20, 1970

TO: Dr. Charles D. Harrington, Chairman, Aerospace Safety Advisory Panel

SUBJECT: Review of Procedures and Findings of Apollo 13 Review Board

Attachment: (a) Memorandum dated April 17, 1970, to Mr. Edgar M. Cortright, subject: Establishment of Apollo 13 Review Board

References: (a) Section 6, National Aeronautics and Space Administration Authorization Act, 1968

(b) NMI 1156.14 - Aerospace Safety Advisory Panel

. In accordance with References (a) and (b), the Aerospace Safety Advisory Panel (hereafter referred to as the Panel) is requested to eview the procedures and findings of the Apollo 13 Review Board (hereafter referred to as the Board) established by Attachment (a).

The procedures established by the Board will be made available to the Panel for review and content as provided in paragraph 4(a) of ttachment (a).

As Chairman of the Panel, you are designated an Observer on the Board. In this capacity, you, or another member of the Panel esignated by you, are authorized to be present at those regular meetings of the Board you desire to attend. You are also authorized to eceive oral progress reports from the Chairman of the Board or his designee from time to time to enable you to keep the Panel fully formed on the work of the Board.

The final report and any interim reports of the Board will be made available promptly to the Panel for its review.

The Panel is requested to report to us on the procedures and findings of the Board at such times and in such form as you consider ppropriate, but no later than 10 days after the submission to us of the final report of the Board.

George M. Low — Deputy Administrator
T. O. Paine — Administrator

nclosure

: Mr. Edgar M. Cortright, Chairman, Apollo 13 Review Board
M/Mr. Dale Myers

NATIONAL AERONAUTICS AND SPACE ADMINISTRATION, WASHINGTON D.C 20546
OFFICE OF THE ADMINISTRATOR April 20, 1970

TO: Mr. Dale D. Myers Associate Administrator for Manned Space Flight

SUBJECT: Apollo 13 Review

References: (a) Memorandum dated April 17, 1970, to Mr. Edgar M. Cortright, subject: Establishment of Apollo 13 Review Board

(b) Memorandum dated April 20, 1970, to Dr. Charles D. Harrington, subject: Review of Procedures and Findings of Apollo 13 Review Board

1. As indicated in paragraph 6 of Reference (a), you are directed to;

(a) Assure that all elements of the Office of Manned Space Flight cooperate fully with the Board in providing records, data and technical support as requested.

(b) Undertake through the regular OMSF organization such review , studies, and supporting actions as are required to develop timely recommendations to us on corrective measures to be taken prior to the Apollo 14 mission with respect t hardware, operational procedures, flight crews, and other aspects of the Apollo program.

2. The recommendations referred to in paragraph l(b) above should be submitted to us in such form and at such time as you deer appropriate, but a report should be submitted no later than ten days after the Apollo 13 Review Board submits its final report.

3. The assignments to the Apollo 13 Review Board and to the space Safety Advisory Panel by References (a) and (b), respectively, in n way relieve you of your continuing full responsibility for the conduct of the Apollo and other OMSF programs .

George M. Low
Deputy Administrator

T. Paine
Administrator

cc: Mr. Edgar M. Cortright, Chairman, Apollo 13 Review Board
Mr. Charles D. Harrington, Chairman , Aerospace Safety Advisory Panel

---

Management Instruction

NMI 8621.1 April 14 1966

SUBJECT: MISSION FAILURE INVESTIGATION POLICY AND PROCEDURES

1. PURPOSE

This Instruction establishes the policy and procedures for investigating and documenting the causes of all major mission failures whic occur in the conduct of NASA space and aeronautical activities.

2. APPLICABILITY

This Instruction is applicable to NASA Headquarters and field installations.

3. DEFINITION

For the purpose of this Instruction, the following term shall apply:

In general, a failure is defined as not achieving a major mission objective.

4. POLICY

a). It is NASA policy to investigate and document the causes of all major mission failures which occur in the conduct of i space and aeronautical activities and to take appropriate corrective actions as a result of the findings and recommendatio

b). The Deputy Administrator may conduct independent investigations of major failures in addition to those investigatio required of the Officials-in-Charge of headquarters Program Offices as set forth in paragraph 5a.

5. PROCEDURES

a). Officials-in-Charge of Headquarters Program Offices are responsible, within their assigned areas, for:

(1) Informing promptly the Deputy Administrator of each major failure and apprising him of the nature of the failur status of investigations, and corrective or other actions which are or will be taken.

(2) Determining the causes or probable causes of all failures, taking corrective or other actions, and submitti written reports of such determinations and actions to the Deputy Administrator.

b). When the Deputy Administrator decides to conduct an independent investigation, he will:

(1) Establish a (name of project) Review Board, comprised of appropriate NASA officials;

(2) Define the specific responsibilities of each Board, encompassing such tasks as:

(a) Reviewing the findings, determinations and corrective or other actions which have been developed by contractors, field installations and the Official-in-Charge of cognizant Headquarters Program Office and presenting the Board's conclusions as to their adequacy to the Deputy Administrator.

(b) Reviewing the findings during the course of investigations with cognizant field installation and Headquarters officials.

(c) Recommending such additional steps (for example additional tests) as are considered desirable, to determine the technical and operational causes or probable causes of failure, and to obtain evidence of non-technical contributing factors.

(d) Developing recommendations for corrective and other actions, based on all information available to the Board.

(e) Documenting findings, determination and recommendations for corrective or other actions and submitting such documentation to the Deputy Administrator.

c). Procedures for implementing the Board's recommendations shall be determined by the Deputy Administrator.

6. CANCELLATION

NASA Management Manual Instruction 4-1-7 (T.S. 760). March 24, 1964.

Deputy Administrator

Management Instruction

NMI 1156.14 December 7, 1967

SUBJECT: AEROSPACE SAFETY ADVISORY PANEL

1. PURPOSE

This Instruction sets forth the authority for, and the duties, procedures, organization, and support of the Aerospace Safety Advisory Panel.

2. AUTHORITY

The Aerospace Safety Advisory Panel (hereafter called the "Panel") was established under Section 6 of the National Aeronautics and Space Administration Authorization Act, 1968 (PL 90-67, 90th Congress, 81 Stat. 168, 170). Since the Panel was established by statute, its formation and use are not subject to the provisions of Executive Order 11007 or of NMI 1150.2, except to the extent that such provisions are made applicable to the Panel under this Instruction.

3. DUTIES

a). The duties of the Panel are set forth in Section 6 of the National Aeronautics and Space Administration Authorization Act, 1968, as follows:

"The Panel shall review safety studies and operations plans referred to it and shall make reports thereon, shall advise the Administrator with respect to the hazards of proposed or existing facilities and proposed operations and with respect to the adequacy of proposed or existing safety standards, and shall perform such other duties as the Administrator may request."

b). Pursuant to carrying out its statutory duties, the Panel will review, evaluate, and advise on all elements of NASA's safety system, including especially the industrial safety, systems safety, and public safety activities, and the management of these activities. These key elements of NASA's safety system are identified and delineated as follows:

(1) Industrial Safety. This element includes those activities which, on a continuing basis, provide protection for the well being of personnel and prevention of damage to property involved in NASA's business and exposed to potential hazards associated with carrying out this business. Industrial safety relates especially to the operation of facilities in the many programs of research, development, manufacture, test, operation, and maintenance. Industrial safety activities include, but are not limited to, such functions as:

(a) Determination of industrial safety criteria.

(b) Establishment and implementation of safety standards and procedures for operation and maintenance of facilities, especially test and hazardous environment facilities.

(c) Development of safety requirements for the design of new facilities.

(d) Establishment and implementation of safety standards and procedures for operation of program support and administrative aircraft.

(2) Systems Safety. This element includes those activities specifically organized to deal with the potential hazards of comple R&D systems that involve many highly specialized areas of technology. It places particular emphasis on achieving saf operation of these systems over their life cycles, and it covers major systems for aeronautical and space flight activities manned or unmanned, including associated ground based research, development, manufacturing, and test activities. System safety activities include, but are not limited to, such functions as:

(a) Determination of systems safety criteria, including criteria for crew safety.

(b) Determination of safety data requirements.

(c) Performance of systems safety analyses.

(d) Establishment and implementation of systems safety plans.

(3) Public Safety. This element includes those activities which, on a continuing basis, provide protection for the well being c people and prevention of damage to property not involved in NASA's business, but which may nevertheless be exposed t potential hazards associated with carrying out this business. Public safety activities include, but are not limited to, suc functions as:

(a) Determination of public safety criteria.

(b) Establishment and control of public safety hazards associated with facility and systems tests and operations.

(c) Establishment and implementation, as required, of emergency or catastrophe control plans.

(4) Safety Management. This element includes both the program and functional organizations of NASA and its contractor involved in the identification of potential hazards and their elimination or control as set forth in the foregoing description c safety activities. It also includes the management systems for planning, implementing, coordinating, and controlling thes activities. These management systems include, but are not limited to, the following:

(a) The authorities, responsibilities, and working relationships of the organizations involved in safety activities, and th assessment of their effectiveness.

(b) The procedures for insuring the currency and continuity of safety activities, especially systems safety activities whic may extend over long periods of time and where management responsibilities are transferred during the life cycles of th systems.

(c) The plans and procedures for accident/incident investigations, including those for the follow-up on corrective action and the feedback of accident/incident information to other involved or interested organizations.

(d) The analysis and dissemination of safety data.

4. PROCEDURES

(a) The Panel will function in an advisory capacity to the Administrator, and, through him, to those organizational elements responsible for management of the NASA safety activities.
(b) The Panel will be provided with all information required to discharge its advisory responsibilities as they pertain t both NASA and its contractors' safety activities. This information will be made available through the mechanism c appropriate reports, and by means of in situ reviews of safety activities at the various NASA and contractor sites, a deemed necessary by the Panel and arranged through the Administrator. The Panel will thus be enabled to examine an evaluate not only the general status of the NASA safety system, but also the key elements of the planned and ongoin activities in this system.

5. ORGANIZATION

(a) Membership

(1) The Panel will consist of a maximum of nine members, who will be appointed by the Administrator. Appointments will be f a term of six years, except that, in order to provide continuity of membership, one-third of the members appointed originally t the Panel will be appointed for a term of two years, one-third for a term of four years, and one-third for a term of six years.

(2) Not more than four members of the Panel shall be employees of NASA, nor shall such NASA members constitute a majori of the composition of the Panel at any given time.

(3) Compensation and travel allowances for Panel members shall be as specified in Section 6 of the NASA Authorization Act, 196

(b) Officers

(1) The Officers of the Panel shall be a Chairman and a Vice Chairman, who shall be selected by the Panel from their membersh to serve for one-year terms.

(2) The Chairman, or Vice Chairman in his absence, shall preside at all meetings of the Panel and shall have the usual powers of presiding officer.

(c) Committees

(1) The Panel is authorized to establish special committees, as necessary and as approved by the Administrator, to carry o specified tasks within the scope of duties of the Panel.

(2) All such committee activities will be considered an inseparable extension of Panel activities, and will be in accordance with applicable procedures and regulations set forth in this Instruction.

(3) The Chairman of each special committee shall be a member of the Aerospace Safety Advisory Panel. The other committ

members may or may not be members of the Panel, as recommended by the Panel and approved by the Administrator.

(4) Appointment of Panel members to committees as officers or members will be either for one year, for the duration of their term as Panel members, or for the lifetime of the committee, whichever is the shortest. Appointments of non-Panel members to committees will be for a period of one year or for the lifetime of the committee, whichever is shorter.

(5) Compensation and travel allowances for committee members who are not members of the Panel shall be the same as for members of the Panel itself, except that compensation for such committee members appointed from outside the Federal Government shall be at the rate prescribed by the Administrator for comparable services.

(d) Meetings

(1) Regular meetings of the Panel will be held as often as necessary and at least twice a year. One meeting each year shall be an Annual Meeting. Business conducted at this meeting will include selecting the Chairman and the Vice Chairman of the Panel, recommending new committees and committee members as required or desired, approving the Panel's annual report to the Administrator, and such other business as may be required.

(2) Special meetings of the Panel may be called by the Chairman, by notice served personally upon or by mail or telegraph to the usual address of each member at least five days prior to the meeting.

(3) Special meetings shall be called in the same manner by the Chairman, upon the written request of three members of the Panel.

(4) If practicable, the object of a special meeting should be sent in writing to all members, and if possible a special meeting should be avoided by obtaining the views of members by mail or otherwise, both on the question requiring the meeting and on the question of calling a special meeting.

(5) All meetings of special committees will be called by their respective chairman pursuant to and in accordance with performing their specified tasks.

(6) Minutes of all meetings of the Panel, and of special committees established by the Panel, will be kept. Such minutes shall, at a minimum, contain a record of persons present, a description of matters discussed and conclusions reached, and copies of all reports received, issued, or approved by the Panel or committee. The accuracy of all minutes will be certified to by the Chairman of the Panel (or by the Vice Chairman in his absence) or of the committee.

(e) Reports and Records

(1) The Panel shall submit an annual report to the Administrator.

(2) The Panel will submit to the Administrator reports on all safety reviews and evaluations with comments and recommendations as deemed appropriate by the Panel.

(3) All records and files of the Panel, including agendas, minutes of Panel and committee meetings, studies, analyses, reports, or other data compilations or work papers, made available to or prepared by or for the Panel, will be retained by the Panel.

(f) Avoidance of Conflicts of Interest

(1) Non-governmental members of the Panel, and of special committees established by the Panel, are "Special Government Employees" within the meaning of NHB 1900.2A, which sets forth guidance to NASA Special Government Employees regarding the avoidance of conflicts of interest and the observance of ethical standards of conduct. A copy of NHB 1900.2A and related NASA instructions on conflicts of interest will be furnished to each Panel or committee member at the time of his appointment as a NASA consultant or expert.

(2) Non-governmental members of the Panel or a special committee will submit a "NASA Special Government Employees Confidential Statement of Employment and Financial Interests" (NASA Form 1271) prior to participating in the activities of the Panel or a special committee.

. SUPPORT

a). A staff, to be comprised of full-time NASA employees, shall be established to support the Panel. The members of this staff will be fully responsive to direction from the Chairman of the Panel.

b). The director of this staff will serve as Executive Secretary to the Panel. The Executive Secretary of the Panel, in accordance with the specific instructions from the Chairman of the Panel, shall:

(1) Administer the affairs of the Panel and have general supervision of all arrangements for safety reviews and evaluations, and other matters undertaken by the Panel.

(2) Insure that a written record is kept of all transactions, and submit the same to the Panel for approval at each subsequent meeting.

(3) Insure that the same service is provided for all special committees of the Panel.

James E. Webb - Administrator

CHAPTER 2

BOARD HISTORY AND PROCEDURES

PART 1. SUMMARY OF BOARD HISTORY AND PROCEDURES

The Apollo 13 Review Board was established on April 17, 1970, by the NASA Administrator and Deputy Administrator under the authority of NASA Management Instruction 8621.1, dated April 14, 1966. In the letter establishing the Board, Mr. Edgar M. Cortright, Director of Langley Research Center, was appointed as Chairman and the general responsibilities of the Board were set forth. The seven additional members of the Board were named in a letter from the Administrator and the Deputy Administrator to the Chairman, dated April 21, 1970. This letter also designated a Manned Space Flight Technical Support official, a Counsel to the Board, several other supporting officials, and several observers from various organizations. In addition, in a letter dated April 20, 1970, to Dr. Charles D. Harrington, Chairman of the NASA Aerospace Safety Advisory Panel, that Panel was requested to review the Board's procedures and findings.

The Review Board convened at the Manned Spacecraft Center, Houston, Texas, on Tuesday, April 21, 1970. Four Panels of the Board were formed, each under the overview of a member of the Board. Each of the Panels was chaired by a senior official experienced in the area of review assigned to the Panel. In addition, each Panel was manned by a number of specialists, thereby providing a nucleus of expertise for the review activity. During the period of the Board's review activities, the Chairmen of the four Panels were responsible for the conduct of evaluations, analyses, and other studies bearing on their Panel assignments, for preparing preliminary findings and recommendations, and for developing other information for the Board's consideration. To overview these Panel efforts, each member of the Board assumed specific responsibilities related to the overall review.

In addition to the direct participants in the Board activity, a number of observers and consultants also attended various meetings of the Board or its constituent Panels. These individuals assisted the Review Board participants with advice and counsel in their areas of expertise and responsibilities.

While the Board's intensive review activities were underway, the Manned Spacecraft Center Apollo 13 Investigation Team, under James A. McDivitt, Colonel, USAF, was also conducting its own analysis of the accident on Apollo 13. Coordination between the Investigation Team work and the Apollo 13 Review Board activities was effected through the MSF Technical Support official and by maintaining a close and continuing working relationship between the Panel Chairmen and officials of the MSC Investigation Team.

The Board Chairman established a series of administrative procedures to guide the Board's activities. In addition, specific assignments of responsibility were made to all individuals involved in the Board's activities so as to insure an efficient review activity. Overall logistic and administrative support was provided by MSC.

The Board conducted both Executive and General Sessions. During the Executive Sessions, plans were agreed upon for guiding the Board's activities and for establishing priorities for tests, analyses, studies, and other Board efforts. At the General Sessions, status of Panel activities was reviewed by the Board with a view towards coordination and integration of all review activities. In addition, Board members regularly attended daily status meetings of the Manned Spacecraft Center Investigation Team.

In general, the Board relied on Manned Spacecraft Center post-mission evaluation activities to provide the factual data upon which evaluation, assessment, and analysis efforts could be based. However, the Board, through a regular procedure also levied specific data collection, reduction, and analysis requirements on MSC. Test support for the Board was conducted primarily at MSC but also included tests run at other NASA Centers. Members of the Board and its Panels also visited a number of contractor facilities to review manufacturing, assembly, and test procedures applicable to the Apollo 13 mission.

The Chairman of the Board provided the NASA Deputy Administrator with oral progress reports. These reports summarized the status of Review Board activities at the time and outlined the tasks still ahead. All material used in these interim briefings was incorporated into the Board's official files.

As a means of formally transmitting its findings, determinations, and recommendations, the Board chose the format of this Final Report which includes both the Board's judgments as well as the reports of the individual Panels.

A general file of all the data and information collected and examined by the Board has been established at the Langley Research Center, Hampton, Virginia. In addition, the MSC Investigation Team established a file of data at MSC.

PART 2. BIOGRAPHIES OF BOARD MEMBERS, OBSERVERS, AND PANEL CHAIRMEN

CHAIRMAN OF THE APOLLO 13 REVIEW BOARD

EDGAR M. CORTRIGHT
NASA Langley Research Center

Edgar M. Cortright, 46, Director of the NASA Langley Research Center, Hampton, Virginia, is Chairman of the Apollo 13 Review Board.

Mr. Cortright has been an aerospace scientist and administrator for 22 years. He began his career at NASA's Lewis Research Center, Cleveland, Ohio, in 1948 and for the next 10 years specialized in research on high speed aerodynamics there.

In October 1958, Mr. Cortright was named Chief of Advanced Technology Programs at NASA Headquarters, Washington, D. C., where he directed initial formulation of NASA's Meteorological Satellite Program. In 1960, he became Assistant Director for Lunar and Planetary Programs and directed the planning and implementation of such projects as Mariner, Ranger, and Surveyor.

Mr. Cortright became Deputy Director of the Office of Space Sciences in 1961, and Deputy Associate Administrator for Space Science and Applications in 1963, in which capacities he served as General Manager of NASA's space flight program using automated spacecraft. He joined the Office of Manned Space Flight as Deputy Associate Administrator in 1967 and served in a similar capacity until he was appointed Director of the Langley Research Center in 1968.

He is a Fellow of the American Institute of Aeronautics and Astronautics and of the American Astronautical Society. He has received the Arthur S. Fleming Award, the NASA Medal for Outstanding Leadership, and the NASA Medal for Distinguished Service.

Mr. Cortright is the author of numerous technical reports and articles, and compiled and edited the book, "Exploring Space With a Camera."

He is a native of Hastings, Pennsylvania, and served as a U.S. Navy officer in World War II. He received Bachelor and Master of Science degrees in aeronautical engineering from the Rensselaer Polytechnic Institute.

Mr. and Mrs. Cortright are the parents of two children.

MEMBERS OF THE APOLLO 13 REVIEW BOARD

ROBERT F. ALLNUTT
NASA Headquarters

Robert F. Allnutt, 34, Assistant to the NASA Administrator, Washington, D. C., is a member of the Apollo 13 Review Board.

Mr. Allnutt was named to his present position this year. Prior to that, he had been Assistant Administrator for Legislative Affairs since 1967.

He joined NASA in 1960 as a patent attorney at the Langley Research Center, Hampton, Virginia. In 1961, he was transferred to NASA Headquarters, Washington, D.C.

Mr. Allnutt served as Patent Counsel for Communications Satellite Corporation from January to September 1965, when he returned to NASA Headquarters as Assistant General Counsel for Patent Matters.

He is admitted to the practice of law in the District of Columbia and the state of Virginia and is a member of the American Bar Association and the Federal Bar Association.

Mr. Allnutt was graduated from Virginia Polytechnic Institute with a B.S. degree in industrial engineering. He received Juris Doctor and Master of Laws degrees from George Washington University Law School.

Mr. and Mrs. Allnutt are the parents of two sons. The family lives in Washington, D. C.

NEIL A. ARMSTRONG
NASA Astronaut

Neil A. Armstrong, 39, NASA astronaut, is a member of the Apollo 13 Review Board.

Commander of the Apollo 11 mission and the first man on the moon, Mr. Armstrong has distinguished himself as an astronaut and an engineering test pilot.

Prior to joining the astronaut team at the Manned Spacecraft Center, Houston, Texas, in 1962, Mr. Armstrong was an X-15 rocket aircraft project pilot at the NASA Flight Research Center, Edwards, California.

Mr. Armstrong joined NASA at the Lewis Research Center, Cleveland, Ohio, in 1955, and later transferred to the Flight Research Center as aeronautical research pilot.

His initial space flight was as command pilot of Gemini VIII, launched March 16, 1966. He performed the first successful docking of two vehicles in space. The flight was terminated early due to a malfunctioning thruster, and the crew was cited for exceptional piloting skill in overcoming the problem and accomplishing a safe landing. He has served on backup crews for both Gemini and Apollo.

Mr. Armstrong is a Fellow of the Society of Experimental Test Pilots, Associate Fellow of the American Institute of Aeronautics and Astronautics, and member of the Soaring Society of America. He has received the Institute of Aerospace Sciences Octave Chanute Award, the AIAA Astronautics Award, the NASA Exceptional Service Medal, the John F. Montgomery Award, and the Presidential Medal of Freedom.

He is a native of Wapakoneta, Ohio, and received a B.S. degree in aeronautical engineering from Purdue University and a M.S. degree from the University of Southern California. He was a naval aviator from 1949 to 1952 and flew 78 combat missions during the Korean Action.

Mr. and Mrs. Armstrong have two sons.

JOHN F. CLARK
NASA Goddard Space Flight Center

Dr. John F. Clark, 49, Director of the NASA Goddard Space Flight Center, Greenbelt, Maryland, is a member of the Apollo 13 Review Board.

He is an internationally known authority on atmospheric and space sciences, holds four patents in electronic circuits and systems, and has written many scientific papers on atmospheric physics, electronics, and mathematics.

Dr. Clark joined NASA in 1958 and served in the Office of Space Flight Programs at NASA Headquarters until 1961 when he was named Director of Geophysics and Astronomy Programs, Office of Space Sciences. From 1962 until 1965, he was Director of Sciences and Chairman of the Space Science Steering Committee, Office of Space Science and Applications.

In 1965, Dr. Clark was appointed Deputy Associate Administrator for Space Science and Applications (Sciences), and later that year, Acting Director of Goddard. He was named director of the center in 1966.

Dr. Clark began his career in 1942 as an electronics engineer at the Naval Research Laboratory, Washington, D.C. From 1947 to 1948 he was Assistant Professor of Electronic Engineering at Lehigh University, Bethlehem, Pennsylvania. He returned to NRL in 1948; an prior to joining NASA, served as head of the Atmospheric Electricity Branch there.

He is a member of the American Association of Physics Teachers, American Geophysical Union, Scientific Research Society of Americ Philosophical Society of Washington, the International Scientific Radio Union, and the Visiting Committee on Physics, Lehigh Universit He received the NASA Medals for Exceptional Service, Outstanding Leadership, and Distinguished Service.

Dr. Clark was born in Reading, Pennsylvania. He received a B.S. degree in electrical engineering from Lehigh University, M.S. degree i mathematics from George Washington University, and Ph. D. in physics from the University of Maryland.

Dr. and Mrs. Clark have two children and live in Silver Springs, Maryland.

WALTER R. HEDRICK, JR.
Headquarters, USAF

Brig. Gen. Walter R. Hedrick, Jr., 48, Director of Space, Office of the Deputy Chief of Staff for Research and Development, Headquarter USAF, Washington, D.C., is a member of the Apollo 13 Review Board.

He has participated in most of the Air Force's major nuclear test projects and has extensive experience as a technical project office and administrator.

General Hedrick joined the Army Air Corps as an aviation cadet in 1941 and flew in combat with the 86th Fighter Bomber Group durin World War II. After the War, he was assigned to the 19th Air Force, the 14th Air Force, and as a project officer under Air Force Secretar Stuart Symington. From 1952 to 1955, he was assigned to the Air Force Office of Atomic Energy.

In 1955, he was assigned to the Technical Operations Division, Air Force Special Weapons Command, Kirtland Air Force Base, Ne Mexico. In 1957, he was named Commander of the 4951st Support Squadron, Eniwetok; and the following year, he was reassigned t Kirtland AFB as Assistant to the Group Commander and later as Air Commander of the 4925th Test Group.

General Hedrick joined the Special System Office, Air Force Ballistics Division, Los Angeles, in 1960. He was named Commander of th Satellite Control Facility in 1965, and in 1966, he was appointed Deputy Commander, Air Force Systems Command. He received h present assignment in 1967.

General Hedrick is a Command Pilot and has received numerous Air Force awards.

His home town is Fort Worth, Texas, and he attended Texas Technological College, Lubbock, prior to joining the service. He receive B.S. and M.S. degrees in physics from the University of Maryland.

General and Mrs. Hedrick are the parents of two sons.

VINCENT L. JOHNSON
NASA Headquarters

Vincent L. Johnson, 51, Deputy Associate Administrator for Space Science and Applications (Engineering), NASA Headquarters, is member of the Apollo 13 Review Board.

Mr. Johnson was appointed to his present position in 1967. Prior to that time he had been Director of the Launch Vehicle and Propulsic Programs Division, Office of Space Science and Applications, since 1964. He was responsible for the management and development the light and medium launch vehicles used for NASA's unmanned earth orbital and deep space programs. His division also directe studies of future unmanned launch vehicle and propulsion system requirements.

Mr. Johnson joined NASA in 1960, coming from the Navy Department where he had been an engineer with the Bureau of Weapons. H first assignments with NASA were as Program Manager for the Scout, Delta, and Centaur launch vehicles.

He was a naval officer during World War II, serving with the Bureau of Ordnance. Prior to that, he was a physicist with the Nav Ordnance Laboratory.

Mr. Johnson was born in Red Wing, Minnesota, and attended the University of Minnesota.

He and Mrs. Johnson live in Bethesda, Maryland. They are the parents of two children.

MILTON KLEIN
NASA Headquarters

Milton Klein, 46, Manager, Space Nuclear Propulsion Office, NASA Headquarters, is a member of the Apollo 13 Review Board.

Mr. Klein has been in his present position since 1967. Prior to that he had been Deputy Manager since 1960. The Space Nucle Propulsion Office is a joint activity of the Atomic Energy Commission (AEC) and the National Aeronautics and Space Administratio The office conducts the national nuclear rocket program . He is also Director of the Division of Space Nuclear Systems of the AE responsible for space nuclear electric power activities.

Mr. Klein became associated with atomic energy work in 1946, when he was employed by the Argonne National laboratory. In 1950, joined the AEC's Chicago Operations Office as staff chemical engineer. Later, he was promoted to Assistant Manager for Technic Operations. Generally engaged in reactor development work for stationary power plants, he had a primary role in the power react demonstration program.

Mr. Klein was born in St. Louis, Missouri. He served in the U.S. Navy during World War II.

He has a B.S. degree in chemical engineering from Washington University and a Master of Business Administration degree from Harva University.

Mr. and Mrs. Klein and their three children live in Bethesda, Maryland.

HANS M. MARK
NASA Ames Research Center

)r. Hans M. Mark, 40, Director of the NASA Ames Research Center, Moffett Field, California, is a member of the Apollo 13 Review oard.

rior to being appointed Director of the Ames Research Center he was, from 1964 to 1969, Chairman of the Department of Nuclear ngineering at the University of California, Berkeley, California.

n expert in nuclear and atomic physics, he served as Reactor Administrator of the University of California's Berkeley Research Reactor, rofessor of nuclear engineering and a research physicist at the University's Lawrence Radiation Laboratory, Livermore, California, and onsultant to the U.S. Army and the National Science Foundation. He has written many scientific papers.

xcept for 2 years as an Assistant Professor of Physics at the Massachusetts Institute of Technology from 1958 to 1960, Dr. Mark's dministrative, academic, and research career has been centered at the University of California (Berkeley).

)r. Mark received his A.B. degree in physics from the University of California, Berkeley, in 1951, and returned there as a research physicist 1955, one year after receiving his Ph. D. in physics from M. I. T.

le is a Fellow of the American Physical Society and a member of the American Geophysical Union, the American Society for Engineering ducation and the American Nuclear Society.

)r. Mark was born in Mannheim, Germany, and came to the United States when he was 11 years old. He became a naturalized U.S. tizen in 1945.

)r. and Mrs. Mark are the parents of two children.

COUNSEL TO THE APOLLO 13 REVIEW BOARD

GEORGE T. MALLEY
NASA Langley Research Center

ieorge T. Malley, 57, Chief Counsel, Langley Research Center, Hampton, Virginia, is the legal Counsel to the Apollo 13 Review Board. He lso served as Counsel to the Apollo 204 Review Board.

1r. Malley is the Senior Field Counsel of NASA, and has been assigned to Langley since 1959. He was with the Office of the General Counsel, Department of the Navy, from 1950 to 1959, where he specialized in admiralty and international law.

le is a retired Navy officer and served on active duty from 1939 to 1946, mainly in the South Pacific. His last assignment was ommanding officer of the U.S.S. Fentress.

1r. Malley has an A.B. degree from the University of Rochester and an LL.B. degree from Cornell University Law School. He is a native f Rochester, New York, and is a member of the New York Bar and the Federal Bar Association.

1r. and Mrs. Malley and their two children live in Newport News, Virginia.

MANNED SPACE FLIGHT TECHNICAL SUPPORT

CHARLES W. MATHEWS
NASA Headquarters

Charles W. Mathews, 49, Deputy Associate Administrator for Manned Space Flight, NASA Headquarters, Washington, D. C., directs the)ffice of manned Space Flight technical support to the Apollo 13 Review Board.

1r. Mathews has been a research engineer and project manager for NASA and its predecessor, the National Advisory Committee for eronautics (NACA), since 1943. In his present assignment, he serves as general manager of manned space flight.

rior to his appointment to this position in 1968, he had been Director, Apollo Applications Program, NASA Headquarters, since January 967.

1r. Mathews was Gemini Program Manager at the Manned Spacecraft Center, Houston, Texas, from 1963 until 1967. Prior to that time, e was Deputy Assistant Director for Engineering and Development and Chief of the Spacecraft Technology Division at MSC.

1r. Mathews transferred to MSC (then the Space Task Group) when Project Mercury became an official national program in 1958. He erved as Chief of the Operation Division. He had been at the Langley Research Center, Hampton, Virginia, since 1943 engaged in aircraft ght research and automatic control of airplanes. He became involved in manned spacecraft studies prior to the first Sputnik flights, and e conducted early studies on reentry. Mr. Mathews was chairman of the group which developed detailed specifications for the Mercury acecraft.

1r. Mathews has been awarded the NASA Distinguished Service Medal and the NASA Outstanding Leadership Medal. He has received e NASA Group Achievement Award Gemini Program Team.

e is a Fellow of the American Astronautical Society and an Associate Fellow of the American Institute of Aeronautics and Astronautics. e is the author of numerous technical articles published by NASA.

r. Mathews, a native of Duluth, Minnesota, has a B.S. degree in aeronautical engineering from Rensselaer Polytechnic Institute, Troy, New ork.

. and Mrs. Mathews live in Vienna, Virginia. They have two children.

APOLLO 13 REVIEW BOARD OBSERVERS

WILLIAM A. ANDERS
National Aeronautics and Space Council

William A. Anders, 36, Executive Secretary, National Aeronautics and Space Council, Washington, D.C., is an official observer of the Apo
13 Review Board.

Prior to being appointed to his present position in 1969, Mr. Anders was a NASA astronaut and an Air Force lieutenant colonel. He w
lunar module pilot on the Apollo 8 lunar orbital mission, man's first visit to the vicinity of another celestial body.

Mr. Anders joined the NASA astronaut team at the Manned Spacecraft Center, Houston, Texas, in 1963. In addition to his Apollo 8 flig
he served as backup pilot for Gemini 11 and backup command module pilot for Apollo 11, the first lunar landing mission.

Mr. Anders was commissioned a second lieutenant in the Air Force upon graduation from the U.S. Naval Academy. After flight trainir
he served as a pilot in all-weather interceptor squadrons of the Air Defense Command. Prior to becoming an astronaut, he was a nucle
engineer and instructor pilot at the Air Force Weapons Laboratory, Kirtland Air Force Base, New Mexico.

He is a member of the American Nuclear Society and has been awarded the Air Force Commendation Medal, Air Force Astronaut Win
the NASA Distinguished Service Medal, and the New York State Medal for Valor.

Mr. Anders was born in Hong Kong. He received a B.S. degree from the U.S. Naval Academy and an M.S. degree in nuclear engineeri
from the Air Force Institute of Technology.

Mr. and Mrs. Anders are the parents of five children.

CHARLES D. HARRINGTON
Douglas United Nuclear, Inc.

Dr Charles D Harrington, 59, President and General Manager, Douglas United Nuclear, Inc., Richland, Washington, is an official observ
of the Apollo 13 Review Board.

Dr. Harrington, who has been associated with all phases of the chemical and nuclear industrial fields since 1941, is Chairman of t
Aerospace Safety Advisory Panel, a statutory body created by Congress.

From 1941 to 1961, he was employed by the Mallinckrodt Chemical Works, St. Louis, Missouri. Dr. Harrington started with the compa
as a research chemist and in 1960, after a procession of research and management positions, was appointed Vice President, Mallinckro
Nuclear Corporation and Vice President, Mallinckrodt Chemical Works.

In 1961, when the fuel material processing plant of Mallinckrodt became the Chemicals Division of United Nuclear Corporation, D
Harrington was named Vice President of that division.

He became Senior Vice President, United Nuclear Corporation, Centreville, Maryland, in 1965.

In 1965, Dr. Harrington was appointed President and General Manager, Douglas United Nuclear, Inc. The company manages producti
reactors and fuels fabrication facilities at Hanford, Washington, for the Atomic Energy Commission.

He is the co-author of a book, "Uranium Production Technology," and has written numerous technical papers. He has received t
Mid-West Award of the American Chemical Society for contributions to technology in the nuclear energy field.

He is director of several corporations, including United Nuclear, as sell as professional councils and societies.

Dr. Harrington has M.S., M.A., and Ph. D. degrees in chemistry from Harvard University.

I. IRVING PINKEL
NASA Lewis Research Center

I. Irving Pinkel, 57, Director, Aerospace Safety Research and Data Institute at the NASA Lewis Research Center, Cleveland, Ohio, is offic
observer of the Apollo 13 Review Board.

Until recently, he directed research at Lewis Research Center on rocket propellant and electric power generation systems for spa
vehicles, compressors and turbines for advanced aircraft engines, and lubrication systems for rotating machines for these systems.

Mr. Pinkel entered Government scientific service in 1935 as a physicist with the U.S. Bureau of Mines, Pittsburgh, Pennsylvania. In 19
he joined the staff of the Langley Research Center, Hampton, Virginia, as a physicist. When the Lewis Research Center was built in 19
he transferred there.

He has been elected to Phi Beta Kappa, Sigma Xi, honorary scientific society, and Pi Mu Epsilon, honorary mathematics fraternity. He
an Ohio Professional Engineer, served on the former NACA subcommittees on Meteorological Problems, Icing Problems, Aircraft F
Prevention and Flight Safety, and is a member of the NASA Research and Technology Advisory Subcommittee on Aircraft Operat
Problems. He has been a Special Lecturer, Case Institute of Technology Graduate School.

Mr. Pinkel has received the Flight Safety Foundation Award for contributions to the safe utilization of aircraft, the Laura Taber Barbo
Award for development of a system for suppressing aircraft crash fires, the NACA Distinguished Service Medal, and the NASA Sustain
Superior Performance Award.

He was born in Gloversville, New York, and was graduated from the University of Pennsylvania.

Mr. and Mrs. Pinkel live in Fairview Park, Ohio. They are the parents of two sons.

JAMES E. WILSON, JR.
Committee on Science and Astronautics
United States House of Representatives

ames E. Wilson, Jr., 39, Technical Consultant, United States House of Representatives Committee on Science and Astronautics, is an fficial observer of the Apollo 13 Review Board.

1r. Wilson has been technical consultant to the Committee since 1963. From 1961 to 1963, he was Director of Research and)evelopment, U.S. Naval Propellant Plant, Indian Head, Maryland. Mr. Wilson managed the Polaris Program at Indian Head from 1956 to 961.

rom 1954 to 1956, Mr. Wilson served as an officer in the U.S. Army Signal Corps. He was a development engineer with E.I. DuPont, Vilmington, Delaware , from 1953 to 1954.

1r. Wilson is a member of Phi Sigma Alpha, a National Honor Society; American Institute of Chemical Engineers; American Chemical ociety; and American Ordnance Association.

Ir. Wilson is co-author of several Publications of the House Committee on Science and Astronautics.

le received a B.S. degree in chemical engineering from the University of Maine and a Master of Engineering Administration degree from ieorge Washington University.

Ir. and Mrs. Wilson live in LaPlata Maryland. They have two children.

APOLLO 13 REVIEW BOARD PANEL CHAIRMEN

SEYMOUR C. HIMMEL
NASA Lewis Research Center

r. Seymour C. Himmel, Assistant Director for Rockets and Vehicles, Lewis Research Center, Cleveland, Ohio, heads the Design Panel f the Apollo 13 Review Board.

r. Himmel joined Lewis in 1948 as an aeronautical research scientist. He has occupied supervisory positions since 1953.

e has been awarded the NASA Exceptional Service Medal and the NASA Group Achievement Award as manager of the Agena Project iroup. Dr. Himmel has served on a number of advisory committees. He is an Associate Fellow of the American Institute of Aeronautics id Astronautics, and a member of Tau Beta Pi and Pi Tau Sigma. He is the author of more than 25 technical papers.

r. Himmel has a Bachelor of Mechanical Engineering degree from the College of the City of New York and M.S. and Ph. D. degrees from ase Institute of Technology.

r. and Mrs. Himmel live in Lakewood, Ohio.

EDWIN C. KILGORE
NASA Langley Research Center

lwin C. Kilgore, 47, Deputy Chief, Engineering and Technical Services, Langley Research Center, Hampton, Virginia, heads the Project anagement Panel of the Apollo 13 Review Board.

r. Kilgore joined the Langley science staff in 1944 and served in a variety of technical and management positions until promotion to s present position in 1968.

e has received the Honorary Group Achievement Award for his role in achieving a record of 97 consecutive successes for solid opellant rocket motors and the NASA-Lunar Orbiter Project Group Achievement Award for outstanding performance. He is a ember of Pi Tau Sigma, honorary mechanical engineering society.

r. Kilgore was born in Coeburn, Virginia. He was graduated from Virginia Polytechnic Institute with a B.S. degree in mechanical igineering.

r. and Mrs. Kilgore and their two daughters live in Hampton.

HARRIS M. SCHURMEIER
California Institute of Technology Jet Propulsion Laboratory

arris M. Schurmeier 45, Deputy Assistant Laboratory Director for Flight Projects, California Institute of Technology Jet Propulsion boratory, Pasadena, California, heads the Manufacturing and Test Panel of the Apollo 13 Review Board.

r. Schurmeier was appointed to his current position in 1969. Prior to that he was Mariner Mars 1969 Project Manager, Voyager Capsule stem Manager and Deputy Manager of the Voyager Project, and Ranger Project Manager at JPL.

e has received the NASA Medals for Exceptional Scientific Achievement and Exceptional Service. In addition, he has received the tronautics Engineer Award, and the NASA Public Service Award.

e was born in St. Paul, Minnesota. He has received a B.S. degree in mechanical engineering, M.S. degree in aeronautical engineering, and rofessional degree in aeronautical engineering from the California Institute of Technology.

. Schurmeier was a naval officer in World War II. He and his wife and four children live in Altadena, California.

FRANCIS B. SMITH
NASA Headquarters

Francis B. Smith, 47, Assistant Administrator for University Affairs, NASA Headquarters, is leader of the Mission Events Panel of th Apollo 13 Review Board.

Mr. Smith has been in his present position since 1967. Prior to that he had been Assistant Director, Langley Research Center, Hampto Virginia, since 1964. He joined the Langley science staff in 1947. He is an expert in several fields, including radio telemetry, rada electronic tracking systems, and missile and range instrumentation.

Mr. Smith was born in Piedmont, South Carolina, and received a B.S. degree in electrical engineering from the University of Sout Carolina, Where he was elected to Phi Beta Kappa. He remained at the University as an instructor from 1945 to 1944 and then serve in the U.S. Navy until 1946.

Mr. and Mrs. Smith and their three children live in Reston, Virginia.

PART 3. BOARD ORGANIZATION AND GENERAL ASSIGNMENTS FOR BOARD PANELS

BOARD ORGANIZATION*

After reviewing the scope of the Board's charter, the Chairman and Board Members agreed upon the Panel and Support Office structur depicted on the following organization chart. Each Panel was assigned specific responsibilities for reviewing major elements of the overa Board task, with particular emphasis upon establishing a sound and independent technical data base upon which findings, determination and recommendations by the Board could be based. The Panels were staffed with individual NASA specialists and established workir arrangements with the Manned Space Flight line organization personnel working in analogous areas.

The Board's support offices were structured to provide necessary staff, logistics, and administrative support without duplication available MSC assistance.

In addition to this structure, the Board and Panels also utilized the special assistance of expert consultants.

Panel assignments, complete Panel membership, and the official Board organization approved by the Chairman are included in this pa of the Board report.

GENERAL ASSIGNMENTS FOR BOARD PANELS (AS DOCUMENTED IN THE BOARD'S ADMINISTRATIVE PROCEDURES)

Panel 1 - Mission Events Panel

It shall be the task of the Mission Events Panel to provide a detailed and accurate chronology of all pertinent events and actions leadi to, during, and subsequent to the Apollo 13 incident. This information, in narrative and graphical time history form, will provide the Apol 13 Review Board an official events record on which their analysis and conclusions may be based. This record will be published in a for suitable for inclusion in the Review Board's official report.

The Panel will report all significant events derived from telemetry records, air-to-ground communications transcripts, crew and contr center observations, and appropriate documents such as the flight plan, mission technique description, Apollo Operation Handbook, a crew checklists. Correlation between various events and other observations related to the failure will be noted. Where telemetry da are referenced, the Panel will comment as appropriate on its significance, reliability, accuracy, and on spacecraft conditions which mig have generated the data.

The chronology will consist of three major sections: Pre-incident Events, Incident Events, and Post-incident Events. The decision-maki process leading to the safe recovery, referencing the relevant contingency plans and available alternates, will be included.

Pre-incident Events — This section will chronicle the progress of the flight from the countdown to the time of the incident. All acti and data relevant to the subsequent incident will be included.

Incident Events — This section will cover that period of time beginning at 55 hours and 52 minutes after lift-off and continuing so lo as abnormal system behavior is relevant to the failure.

Post-incident Events — This section will document the events and activities subsequent to the incident and continuing to missi termination (Splash). Emphasis will be placed on the rationale used on mission completion strategy.

Panel 1 Membership

Mr. F. B. Smith, Panel Chairman
Assistant Administrator for University Affairs, NASA Headquarters Washington, D. C.

Dr. Tom B. Ballard
Aerospace Technologist, Flight Instrument Division, Langley Research Center, Hampton, Virginia

Mr. M. P. Frank
Flight Director, Flight Control Division, Manned Spacecraft Center, Houston, Texas

Mr. John J. Williams
Director, Spacecraft Operations, Kennedy Space Center, Florida

Mr. Neil Armstrong
Board Member and Panel Monitor Astronaut Manned Spacecraft Center Houston, Texas

(Editor's note - See slide 1.)

Panel 2 - Manufacturing and Test Panel

The Manufacturing and Test Panel shall review the manufacturing and testing, including the associated reliability and quality assurance activities, of the flight hardware components involved in the flight failure as determined from the review of the flight data and the analysis of the design. The purpose of this review is to ascertain the adequacy of the manufacturing procedures, including any modifications, and the preflight test and checkout program, and any possible correlation of these activities with the inflight events.

The Panel shall consist of three activities:

Fabrication and Acceptance Testing — This will consist of reviewing the fabrication, assembly, and acceptance testing steps actually used during the manufacturing of the specific flight hardware elements involved. Fabrication, assembly, and acceptance testing procedures and records will be reviewed, as well as observation of actual operations when appropriate.

Subsystem and System Testing — This will consist of reviewing all the flight qualification testing from the completion of the component level acceptance testing up through the countdown to lift-off for the specific hardware involved. Test procedures and results will be reviewed as well as observing specific tests where appropriate. Results of tests on other serial number units will also be reviewed when appropriate.

Reliability and Quality Assurance — This will be an overview of both the manufacturing and testing, covering such things as parts and material qualification and control, assembly and testing procedures, and inspection and problem/failure reporting and closeout.

Panel 2 Membership

Mr. Harris M. Schurmeier Panel Chairman
Deputy Assistant Laboratory Director for Flight Projects, Jet Propulsion Laboratory, Pasadena, California

Mr. Edward F. Baehr
Assistant Chief, Launch Vehicles Division, Deputy Manager, Titan Project, Lewis Research Center, Cleveland, Ohio

Mr. Karl L. Heimburg
Director, Astronautics Laboratory, Marshall Space Flight Center, Huntsville, Alabama

Mr. Brooks T. Morris
Manager, Quality Assurance add Reliability Office, Jet Propulsion Laboratory, Pasadena, California

Dr. John F. Clark, Board Member and Panel Monitor
Director, Goddard Space Flight Center, Greenbelt, Maryland

Panel 3 - Design Panel

The Design Panel shall examine the design of the oxygen and associated system to the extent necessary to support the theory of failure. After such review the Panel shall indicate a course of corrective action which shall include requirements for further investigations and/or redesign. In addition, the Panel shall establish requirements for review of other Apollo spacecraft systems of similar design.

The Panel shall consist of four subdivisions

Design Evaluation — This activity shall review the requirements and specifications governing the design of the systems, subsystem and components, their derivation, changes thereto and the reasons therefore and the design of the system in response to the requirements, including such elements as design approach, material selection, stress analysis, development and qualification test program, and results. This activity shall also review and evaluate proposed design modifications, including changes in operating procedures required by such modifications.

Failure Modes and Mechanism — This activity shall review the design of the system to ascertain the possible sources of failure and the manner in which failures may occur. In this process, they shall attempt to correlate such modes with the evidence from flight and ground test data. This shall include considerations such as: energy sources, materials compatibility, nature of pressure vessel failure, effects of environment and service, the service history of any suspect system and components, and any degradation that may have occurred.

Electrical — This activity shall review the design of all electrical components associated with the theory of failure to ascertain their adequacy. This activity shall also review and evaluate proposed design modifications including changes in operating procedures required by such modifications.

Related System — This activity shall review the design of all systems similar to that involved in the Apollo 13 incident with the view to establishing any commonality of design that may indicate a need for redesign. They shall also consider the possibility of design modifications to permit damage containment in the event of a failure.

Panel 3 Membership

Dr. Seymour C. Himmel, Panel Chairman
Assistant Director for Rockets and Vehicles Lewis Research Center Cleveland, Ohio

Mr. William F. Brown, Jr.
Chief Strength of Materials Branch Materials & Structures Division Administration Directorate Lewis Research Center Cleveland, Ohio

Mr. R. N. Lindley
Special Assistant to the Associate Administrator for Manned Space Flight NASA Headquarters Washington, D. C.

Dr. William R. Lucas
Director, Program Development, Marshall Space Flight Center, Huntsville, Alabama

Mr. J. F. Saunders, Jr.
Project Officer for Command and Service Module Office of Manned Space Flight NASA Headquarters Washington, D. C.

Mr. Robert C. Wells
Head, Electric Flight System Section Vehicles Branch Flight Vehicles and Systems Division Office of Engineering and Technical Service
Langley Research Center Hampton, Virginia

Mr. Vincent L. Johnson, Board Member and Panel Monitor Deputy Associate Administrator for Engineering Office of Space Science an
Applications NASA Headquarters Washington, D. C.

Panel 4 - Project Management Panel

The Project Management Panel will undertake the following tasks:

1. Review and assess the effectiveness of the management structure employed in Apollo 13 in all areas pertinent to the Apollo 1 incident. This review will encompass the organization, the responsibilities of organizational elements, and the adequacy of the staffin

2. Review and assess the effectiveness of the management system employed on Apollo 13 in all areas pertinent to the Apollo 1 incident. This task will include the management system employed to control the appropriate design, manufacturing, and tes operations; the processes used to assure adequate communications between organizational elements; the processes used to contr hardware and functional interfaces; the safety processes involved; and protective security.

3. Review the project management lessons learned from the Apollo 13 mission from the standpoint of their applicability to subseque Apollo missions.

Tasks 1 and 2, above, should encompass both the general review of the processes used in Apollo 13 and specific applicability to th possible cause or causes of the mission incident as identified by the Board.

Panel 4 Membership

E. C. Kilgore, Panel Chairman
Deputy Chief, Office of Engineering and Technical Services, Langley Research Center, Hampton, Virginia

R. D. Ginter
Director of Special Programs Office, Office of Advanced Research and Technology NASA Headquarters Washington, D.C.

Merrill H. Mead
Chief of Programs and Resources Office, Ames Research Center, Moffett Field, California

James B. Whitten
Assistant Chief, Aeronautical and Space Mechanics Division, Langley Research Center, Hampton, Virginia

Milton Klein
Board Member and Panel Monitor Manager, AEC-NASA Space Nuclear Propulsion Office Washington, D.C.

Board Observers

William A. Anders
Executive Secretary National Aeronautics and Space Council Washington, D.C.

Dr. Charles D. Harrington
Chairman NASA Aerospace Safety Advisory Panel Washington, D.C.

I. Irving Pinkel
Director, Aerospace Safety Research and Data Institute, Lewis Research Center, Cleveland, Ohio

Mr. James E. Wilson
Technical Consultant to the Committee on Science and Astronautics United States House of Representatives Washington, D.C.

Apollo 13 Review Board Support Staff

Brian M. Duff
Public Affairs Officer, Manned Spacecraft Center, Houston, Texas

Gerald J. Mossinghoff
Director of Congressional Liaison NASA Headquarters Washington, D.C.

Edward F. Parry
Counsel to Office of Manned Space Flight NASA Headquarters Washington, D.C.

Raymond G. Romatowski
Deputy Assistant Director for Administration, Langley Research Center, Hampton, Virginia

Ernest P. Swieda
Deputy Chief, Skylab Program Control Office, Kennedy Space Center, Florida

Consultants to the Board

Dr. Wayne D. Erickson, Head
Aerothemochemistry Branch
Langley Research Center
Hampton, Virginia

Dr. Robert Van Dolah
Acting Research Director
Safety Research Center
Bureau of Mines
Pittsburgh, Pennsylvania

MSC Support to the Board

These persons were detailed by MSC to support the Apollo 13 Review Board during its review activity at MSC. They are identified by MSC position title.

Roy C. Aldridge
Assistant to the Director of Administration

Mary Chandler
Secretary

Rex Cline
Technical Writer/Editor

Evon Collins
Program Analyst

Leroy Cotton
Equipment Specialist

Maureen Cruz
Travel Clerk

Janet Harris
Clerk Stenographer

Marjorie Harrison
Secretary

Phyllis Hayes
Secretary

William N. Henderson
Management Analyst

Sharon Laws
Secretary

Carolyn Lisenbee
Secretary

Judy Miller
Secretary

Jamie Moon
Technical Editor

Dorothy Newberry
Administrative Assistant

Lettie Reed
Editorial Assistant

Charlene Rogozinski
Secretary

Joanne Sanchez
Secretary

Billie Schmidt
Employee Development Specialist

Frances Smith
Secretary

George Sowers
Management Presentations Officer

Elaine Stemerick
Secretary

Mary Thompson
Administrative Assistant

Alvin C. Zuehlke
Electrical Engineer

PART 4. SUMMARY OF BOARD ACTIVITIES

APRIL 19, 1970

Chairman E. M. Cortright met with Langley officials to begin planning the Apollo 13 Review Board approach. Tentative list of Panel Members and other specialists were developed for consideration.

APRIL 20, 1970

Chairman Cortright met with the NASA Administrator, Deputy Administrator, and key NASA officials in Washington, D.C., to discuss Board membership.

The Chairman met with NASA Office of Manned Space Flight top officials while en route to MSC on NASA aircraft and discussed program organization plans for review of the accident, and coordination with Apollo 13 Review Board activity.

APRIL 21, 1970

Chairman Cortright met with MSC officials to discuss Apollo 13 Review Board support.

A formal MSC debriefing of the Apollo 13 crew was conducted for MSC officials and Apollo 13 Review Board personnel already at MSC.

Detailed discussions between early arrivals on the Review Board and the MSC Investigation Team were held to provide quick-look data on the Apollo 13 accident and to develop detailed procedures for MSC support of the Apollo 13 Board.

Chairman Cortright met with members of the Press to report on early activity of the Board and to inform them of plans for keeping the Press current on Board activities.

The first meeting of the Board was held at 8 p.m. to discuss Board composition, structure, assignments, and scope of review. Preliminary plans were developed for appointing various specialists to assist the Board in its analysis and evaluation.

APRIL 22, 1970

The Board met with Colonel McDivitt's MSC Investigation Team to review the progress made by MSC in identifying causes of the accident and in developing an understanding of sequences and relationships between known inflight events. In addition, MSC officials briefed the Board on MSC Investigation Team structure and assignments.

The Board met with Panel 1 of the MSC Investigation Team for detailed discussion of inflight events and consideration of early conclusions on implications of preliminary data analysis.

The Board held its second meeting to discuss MSC investigative efforts and additional appointments of Panel specialists.

Board members attended Panel I evening roundup of day's evaluation activities, which included detailed discussions of specific studies data reductions, and support test activities already underway.

APRIL 23, 1970

The Apollo 13 Review Board established itself in proximity to the MSC Investigation Team in Building 45, and arranged for a administrative and logistics support to the Board.

A daily schedule of meetings, reviews, briefings, and discussions was established, including preliminary plans for contractor meetings special support tests, and accumulation of accident-related information.

Initial task assignments and responsibilities were made to Board Panels as guidance for detailed review work. Individual Board member were assigned Panel overview responsibilities or other special tasks.

Administrative procedures were developed for Board activity, particularly to provide efficient interface with MSC personnel.

Board and Panel Members again met with MSC officials to further review the sequence of events in the Apollo 13 mission and to examine early hypotheses concerning causes of these events.

The Board convened for an evening meeting to discuss the progress to date and to coordinate Panel activities for the next few days Discussion centered upon immediate requirements for data collection and analysis.
Chairman Cortright appointed additional NASA specialists in order to bring Panels up to strength.

APRIL 24, 1970

Board Members, Panel Chairmen, and MSC officials reviewed additional data analysis made by MSC and contractor personnel with particular emphasis upon the service module (SM) cryogenic system.

The Board convened and reviewed the progress to date. Tentative approvals were given for Board trips to North American Rockwell (NR), Downey, California, Beech Aircraft, Boulder, Colorado, and other locations.

Chairman Cortright briefed the Press on progress to date.

Panel Chairmen and Members continued their detailed analysis of failure modes, test histories, mission events, and other data bearing upon the accident.

Board Members and Panel Chairmen met with Mr. Norman Ryker of NR on NR's activities involving design, qualification, and tests of SM cryogenic oxygen tanks.

APRIL 25, 1970

The Board met to discuss details of onsite inspections of command service module (CSM) flight hardware at principal contractor installations.

Panels examined in detail probable failure modes based on data analyzed at that time.

Specific plans were discussed by the Board relating to evaluation of oxygen tank assembly and checkout operations, including review of component histories.

The MSC Investigation Team members briefed Board personnel on Kennedy Space Center checkout operations of the service module cryogenic and electric power systems, including a detailed briefing covering oxygen tank detanking operations.

APRIL 26, 1970

Board and Panel Members traveled to North American Rockwell, Downey, for detailed briefings by NR engineers and management. NR reviewed its

progress in an intensive analysis of the Apollo 13 malfunction, including a review of approved special tests. Oxygen tank, fuel cell components, assemblies, and other hardware were also inspected.

APRIL 27, 1970

An Executive Session of the Board met to discuss progress of specific analyses required to verify tentative conclusions on oxygen tank failure and service module EPS failure .

Additional Board specialists arrived at MSC and received detailed briefings by MSC and Board personnel on selected aspects of the Apollo 13 data.

Panel Members received and assessed a preliminary MSC evaluation of the Apollo 13 accident, including tentative conclusions on the most probable failure modes.

Procedures were established to provide information flow on the status of review to Board observers.

The Board reviewed work plans for the coming week with each Panel and established review priorities and special task assignments.

APRIL 28, 1970

Chairman Cortright outlined a plan for the Board's preliminary report scheduled for presentation to the Deputy Administrator during his visit to MSC on May 1. Each Panel Chairman was to summarize the status of his Panel's activities for Dr. George Low on Friday, April 29, 1970.

Board Member Neil Armstrong completed arrangements to provide each Board Member and Panel Chairman an opportunity for detailed simulation of the Apollo 13 inflight accident using MSC's CSM simulation equipment.

Board and Panel Members reviewed enhanced photographs of the Apollo 13 service module at the MSC Photographic Laboratory.

Dr. von Elbe of Atlantic Research Company briefed Board and Panel Members on cryogenics and combustion phenomena.

A representative of the Manufacturing and Test Panel performed onsite inspection at Beech Aircraft, Boulder.

Manufacture and Test Panel personnel reviewed detanking procedures followed at KSC during the Apollo 13 countdown demonstration test (CDDT).

Board and Panel personnel reviewed progress to date at a general Board meeting involving all Review Board personnel.

APRIL 29, 1970

Dr. Charles Herrington, Board Observer and Chairman of the Aerospace Safety Advisory Panel, arrived for a 2-day detailed review of Board procedures and progress in the accident review.

The Board reviewed North American Rockwell preliminary recommendations involving oxygen tank redesign.

The Board continued to review and examine oxygen tank ignition sources and combustion propagation processes with specialists from MSC, other NASA Centers, and contractor Personnel.
The Mission Events Panel continued to examine and record details of all significant mission events as a basis for other Panel evaluations and study.

Chairman Cortright convened two Board meetings to review Panel progress to date and to discuss work plans for the next several days.

The Project Management Panel visited North American Rockwell at Downey to review detailed procedures for acceptance tests, subcontractor inspections, project documentation, and other management interface areas.

APRIL 30, 1970

The Safety Advisory Panel continued discussions with Board Chairman and MSC officials on progress of total Apollo 13 review efforts.

Panel Members reviewed instrumentation used in Apollo 13 spacecraft in order to establish the validity of telemetry data being used in Board analysis.

Chairman Cortright convened two Board meetings to review progress of the work and to discuss preliminary findings of the Board.

Project Management personnel visited Beech Aircraft Corporation to review procedures used for assembly of cryogenic oxygen tanks and to discuss communication and information systems within the Apollo Program.

Panels continued to review detailed data in their respective areas.

MAY 1, 1970

Board and Panel personnel participated in a joint MSC/Apollo 13 Review Board status presentation to the NASA Deputy Administrator. The meeting covered all significant Apollo 13 findings and early conclusions on the cause of the accident and appropriate remedial actions.

The MSC staff briefed Board Members on initial evaluations of proposed design changes in oxygen tank system.

Panel Members continued to assess data accumulated from the Apollo 13 mission with particular emphasis upon the design and performance of electric power systems used in the service module.

Board Members and Panel Chairmen reviewed specific test matrix being proposed by Apollo 13 Review Board specialists covering most significant unknowns involved in understanding failure mechanisms.

MAY 2, 1970

Board Members met in General Session to discuss preparation of a complete "failure tree" as an additional guide in conducting a complete review and investigation. Specific aspects of this approach were reviewed.

The Project Management Panel reviewed oxygen tank reliability history and quality assurance criteria used in assembly, test, and checkout of these systems.

Panel specialists continued reviewing data from the mission with emphasis upon integrating various data points into logical failure mode patterns established by MSC and Board personnel.

MAY 3, 1970

Chairman Cortright and Board Members conducted a detailed review of individual Panel status and progress and established milestones for additional analytical work and preparation of preliminary findings.

The Board and Panel agreed to tentative report structure, including required exhibits, tables, drawings, and other reference data.

The Board established a system for tabulating all significant mission events and explanatory data, including the support tests required to clarify questions raised by events.

Panel Members worked on individual analyses with particular attention to developing requirements for additional test activity in support of tentative conclusions.

The Board agreed to strengthen its technical reviews of combustion propagation and electrical design by adding specialists in these areas.

MAY 4, 1970

The Design Panel continued its intensive review of the "shelf drop" incident at NR involving the cryogenic oxygen flight tank used in Apollo 13 in order to understand possible results of this event.

The Mission Events Panel continued to analyze telemetry data received by MSC, with particular attention on data received in proximity to the data dropout period during the Apollo 13 mission and on fan turn-ons during the flight.

The Board transmitted a formal listing of 62 requests for data, analyses, and support tests required for Board review activity.

The Board continued to meet with individual Panels and support offices to review the status of preliminary findings and work completed.

MAY 5, 1970

The Board met in General Session to discuss the scope and conduct of support test activity, including careful documentation of test methods and application of test results.

MSC personnel briefed Panel Members on availability of additional telemetry data in the MSC data bank in order to insure Board consideration of all possible useful data.

Panels commenced initial drafting of preliminary findings in specific areas, including summary descriptions of system performance during the Apollo 13 flight.

The Board met with the MSC Investigation Team for complete review of the proposed test program.

MAY 6, 1970

Board Members, MSC personnel, and Members of NASA's Aerospace Safety Advisory Panel met for detailed discussions and evaluation of accident review status and progress. The review covered Oxygen tank questions, recovery operations, and a mission simulation by MSC astronauts.

Panel Members continued to Work on the preparation of preliminary Panel drafts.

Chairman Cortright transmitted additional requests for tests to MSC and modified procedures for control of overall test activity relating to the Apollo 13 accident.

MAY 7, 1970

The General Board Session reviewed complete analysis and test support activities being conducted for the Board and MSC at various governmental and contractor installations.

Board and Panel Members met to discuss Ames laboratory tests concerning liquid oxygen combustion initiation energies required in the cryogenic oxygen tank used in the Apollo 13 SM.

Panel 1 Members reviewed mission control equipment and operating procedures used during the Apollo 13 mission and reviewed actual mission events in detail.

The Panels continued to develop preliminary drafts of their reviews and analyses for consideration by the Board.

MAY 8, 1970

Dr. Robert Van Dolah, Bureau of Mines, joined the Board as a consultant on combustion propagation and reviewed Apollo 13 Review Board data developed to date.

The General Board Session convened to review proposed report format and scope. An agreement was reached on appendices, on the structure of the report, and on the degree of detail to be included in individual Panel reports.

Chairman Cortright assigned additional specific test overview responsibilities to members of the Apollo 13 Review activity.

Panel 1 conducted a formal interview with the MSC Flight Director covering all significant mission events from the standpoint of ground controllers.

Panels 2 through 4 continued developing preliminary reports. Panel 4 announced a formal schedule of interviews of MSC, contractors, and NASA Headquarters personnel.

Board Members explored in detail possible failure mode sequences developed by MSC personnel involving ignition and combustion within the SM cryogenic oxygen tank

The Board recessed for 3 days, leaving a cadre of personnel at MSC to edit preliminary drafts developed by the Panels and to schedule further activity for the week of May 11.

MAY 9, 1970

Board in recess.

MAY 10, 1970

Board in recess.

MAY 11, 1970

Board in recess. MSC support personnel continued work obtaining additional technical data for Board review.

MAY 12, 1970

Board Members returned to MSC.

Board Members attended a General Session to review progress and status of the report.

Panel Chairmen reported on individual progress of work and established schedules for completion of analyses and evaluations.

Chairman Cortright reported on the Langley Research Center support test program aimed at simulation of SM panel ejection energy pulses.

MAY 13, 1970

Board Members reviewed preliminary drafts of report chapter on Review and Analysis and Panel 1 report on Mission Events.

Mission Events Panel Members interviewed Electrical, Electronic, and Communications Engineer (EECOM) and one of the Apollo 13 Flight Directors on activities which took place in the Mission Control Center (MCC) during and after the flight accident period.

Panel 4, Project Management Panel, conducted interviews with principal Apollo 13 program personnel from MSC and contract organizations.
Panel Members continued drafting preliminary versions of Panel reports for review by the Board.

Manufacturing and Test Panel representatives discussed program for oxygen tank testing to be conducted at Beech Aircraft.

Board Members met in General Session to review report milestones and required test data for the week ahead.

MAY 14, 1970

Board met in General Session to review Panel report progress and to agree to firm schedules for completion of all Review Board assignments.

Project Management Panel continued to interview key Apollo project personnel from NASA Centers and contractors.

Panel Members circulated first drafts of all Panel reports to Board Members for review and correction.

MAY 15, 1970

Mission Events Panel personnel interviewed Apollo 13 Command Module Pilot John Swigert to verify event chronology compiled by the Panel and to review crew responses during Apollo 13 mission.

Project Management Panel continued interviewing key project personnel with NASA Centers and contractors.

MSC personnel provide Board Members and Panel Chairmen with a detailed briefing on all support tests and analyses being performed in connection with the MSC and Board reviews.

Board Members met in Executive Session to review preliminary drafts of Panel reports and findings and determinations and to provide additional instructions and guidance to Panel Chairmen.

Panel Members continued to review and edit early Panel drafts and to compile reference data in support of findings.

MAY 16, 1970

Board met in General Session to review further revisions of preliminary findings and determinations and to establish working schedules for completion of the Board report.

Panel Members continued to edit and refine Panel reports on basis of discussions with MSC personnel and further analysis of Apollo 13 documentation.

MAY 17, 1970

Draft material for all parts of Board report was reviewed by Panel Members and staff. Changes were incorporated in all draft material and re-circulated for additional review and comment.

Board Members met in General Session to review report progress and to examine results from recent support tests and analyses being conducted at various Government and contractor installations.

The Apollo 13 Review Board discussed a continuing series of support tests for recommendation to MSC following presentation of report and recess of the Board.

MAY 18, 1970

Board Members reviewed Special Tests and Analyses Appendix of the report and examined results of completed tests.

Board met in General Session to discuss control procedures for reproduction and distribution of Board report.

Mission Events Panel distributed a final draft of their report for review by Board Members.

Board reviewed a preliminary draft of findings and determinations prepared by Panel Chairmen, Board Members, and Board Chairman.

A Manufacture and Test Panel representative reviewed special oxygen tank test programs at Beech Aircraft.

MAY 19, 1970

Board Members met in Executive Session to continue evaluation and assessment of preliminary findings, determinations, and recommendations prepared by individual Board Members and Panel Chairmen.

Board met in General Session to review final draft of Mission Events Panel report.

Manufacture and Test Panel preliminary report was distributed to Board Members for review and comment.

Design Panel preliminary report was distributed to Board Members for review and comment.

Design Panel Members met with MSC Team officials to discuss further test and analyses support for the Board.

MAY 20, 1970

Board Members met in Executive Session to review and evaluate reports from the Design Panel and from the Manufacturing and Test Panel.

Project Management Panel distributed final draft of its report to Board Members for review and comment.

Chairman Cortright met with Mr. Bruce Lundin of the Aerospace Safety Advisory Panel to discuss progress of Board review and analysis.

MAY 21, 1970

Board Members met in Executive Session for final review of Project Management Panel report.

Board Members and others met with MSC officials to review in detail the activities and actions taken after the Apollo 204 accident concerning ignition flammability for materials and control in the CSM.

A third draft of preliminary findings, determinations, and recommendations was developed and circulated by the Chairman for review and comment.

Arrangements were made with NASA Headquarters officials for packaging, delivery, and distribution of the Board's final report.

Mission Events Panel conducted an interview with Lunar Module Pilot Haise to review selected mission events bearing on the accident.

MAY 22, 1970

Mission Events Panel representatives met with MSC officials to review in detail several events which occurred during later flight stages.

Board met in Executive Session to assess latest drafts of findings, determinations, and recommendations circulated by the Chairman.

Board met in General Session to review total progress in all report areas and to establish final schedule for preparation of Board report.

Langley Research Center representative M. Ellis briefed the Board on ignition and combustion of materials in oxygen atmosphere tests being conducted in support of the Apollo 13 Review.

Board Observer I. I. Pinkel briefed the Board on Lewis Research Center fire propagation tests involving Teflon.

MAY 23, 1970

Board Members reviewed Chapter 4 of Board report entitled "Review and Analysis."

Panel Chairmen reviewed draft findings and determinations prepared by the Board.

MAY 24, 1970

Board Members reviewed NASA Aerospace Safety Panel report covering Apollo activities during the period of 1968-69.

Board met in Executive Session for detailed review of support test status and progress and of documentation describing the results of test activity.

Board met in Executive Session for further review of findings, determinations, and recommendations.

MAY 25, 1970

Board met in Executive Session to review test progress and decided to postpone submittal of final report until June 8 in order to consider results of Langley Research Center panel ejection tests.

Board Members continued to review MSC Investigation team preliminary drafts and refine Apollo 13 data in the various Board appendices.

Board met in Executive Session for further consideration of findings, determinations, and recommendations.

MAY 26, 1970

Board met in General Session and interviewed Astronaut James Lovell regarding crew understanding of inflight accident.

Board Members reviewed proposed MSC tank combustion test and agreed to test methodology and objectives.

Panel Members continued preparation of individual Panel reports.

MAY 27, 1970

Board and Panel Members received a detailed briefing on thermostatic switch failure during MSC heater tube temperature tests.

Aerospace Safety Advisory Panel met with Chairman Cortright, Board Members, and Panel Chairmen to review Board progress and status of findings and conclusions.

Board met in General Session to review status of Panel reports, documentation of test data and results, and plans for report typing and review.

Board agreed to recess for several days to accumulate additional test information on panel separation and full scale tank ignition data.

MAY 28 - MAY 31 1970

Board in recess

JUNE 1, 1970

Board Members returned to MSC.

Board and Panel Members met in General Session to discuss revisions Of Panel reports in light of latest information regarding thermostatic switch failure during CDDT at KSC.

Board approved new schedule for Board report calling for final versions of Panel reports by Monday, June 8.

JUNE 2, 1970

Chairman Cortright briefed the Press on the status of the Board's work and future plans.

Board and Panel Members participated in a detailed interview and discussion with MSC and contractor personnel regarding specific coordination steps taken during oxygen task no. 2 detanking operations at KSC.

Board Members met in Executive Session to review latest test results and to assess status of Board findings and determinations.

JUNE 3, 1970

Board and Panel Members met with MSC Program Office personnel for a detailed update of recent MSC information and analyses stemming from ongoing test programs.

Board Members and Panel Chairmen completed final reviews of Panel reports and also reviewed final draft of findings, determinations, and recommendations.

Board and Panel Members received a detailed briefing on thermostatic switch questions with emphasis upon actions of various organizations during and after detanking operations at KSC.

JUNE 4, 1970

Board Members met in Executive Session and completed final revisions of Chapter 4 of the Board summary.

Board and Panel Members witnessed a special full-scale tank ignition test performed at MSC.

Panel Chairmen completed final revisions of individual Panel reports and submitted copy to the Reports Editorial Office.

Board met in Executive Session and agreed to final schedule for report printing and delivery to the Administrator on June 15, 1970.

JUNE 5, 1970

Board Members met in Executive Session and completed work on Chapter 5 of the Board Summary Report (Findings, Determinations, and Recommendations).

Board Members reviewed final version of Project Management Panel report and authorized printing as Appendix E.

Board Members Hedrick and Mark completed final tabulation of test support activities performed for the Board.

Board Members reviewed films of special test activities performed at various NASA Centers.

JUNE 6, 1970

Board met in Executive Session throughout the day and completed its review of Chapter 5 of its report (Findings, Determinations, and Recommendations).

Board Members completed review of analyses to be incorporated in Appendix F, Special Tests and Analyses.

JUNE 7, 1970

The Board met in Executive Session and approved plans and schedules for final editorial review and publication of the Board report.

The Chairman recessed the Board until June 15 at which time the Board is scheduled to reconvene in Washington, D.C., to present its report to the NASA Administrator and Deputy Administrator.

CHAPTER 3

DESCRIPTION OF APOLLO 13 SPACE VEHICLE AND MISSION SUMMARY

This chapter is extracted from Mission Operation Report No. M-932-70, Revision 3, published by the Program and Special Reports Division (XP), Executive Secretariat, NASA Headquarters, Washington, D.C.

Discussion in this chapter is broken into two parts. Part 1 is designed to acquaint the reader with the flight hardware and with the mission monitoring, support, and control functions and capabilities. Part 2 describes the Apollo 13 mission and gives a mission sequence of events summary.

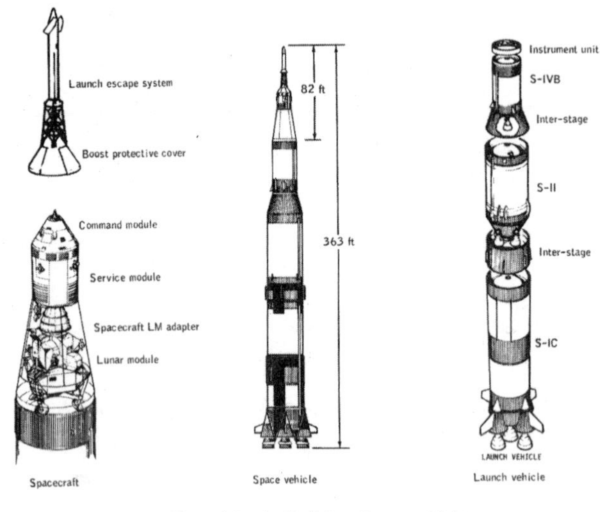

Figure 3-1.- Apollo/Saturn V space vehicle.

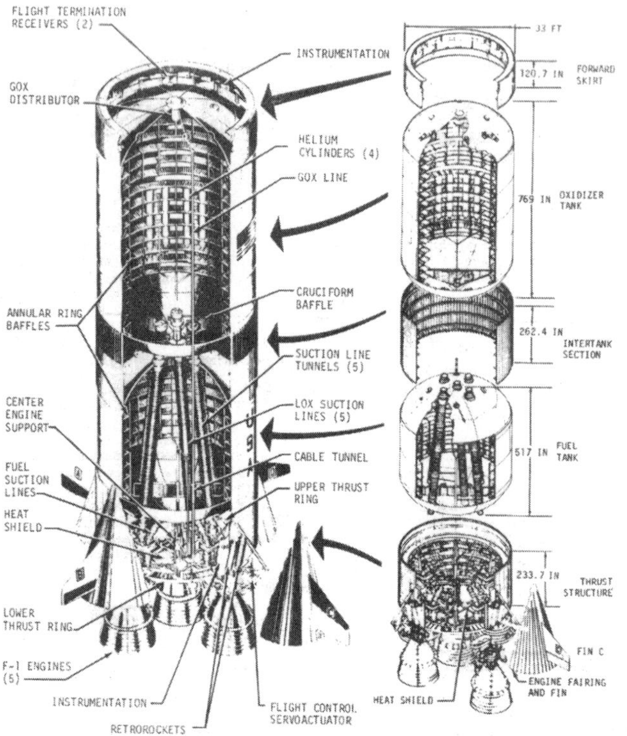

Figure 3-2.- S-IC stage.

PART 1 APOLLO/SATURN V SPACE VEHICLE

The primary flight hardware of the Apollo Program consists of the Saturn V launch vehicle and Apollo spacecraft (fig. 3-1). Collectively, they are designated the Apollo/Saturn V space vehicle (SV). Selected major systems and subsystems of the space vehicle may be summarized as follows.

SATURN V LAUNCH VEHICLE

The Saturn V launch vehicle (LV) is designed to boost up to 300,000 pounds into a 105-nautical mile earth orbit and to provide for lunar payloads of over 100,000 pounds. The Saturn V LV consists of three propulsive stages (S-IC, S-II, S-IVB), two interstages, and instrument unit (IU).

S-IC Stage

The S-IC stage (fig. 3-2) is a large cylindrical booster, 138 feet long and 33 feet in diameter, powered by five liquid propellant F-1 rocket engines. These engines develop a nominal sea level thrust total of approximately 7,650,000 pounds. The stage dry weight is approximately 288,000 pounds and the total loaded stage weight is approximately 5,031,500 pounds. The S-IC stage interfaces structurally and electrically with the S-II stage. It also interfaces structurally, electrically, and pneumatically with ground support equipment (GSE) through two umbilical service arms, three tail service masts, and certain electronic systems by antennas. The S-IC stage is instrumented for operational measurements or signals which are transmitted by its independent telemetry system.

S-II Stage

The S-II stage (fig. 3-3) is a large cylindrical booster, 81.5 feet long and 33 feet in diameter, powered by five liquid propellant J-2 rocket engines which develop a nominal vacuum thrust of 230,000 pounds each for a total of 1,150,000 pounds. Dry weight of the S-II stage is approximately 78,050 pounds. The stage approximate loaded gross weight is 1,075,000 pounds. The S-IC/S-II interstage weighs 10,460 pounds. The S-II stage is instrumented for operational and research and development measurements which are transmitted by its independent telemetry system. The S-II stage has structural and electrical interfaces with the S-IC and S-IVB stages, and electric, pneumatic, and fluid interfaces with GSE through its umbilicals and antennas.

S-IVB Stage

The S-IVB stage (fig. 3-4) is a large cylindrical booster 59 feet long and 21.6 feet in diameter, powered by one J-2 engine. The S-IVB stage is capable of multiple engine starts. Engine thrust is 203,000 pounds. This stage is also unique in that it has an attitude control capability independent of its main engine. Dry weight of the stage is 25,050 pounds. The launch weight of the stage is 261,700 pounds. The interstage weight of 8100 pounds is not included in the stated weights. The stage is instrumented for functional measurements or signals which are transmitted by its independent telemetry system.

The high performance J-2 engine as installed in the S-IVB

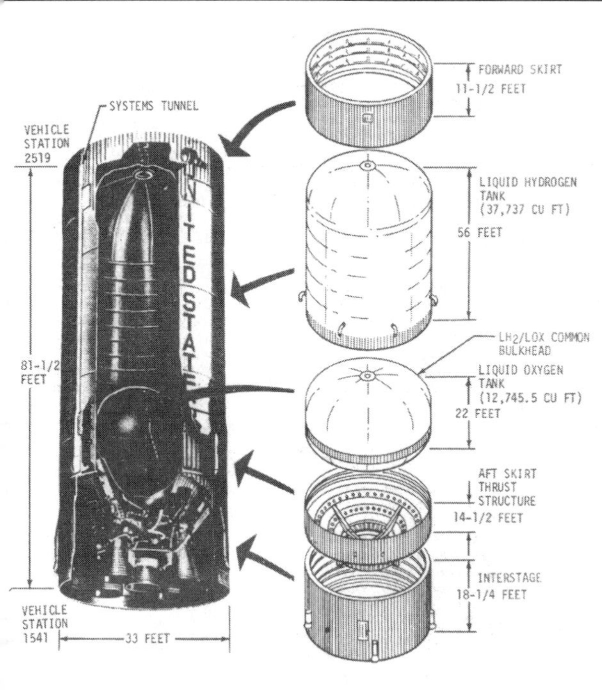

Figure 3-3.- S-II stage.

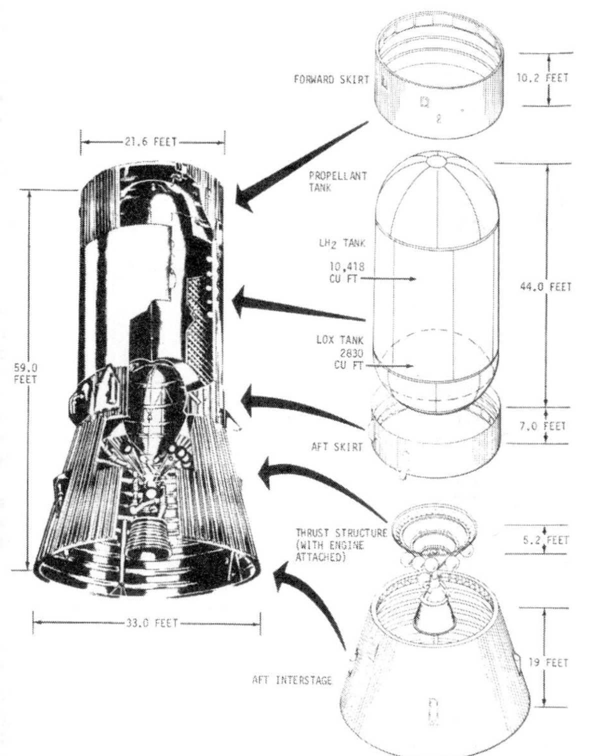

Figure 3-4.- S-IVB stage.

stage has a multiple start capability. The S-IVB J-2 engine is scheduled to produce a thrust of 203,000 pounds during its first burn to earth orbit and a thrust of 178,000 pounds (mixture mass ratio of 4.5:1) during the first 100 seconds of translunar injection. The remaining translunar injection acceleration is provided at a thrust level of 203,000 pounds (mixture mass ratio of 5.0:1). The engine valves are controlled by a pneumatic system powered by gaseous helium which is stored in a sphere inside a start bottle. An electrical control system that uses solid stage logic elements is used to sequence the start and shutdown operations of the engine.

Instrument Unit

The Saturn V launch vehicle is guided from its launch pad into earth orbit primarily by navigation, guidance, and control equipment located in the instrument unit (IU). The instrument unit is a cylindrical structure 21.6 feet in diameter and 3 feet high installed on top of the S-IVB stage. The unit weighs 4310 pounds and contains measurements and telemetry, command communications, tracking, and emergency detection system components along with supporting electrical power and the environmental control system.

APOLLO SPACECRAFT

The Apollo spacecraft (S/C) is designed to support three men in space for periods up to 2 weeks, docking in space, landing on and returning from the lunar surface, and safely entering the earth's atmosphere. The Apollo S/C consists of the spacecraft-to-LM adapter (SLA), the service module (SM), the command module (CM), the launch escape system (LES), and the lunar module (LM) The CM and SM as a unit are referred to as the command and service module (CSM).

Spacecraft-to-LM Adapter

The SLA (fig. 3-5) is a conical structure which provides a structural load path between the LV and SM and also supports the LM. Aerodynamically, the SLA smoothly encloses the irregularly shaped LM and transitions the space vehicle diameter from that of the upper stage of the LV to that of the SM. The SLA also encloses the nozzle of the SM engine and the high gain antenna.

Spring thrusters are used to separate the LM from the SLA. After the CSM has docked with the LM, mild charges are fired to release the four adapters which secure the LM in the SLA. Simultaneously, four spring thrusters mounted on the lower (fixed) SLA panels push against the LM landing gear truss assembly to separate the spacecraft from the launch vehicle.

Service Module

The service module (SM) (fig. 3-6) provides the main spacecraft propulsion and maneuvering capability during a mission. The SM provides most of the spacecraft consumables (oxygen, water, propellant, and hydrogen) and supplements environmental, electrical power, and propulsion requirements of the CM. The SM remains attached to the CM until it is jettisoned just before CM atmospheric entry.

Structure — The basic structural components are forward and aft (upper and lower) bulkheads, six radial beams, four sector honeycomb panels, four reaction control system honeycomb panels, aft heat shield, and a fairing. The forward and aft bulkheads cover the top and bottom of the SM. Radial beam trusses extending above the forward bulkhead support and secure the CM. The radial beams are made of solid aluminum alloy which has been machined and chem-milled to thickness' varying between 2 inches and 0.018 inch. Three of these beams have compression pads and the other three have shear-compression pads and tension ties. Explosive charges in the center sections of these tension ties are used to separate the CM from the SM.

An aft heat shield surrounds the service propulsion engine to protect the SM from the engine's heat during thrusting. The gap between the CM and the forward bulkhead of the SM is closed off with a fairing which is composed of eight electrical power system radiators alternated with eight aluminum honeycomb panels. The sector and reaction control system panels are 1 inch thick and are made of aluminum honeycomb core between two

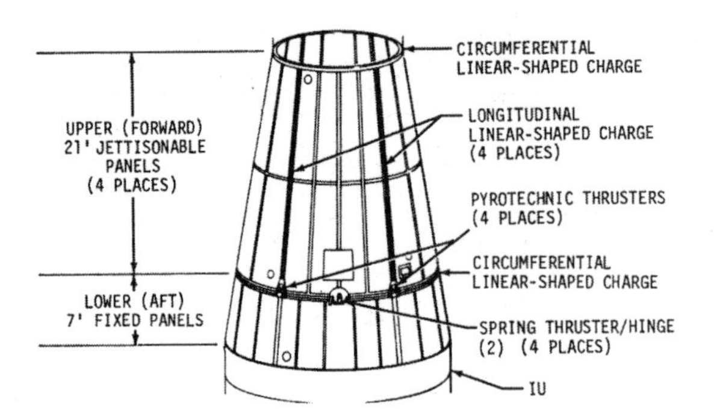

Figure 3-5.- Spacecraft-to-LM adapter.

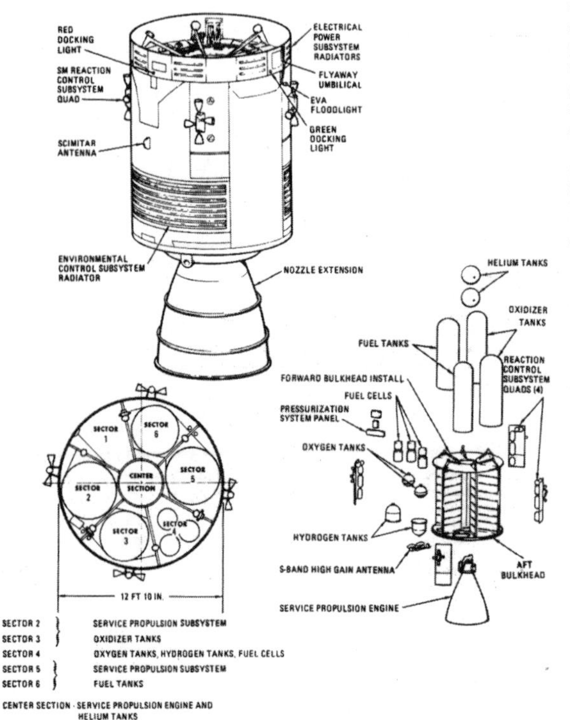

Figure 3-6.- Service module.

aluminum face sheets. The sector panels are bolted to the radial beams. Radiators used to dissipate heat from the environmental control subsystem are bonded to the sector panels on opposite sides of the SM. These radiators are each about 30 square feet in area.

The SM interior is divided into six sectors, or bays, and a center section. Sector one is currently void. It is available for installation of scientific or additional equipment should the need arise. Sector two has part of a space radiator and a reaction control system (RCS) engine quad (module) on its exterior panel and contains the service propulsion system (SPS) oxidizer sump tank. This tank is the larger of the two tanks that hold the oxidizer for the SPS engine. Sector three has the rest of the space radiator and another RCS engine quad on its exterior panel and contains the oxidizer storage tank. This tank is the second of two SPS oxidizer tanks and feeds the oxidizer sump tank in sector two. Sector four contains most of the electrical power generating equipment. It contains three fuel cells, two cryogenic oxygen and two cryogenic hydrogen tanks, and a power control relay box. The cryogenic tanks supply oxygen to the environmental control subsystem and oxygen and hydrogen to the fuel cells. Sector five has part of an environmental control radiator and an RCS engine quad on the exterior panel and contains the SPS engine fuel sump tank. This tank feeds the engine and is also connected by feed lines to the storage tank in sector six. Sector six has the rest of the environmental control radiator and an RCS engine quad on its exterior and contains the SPS engine fuel storage tank which feeds the fuel sump tank in sector five. The center section contains two helium tanks and the SPS engine. The tanks are used to provide helium pressurant for the SPS propellant tanks.

Propulsion — Main spacecraft propulsion is provided by the 20,500-pound thrust SPS. The SPS engine is a re-startable, non-throttleable engine which uses nitrogen tetroxide (N2O4) as an oxidizer and a 50-50 mixture of hydrazine and unsymmetrical-dimethylhydrazine (UDMH) as fuel. (These propellants are hypergolic, i.e., they burn spontaneously when combined without need for an igniter.) This engine is used for major velocity changes during the mission, such as midcourse corrections, lunar orbit insertion, transearth injection, and CSM aborts. The SPS engine responds to automatic firing commands from the guidance and navigation system or to commands from manual controls. The engine assembly is gimbal-mounted to allow engine thrust-vector alignment with the spacecraft center of mass to preclude tumbling. Thrust-vector alignment control is maintained by the crew. The SM RCS provides for maneuvering about and along three axes.

Additional SM systems — In addition to the systems already described, the SM has communication antennas, umbilical connections, and several exterior mounted lights. The four antennas on the outside of the SM are the steerable S-band high-gain antenna, mounted on the aft bulkhead; two VHF omnidirectional antennas, mounted on opposite sides of the module near the top; and the rendezvous radar transponder antenna, mounted in the SM fairing.

Seven lights are mounted in the aluminum panels of the fairing. Four lights (one red, one green, and two amber) are used to aid the astronauts in docking: one is a floodlight which can be turned on to give astronauts visibility during extravehicular activities, one is a flashing beacon used to aid in rendezvous, and one is a spotlight used in rendezvous from 500 feet to docking with the LM.

SM/CM separation — Separation of the SM from the CM occurs shortly before entry. The sequence of events during separation is controlled automatically by two redundant service module jettison controllers (SMJC) located on the forward bulkhead of the SM.

Command Module

The command module (CM) (fig. 3-7) serves as the command, control, and communications center for most of the mission. Supplemented by the SM, it provides all life support elements for three crewmen in the mission environments and for their safe return to the earth's surface. It is capable of attitude control about three axes and some lateral lift translation at high velocities in earth atmosphere. It also permits LM attachment, CM/LM ingress and egress, and serves as a buoyant vessel in open ocean.

Structure — The CM consists of two basic structures joined together: the inner structure (pressure shell) and the outer structure (heat shield). The inner structure, the pressurized crew compartment, is made of aluminum sandwich construction consisting of a welded aluminum inner skin, bonded aluminum honeycomb core, and outer face sheet. The outer structure is basically a heat shield and is made of stainless steel brazed honeycomb brazed between steel alloy face sheets. Parts of the area between the inner and outer sheets are filled with a layer of fibrous insulation as additional heat protection.

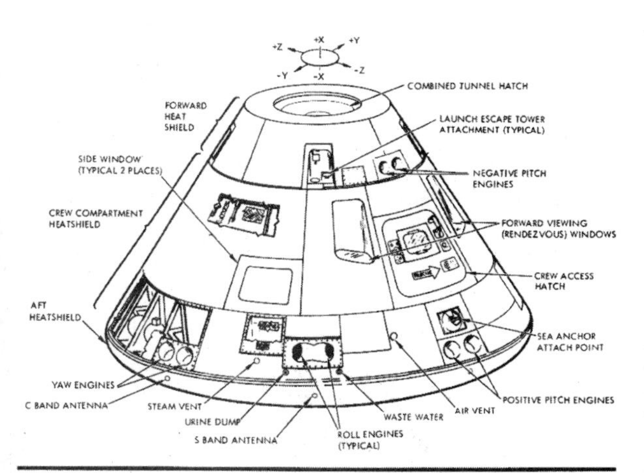

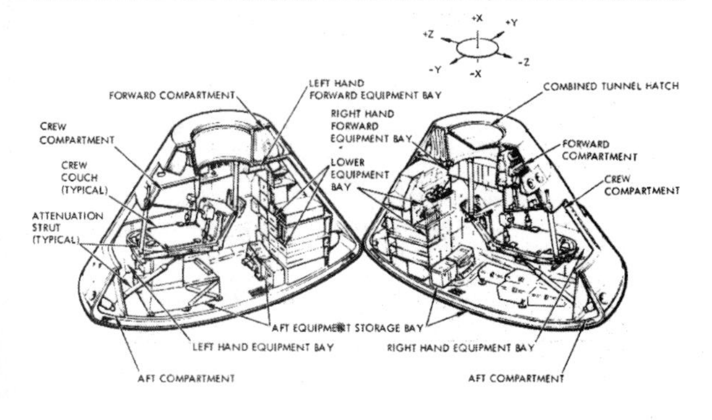

Figure 3-7.- Command module.

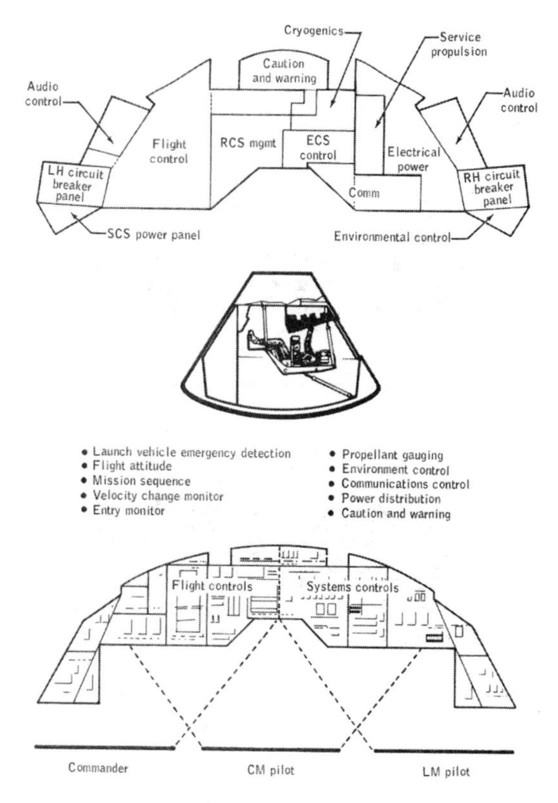

Figure 3-8.- CM main display console.

Display and controls —The main display console (MDC) (fig. 3-8) has been arranged to provide for the expected duties of crew members. These duties fall into the categories of Commander, CM Pilot, and LM Pilot, occupying the left, center, and right couches, respectively. The CM Pilot also acts as the principal navigator. All controls have been designed so they can be operated by astronauts wearing gloves. The controls are predominantly of four basic types: toggle switches, rotary switches with click-stops, thumb-wheels, and push buttons. Critical switches are guarded so that they cannot be thrown inadvertently. In addition, some critical controls have locks that must be released before they can be operated.

Flight controls are located on the left center and left side of the MDC, opposite the Commander. These include controls for such subsystems as stabilization and control, propulsion, crew safety, earth landing, and emergency detection. One of two guidance and navigation computer panels also is located here, as are velocity, attitude, and altitude indicators.

The CM Pilot faces the center of the console, and thus can reach many of the flight controls, as well as the system controls on the right side of the console. Displays and controls directly opposite him include reaction control, propellant management, caution and warning, environmental control, and cryogenic storage systems. The rotation and translation controllers used for attitude, thrust vector, and translation maneuvers are located on the arms of two crew couches. In addition, a rotation controller can be mounted at the navigation position in the lower equipment bay.

Critical conditions of most spacecraft systems are monitored by a caution and warning system. A malfunction or out-of-tolerance condition results in illumination of a status light that identifies the abnormality. It also activates the master alarm circuit, which illuminates two master alarm lights on the MDC and one in the lower equipment bay and sends an alarm tone to the astronauts' headsets. The master alarm lights and tone continue until a crewman resets the master alarm circuit. This can be done before the crewmen deal with the problem indicated. The caution and warning system also contains equipment to sense its malfunctions.

Lunar Module

The lunar module (LM) (fig. 3-9) is designed to transport two men safely from the CSM, in lunar orbit, to the lunar surface, and return them to the orbiting CSM. The LM provides operational capabilities such as communications, telemetry, environmental support, transportation of scientific equipment to the lunar surface and returning surface samples with the crew to the CSM.

The lunar module consists of two stages: the ascent stage and the descent stage. The stages are attached at four fittings by explosive bolts. Separable umbilicals and hard-line connections provide subsystem continuity to operate both stages as a single unit until separate ascent stage operation is desired. The LM is designed to operate for 48 hours after separation from the CSM, with a maximum lunar stay time of 44 hours. Table 3-I is a weight summary of the Apollo/Saturn 5 space vehicle for the Apollo 13 mission.

Main propulsion — Main propulsion is provided by the descent propulsion system (DPS) and the ascent propulsion system (APS). Each system is wholly independent of the other. The DPS provides the thrust to control descent to the lunar surface. The APS can provide the thrust for ascent from the lunar surface. In case of mission abort, the APS and/or DPS can place the LM into a rendezvous trajectory with the CSM from any point in the descent trajectory. The choice of engine to be used depends on the cause for abort, on how long the descent engine has been operating, and on the quantity of propellant remaining in the descent stage. Both propulsion system use identical hypergolic propellants. The fuel is a 50-50 mixture of hydrazine and unsymmetrical dimethylhydrazine and the oxidizer is nitrogen tetroxide. Gaseous helium pressurizes the propellant feed system . Helium storage in the DPS is at cryogenic temperatures in the super-critical state and in the APS it is gaseous at ambient temperatures.

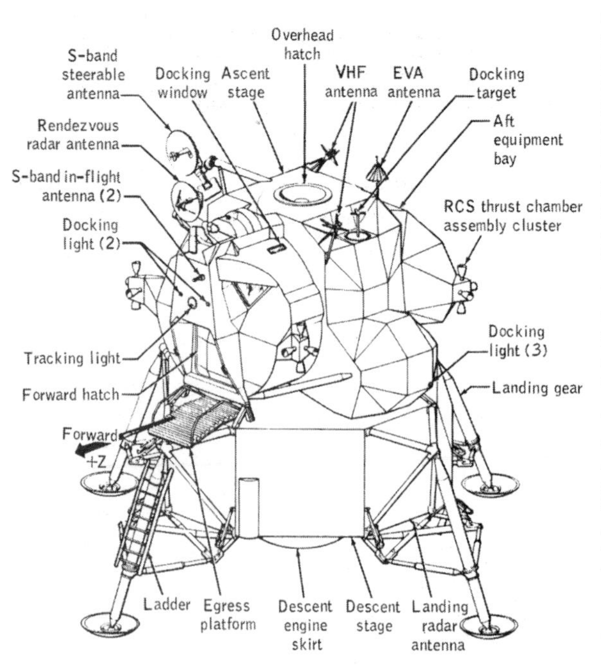

Figure 3-9.- Lunar module.

TABLE 3-I.- APOLLO 13 WEIGHT SUMMARY (WEIGHT IN POUNDS)

Stage/module	Inert weight	Total expendables	Total weight	Final separation weight
S-IC	288000	4746870	5034870	363403
S-IC/S-II interstage	11464	---	11464	---
S-II stage	78050	996960	1075010	92523
S-II/S-IVB interstage	8100	---	8100	---
S-IVB stage	25050	236671	261721	35526
Instrument unit	4482	---	4482	---
Launch vehicle at ignition 6,395,647				
Spacecraft-LM adapter	4044	---	4044	---
Lunar module	9915	23568	33483	*33941
Service module	10532	40567	51099	**14076
Command module	12572	---	12572	**11269 (Landing)
Launch escape system	9012	---	9012	---

* CSM/LM separation
** CM/SM separation

TABLE 3-I.- APOLLO 13 WEIGHT SUMMARY (WEIGHT IN POUNDS) - Concluded

Stage/module	Inert weight	Total expendables	Total weight	Final separation weight
Spacecraft at ignition 110,210				
Space vehicle at ignition			6505857	
S-IC thrust buildup			(-)84598	
Space vehicle at lift-off			6421259	
Space vehicle at orbit insertion			299998	

Ullage for propellant settling is required prior to descent engine start and is provided by the +X axis reaction engines. The descent engine is gimbaled, throttleable, and re-startable. The engine can be throttled from 1050 pounds of thrust to 6300 pounds. Throttle positions above this value automatically produce full thrust to reduce combustion chamber erosion. Nominal full thrust is 9870 pounds. Gimbal trim of the engine compensates for a changing center of gravity of the vehicle and is automatically accomplished by either the primary guidance and navigation system (PGNS) or the abort guidance system (AGS). Automatic throttle and on/off control is available in the PGNS mode of operation.

The AGS commands on/off operation but has no automatic throttle control capability. Manual control capability of engine firing functions has been provided. Manual thrust control override may, at any time, command more thrust than the level commanded by the LM guidance computer (LGC).

The ascent engine is a fixed, non-throttleable engine. The engine develops 3500 pounds of thrust, sufficient to abort the lunar descent, to launch the ascent stage from the lunar surface and place it in the desired lunar orbit. Control modes are similar to those described for the descent engine. The APS propellant is contained in two spherical titanium tanks, one for oxidizer and the other for fuel. Each tank has a volume of 36 cubic feet. Total fuel weight is 2008 pounds, of which 71 pounds are unusable. Oxidizer Weight is 3170 pounds, of which 92 pounds are unusable. The APS has a limit of 35 starts, must have a propellant bulk temperature between 50° F and 90° F prior to start, must not exceed 460 seconds of burn time, and has a system life of 24 hours after pressurization.

Electrical Power system — The electrical power system (EPS) contains six batteries which supply the electrical power requirements of the LM during undocked mission phases. Four batteries are located in the descent stage and two in the ascent stage. Batteries for the explosive devices system are not included in this system description. Postlaunch LM power is supplied by the descent stage batteries until the LM and CSM are docked. While docked, the CSM supplies electrical power to the LM up to 296 watts (peak). During the lunar descent phase, the two ascent stage batteries are paralleled with the descent stage batteries for additional power assurance. The descent stage batteries are utilized for LM lunar surface operations and checkout. The ascent stage batteries are brought on the line just before ascent phase staging. All batteries and busses may, be individually monitored for load, voltage, and failure. Several isolation and combination modes are provided.

Two inverters, each capable of supplying full load, convert the dc to ac for 115-volt, 400-hertz supply. Electrical power is distributed by the following busses: LM Pilot's dc bus, Commander's dc bus, and ac busses A and B.

The four descent stage silver-zinc batteries are identical and have a 400 ampere-hour capacity at 28 volts. Because the batteries do not have a constant voltage at various states of charge/load levels, "high" and "low" voltage taps are provided for selection. The "low voltage" tap is selected to initiate use of a fully charged battery. Cross-tie circuits in the busses facilitate an even discharge of the batteries regardless of distribution combinations. The two silver-zinc ascent stage batteries are identical to each other and have a 296 ampere-hour capacity at 28 volts. The ascent stage batteries are normally connected in parallel for even discharge. Because of design load characteristics, the ascent stage batteries do not have and do not require high and low voltage taps.

Nominal voltage for ascent stage and descent stage batteries is 30.0 volts. Reverse current relays for battery failure are one of many components designed into the EPS to enhance EPS reliability. Cooling of the batteries is provided by the environmental control system cold rail heat sinks. Available ascent electrical energy is 17.8 kilowatt hours at a maximum drain of 50 amps per battery and descent energy is 46.9 kilowatt hours at a maximum drain of 25 amps per battery.

MISSION MONITORING, SUPPORT, AND CONTROL

Mission execution involves the following functions: prelaunch checkout and launch operations; tracking the space vehicle to determine its present and future positions; securing information

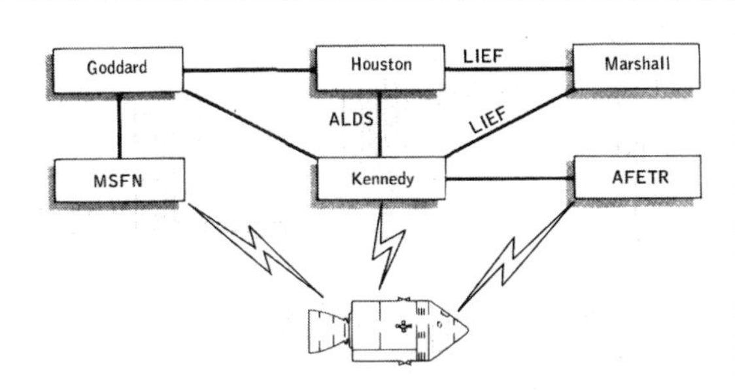

ALDS - Apollo Launch Data System
LIEF - Launch Information Exchange Facility

Figure 3-10.- Basic telemetry, command, and communication interfaces for flight control.

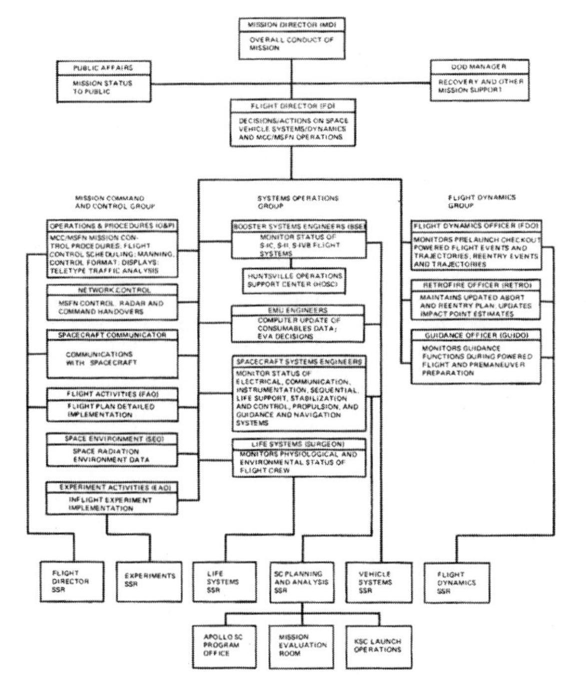

Figure 3-11.- Mission Control Center organization.

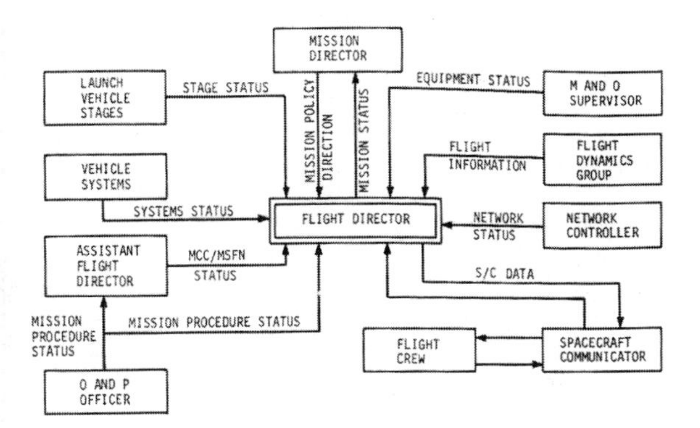

Figure 3-12.- Information flow within the Mission Operations Control Room.

on the status of the flight crew and space vehicle systems (via telemetry); evaluation of telemetry information; commanding the space vehicle by transmitting real-time and updata commands to the onboard computer; and voice communication between flight and ground crews.

These functions require the use of a facility to assemble and launch the space vehicle (see Launch Complex), a central flight control facility, a network of remote stations located strategically around the world, a method of rapidly transmitting and receiving information between the space vehicle and the central flight control facility, and a real-time data display system in which the data are made available and presented in usable form at essentially the same time that the data event occurred.

The flight crew and the following organizations and facilities participate in mission control operations:

a). Mission Control Center (MCC), Manned Spacecraft Center (MSC), Houston, Texas. The MCC contains the communication, computer display, and command systems to enable the flight controllers to effectively monitor and control the space vehicle.

b). Kennedy Space Center (KSC), Cape Kennedy, Florida. The space vehicle is launched from KSC and controlled from the Launch Control Center (LCC). Prelaunch, launch, and powered flight data are collected at the Central Instrumentation Facility (CIF) at KSC from the launch pads, CIF receivers, Merritt Island Launch Area (MILA), and the downrange Air Force Eastern Test Range (AFETR) stations. These data are transmitted to MCC via the Apollo Launch Data System (ALDS). Also located at KSC (AFETR) is the Impact Predictor (IP), for range safety purposes.

c). Goddard Space Flight Center (GSFC), Greenbelt, Maryland. GSFC manages and operates the Manned Space Flight Network (MSFN) and the NASA communications (NASCOM) network. During flight, the MSFN is under the operational control of the MCC.

d). George C. Marshall Space Flight Center (MSFC), Huntsville, Alabama. MSFC, by means of the Launch Information Exchange Facility (LIEF) and the Huntsville Operations Support Center (HOSC) provides launch vehicle systems real-time support to KSC and MCC for preflight, launch, and flight operations.

A block diagram of the basic flight control interfaces is shown in figure 3-10.

Vehicle Flight Control Capability

Flight operations are controlled from the MCC. The MCC has two flight control rooms, but only one control room is used per mission. Each control room, called a Mission Operations Control Room (MOCR), is capable of controlling individual Staff Support Rooms (SSR's) located adjacent to the MOCR. The SSR's are manned by flight control specialists who provide detailed support to the MOCR. Figure 3-11 outlines the organization of the MCC for flight control and briefly describes key responsibilities. Information flow within the MOCR is shown in figure 3-12.

The consoles within the MOCR and SSR's permit the necessary interface between the flight controllers and the spacecraft. The displays and controls on these consoles and other group displays provide the capability to monitor and evaluate data concerning the mission and, based on these evaluations, to recommend or take appropriate action on matters concerning the flight crew and spacecraft.

Problem concerning crew safety and mission success are identified to flight control personnel in the following ways;

a). Flight crew observations
b). Flight controller real-time observations
c). Review of telemetry data received from tape recorder playback
d). Trend analysis of actual and predicted values
e). Review of collected data by systems specialists
f). Correlation and comparison with previous mission data
g). Analysis of recorded data from launch complex testing

PART 2. APOLLO 13 MISSION DESCRIPTION

PRIMARY MISSION OBJECTIVES

The primary mission objectives were as follows;

Perform selenological inspection, survey, and sampling of materials in a pre-selected region of the Fra Mauro Formation.

Deploy and activate an Apollo Lunar Surface Experiments Package (ALSEP).

Develop man's capability to work in the lunar environment.

Obtain photographs of candidate exploration sites.

Table 3-II lists the Apollo 13 mission sequence of major events and the time of occurrence in ground elapsed time .

TABLE 3-II. - APOLLO 13 MISSION SEQUENCE OF EVENTS

Event	Ground elapsed time (hr:min:sec)
Range zero (02:13:00.0 p.m. e.s.t., April 11)	00:00:00
Earth parking orbit insertion	00:12:40
Second S-IVB ignition	02:35:46
Translunar injection	02:41:47
CSM/S-IVB separation	03:06:39
Spacecraft ejection from S-IVB	04:01:03
S-IVB APS evasive maneuver	04:18:01
S-IVB APS maneuver for lunar impact	05:59:59
Midcourse correction - 2 (hybrid transfer)	30:40:50
Cryogenic oxygen tank anomaly	55:54:53
Midcourse correction - 4	61:29:43
S-IVB lunar impact	77:56:40
Pericynthion plus 2-hour maneuver	79:27:39
Midcourse correction - 5	105:18:32
Midcourse correction - 7	137:39:49
Service module jettison	138:02:06
Lunar module jettison	141:30:02
Entry interface	142:40:47
Landing	142:54:41

Launch and Earth Parking Orbit

Apollo 13 was successfully launched on schedule from Launch Complex 39A, Kennedy Space Center, Florida, at 2:13 p.m. e.s.t., April 11, 1970. The launch vehicle stages inserted the S-IVB/instrument unit (IU)/ spacecraft combination into an earth parking orbit with an apogee of 100.2 nautical miles (n. mi.) and a perigee of 98.0 n. mi. (100-n. mi. circular planned). During second stage boost, the center engine of the S-II stage cut off about 132 seconds early, causing the remaining four engines to burn approximately 34 seconds longer than predicted. Space vehicle velocity after S-II boost was 223 feet per second (fps) lower than planned. As a result, the S-IVB orbital insertion burn was approximately 9 seconds longer than predicted with cutoff velocity within about 1.2 fps of planned. Total launch vehicle burn time was about 44 seconds longer than predicted. A greater than 3-sigma probability of meeting translunar injection (TLI) cutoff conditions existed with remaining S-IVB propellants.

After orbital insertion, all launch vehicle and spacecraft systems were verified and preparation was made for translunar injection (TLI). Onboard television was initiated at 01:35 ground elapsed time (g.e.t.) for about 5.5 minutes. The second S-IVB burn was initiated on schedule for TLI. All major systems operated satisfactorily and all end conditions were nominal for a free-return circumlunar trajectory.

Translunar Coast

The CSM separated from the LM/IU/S-IVB at about 03:07 g.e.t. Onboard television was then initiated for about 72 minutes and clearly showed CSM "hard docking," ejection of the CSM/LM from the S-IVB at about 04:01 g.e.t., and the S-IVB auxiliary propulsion system (APS) evasive maneuver as well as spacecraft interior and exterior scenes. The SM RCS propellant usage for the separation, transposition, docking, and ejection was nominal. All launch vehicle safing activities were performed as scheduled.

The S-IVB APS evasive maneuver by an 8-second APS Ullage burn was initiated at 04:18 g.e.t. and was successfully completed. The liquid oxygen dump was initiated at 04:39 g.e.t. and was also successfully accomplished. The first S-IVB APS burn for lunar target point impact was initiated at 06:00 g.e.t. The burn duration was 217 seconds, producing a differential velocity of approximately 28 fps. Tracking information available at 08:00 g.e.t. indicated that the S-IVB/IU would impact at 6°53' S., 30°53' W. versus the targeted 3° S., 30° W. Therefore, the second S-IVB APS (trim) burn was not required. The gaseous nitrogen pressure dropped in the IU ST-124-M3 inertial platform at 18:25 g.e.t. and the S-IVB/IU no longer had attitude control but began tumbling slowly.

At approximately 19:17 g.e.t., a step input in tracking data indicated a velocity increase of approximately 4 to 5 fps. No conclusions have been reached on the reason for this increase. The velocity change altered the lunar impact point closer to the target. The S-IVB/IU impacted the lunar surface at 77:56:40 g.e.t. (08:09:40 p.m. e.s.t. April 14) at 2.4° S., 27.9° W., and the seismometer deployed during the Apollo 12 mission successfully detected the impact. The targeted impact point was 125 n. mi. from the seismometer. The actual impact point was 74 n. mi. from the seismometer, well within the desired 189-n. mi. (350-km) radius.

The accuracy of the TLI maneuver was such that spacecraft midcourse correction No. 1 (MCC-1), scheduled for 11:41 g.e.t., was not required. MCC-2 was performed as planned at 30:41 g.e.t. and resulted in placing the spacecraft on the desired, non-free-return circumlunar trajectory with a predicted closest approach to the moon on 62 n. mi. All SPS burn parameters were normal. The accuracy of MCC-3 was such that MCC-3, scheduled for 55:26 g.e.t., was not performed. Good quality television coverage of the preparations and performance of MCC-2 was received for 49 minutes beginning at 30:13 g.e.t.

At approximately 55:55 g.e.t. (10:08 p.m. e.s.t.), the crew reported an under-voltage alarm on the CSM main bus B. Pressure was rapidly lost in SM oxygen tank no. 2 and fuel cells 1 and 3 current dropped to zero due to loss of their oxygen supply. A decision was made to abort the mission. The increased load on fuel cell 2 and decaying pressure in the remaining oxygen tank led to the decision to activate the LM, power down the CSM, and use the LM systems for life support.

At 61:30 g.e.t., a 38-fps midcourse maneuver (MCC-4) was performed by the LM DPS to place the spacecraft in a free-return trajectory on which the CM would nominally land in the Indian Ocean south of Mauritius at approximately 152:00 g.e.t.

Transearth Coast

At pericynthion plus 2 hours (79:28 g.e.t.), a LM DPS maneuver was performed to shorten the return trip time and move the earth landing point. The 263.4-second burn produced a differential velocity of 860.5 fps and resulted in an initial predicted earth landing point in the mid-Pacific Ocean at 142:53 g.e.t. Both LM guidance systems were powered up and the primary system was used for this maneuver. Following the maneuver, passive thermal control was established and the LM was powered down to conserve consumables; only the LM environmental control system (ECS) and communications and telemetry system were kept powered up.

The LM DPS was used to perform MCC-5 at 105:19 g.e.t. The 15-second burn (at 10-percent throttle) produced a velocity change of about 7.8 fps and successfully raised the entry flight path angle to -6.52°.

The CSM was partially powered up for a check of the thermal conditions of the CM with first reported receipt of S-band signal at 101:53 g.e.t. Thermal conditions an all CSM systems observed appeared to be in order for entry.

Due to the unusual spacecraft configuration, new procedures leading to entry were developed and verified in ground-based simulations. The resulting timeline called for a final midcourse correction (MCC-7) at entry interface (EI) -5 hours, jettison of the SM at EI -4.5 hours, then jettison of the LM at EI -1 hour prior to a normal atmospheric entry by the CM.

MCC-7 was successfully accomplished at 137:40 g.e.t. The 22.4-second LM RCS maneuver resulted in a predicted entry flight path angle of -6.49°. The SM was jettisoned at 138:02 g.e.t. The crew viewed and photographed the SM and reported that an entire panel was missing near the S-band high-gain antenna and a great deal of debris was hanging out . The CM was powered up and then the LM was jettisoned at 141:30 g.e.t. The EI at 400,000 feet was reached at 142:41 g.e.t.

Entry and Recovery

Weather in the prime recovery area was as follows: broken stratus clouds at 2000 feet; visibility 10 miles; 6-knot ENE winds; and wave height 1 to 2 feet. Drogue and main parachutes deployed normally. Visual contact with the spacecraft was reported at 142:50 g.e.t. Landing occurred at 142:54:41 g.e.t. (01:07:41 p.m. e.s.t., April 17). The landing point was in the mid-Pacific Ocean, approximately 21°40' S., 165°22' W. The CM landed in the stable I position about 3.5 n. mi. from the prime recovery ship, USS IWO JIMA. The crew, picked up by a recovery helicopter, was safe aboard the ship at 1:53 p.m. e.s.t., less than an hour after landing.

CHAPTER 4

REVIEW AND ANALYSIS OF APOLLO 13 ACCIDENT

PART 1. INTRODUCTION

It became clear in the course of the Board's review that the accident during the Apollo 13 mission was initiated in the service module cryogenic oxygen tank no. 2. Therefore the following analysis centers on that tank and its history. In addition, the recovery steps taken in the period beginning with the accident and continuing to reentry are discussed.

Two oxygen tanks essentially identical to oxygen tank no. 2 on Apollo 13, and two hydrogen tanks of similar design, operated satisfactorily on several unmanned Apollo flights and on the Apollo 7, 8, 9, 10, 11, and 12 manned missions. With this in mind, the Board placed particular emphasis on each difference in the history of oxygen tank no. 2 from the history of the earlier tanks, in addition to reviewing the design, assembly, and test history.

PART 2. OXYGEN TANK No. 2 HISTORY

DESIGN

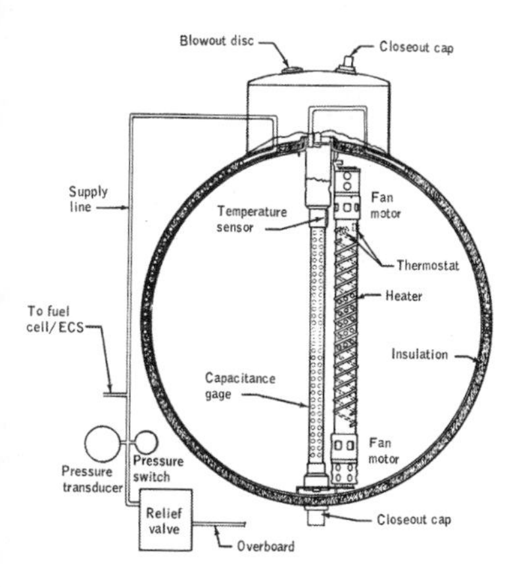

Figure 4-1.- Oxygen tank no. 2 internal components.

On February 26, 1966, the North American Aviation Corporation, now North American Rockwell (NR), prime contractor for the Apollo command and service modules (CSM), awarded a subcontract to the Beech Aircraft Corporation (Beech) to design, develop, fabricate, assemble, test, and deliver the Block II Apollo cryogenic gas storage subsystem. This was a follow-on to an earlier subcontract under which the somewhat different Block I subsystem was procured.

As the simplified drawing in figure 4-1 indicates, each oxygen tank has an outer shell and an inner shell, arranged to provide a vacuum space to reduce heat leak, and a dome enclosing paths into the tank for transmission of fluids and electrical power and signals. The space between the shells and the space in the dome are filled with insulating materials. Mounted in the tank are two tubular assemblies. One, called the heater tube, contains two thermostatically protected heater coils and two small fans driven by 1800 rpm motors to stir the tank contents. The other, called the quantity probe, consists of an upper section which supports a cylindrical capacitance gage used to measure electrically the quantity of fluid in the tank. The inner cylinder of this probe serves both as a fill and drain tube and as one plate of the capacitance gage. In addition, a temperature sensor is mounted on the outside of the quantity probe near the head. Wiring for the gage, the temperature sensor, the fan motors, and the heaters passes through the head of the quantity probe to a conduit in the dome. From there the wiring runs to a connector which ties it electrically to the appropriate external circuits in the CSM. The routing of wiring and lines from the tank through the dome is shown in figure 4-2.

As shown in figure 4-2, the fill line from the exterior of the SM enters the oxygen tank and connects to the inner cylinder of the capacitance gage through a coupling of two Teflon adapters or sleeves and a short length of Inconel tubing. The dimensions and tolerances selected are such that if "worst case" variations in an actual system were to occur, the coupling might not reach from the fill line to the gage cylinder (fig. 4-3). Thus, the variations might be such that a very loose fit would result.

The supply line from the tank leads from the head of the quantity probe to the dome and thence, after passing around the tank between the inner and outer shells, exits through the dome to supply oxygen to the fuel cells in the service module (SM) and the environmental control system (ECS) in the command module (CM). The supply line also connects to a relief valve. Under normal conditions, pressure in the tank is measured by a pressure gage in the supply line and a pressure switch near this gage is provided to turn on the heaters in

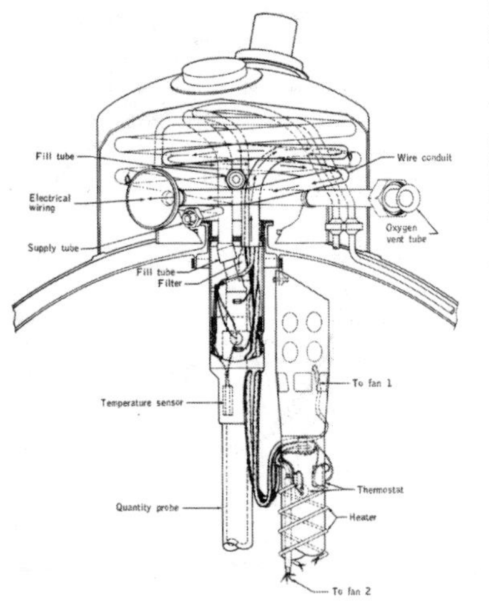

Figure 4-2.- Oxygen tank wiring and lines.

the oxygen tank if the pressure drops below a pre-selected value. This periodic addition of heat to the tank maintains the pressure at a sufficient level to satisfy the demand for oxygen as tank quantity decreases during a flight mission.

The oxygen tank is designed for a capacity of 320 pounds of supercritical oxygen at pressures ranging between 865 to 935 pounds per square inch absolute (psia). The tank is initially filled with liquid oxygen at -297° F and operates over the range from -340° F to +80° F. The term "supercritical" means that the oxygen is maintained at a temperature and pressure which assures that it is a homogeneous, single-phase fluid.

The burst pressure of the oxygen tank is about 2200 psi at -150° F. over twice the normal operating pressure at that temperature. The relief valve is designed to relieve pressure in the oxygen tank overboard at a pressure of approximately 1000 psi. The oxygen tank dome is open to the vacuum between the inner and outer tank shell and contains a rupture disc assigned to blow out at about 75 Psi.

The approximate amounts of principal materials within the oxygen tank are set forth in table 4-1.

Two oxygen tanks are mounted on a shelf in bay 4 of the SM, as shown in figure 4-4. Figures 4-5 through 4-8 are photographs of portions of the Apollo 13 service module (SM 109) at the North American Rockwell plant prior to shipment to KSC. Figure 4-5 shows the fuel cell shelf, with fuel cell 1 on the right, fuel cell 3 on the left, and fuel cell 2 behind cells 1 and 3. The top of oxygen tank no. 2 can be seen at the lower left. Figure 4-6 shows the oxygen tank shelf, with oxygen tank no. 2 at left center. Figure 4-7 shows the hydrogen tank shelf with hydrogen tank no. 1 on top and hydrogen tank no. 2 below. The bottom of the oxygen shelf shows some of the oxygen system instrumentation and wiring, largely covered by insulation. Figure 4-8 is a photograph of the bay 4 panel, which was missing from the service module after the accident.

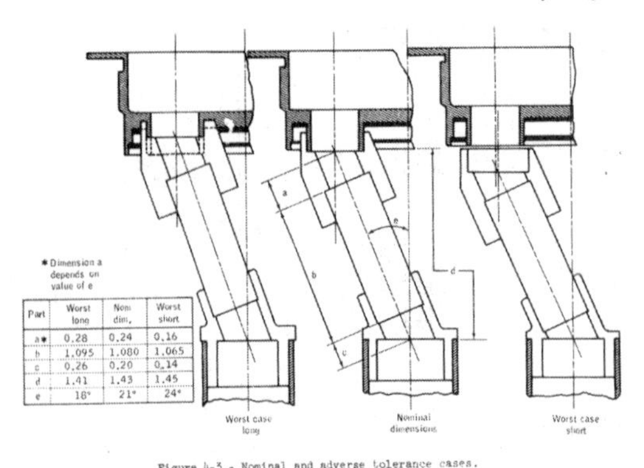

Figure 4-3.- Nominal and adverse tolerance cases.

A more detailed description of the oxygen tank design is contained in Appendix D to this report.

MANUFACTURE

The manufacture of oxygen tank no. 2 began in 1966. Under subcontracts with Beech, the inner shell of the tank was manufactured by the Airite Products Division of Electrada Corporation; the quantity probe was made by Simonds Precision Products, Inc.; and the fans and fan motors were produced by Globe Industries, Inc.

The Beech serial number assigned to the oxygen tank no. 2 flown in the Apollo 13 was 10024XTA0008. It was the eighth Block II oxygen tank built. Twenty-eight Block I oxygen tanks had previously been built by Beech.

TABLE 4-I.- MATERIALS WITHIN OXYGEN TANK

Material	Approximate quantity, lb	Available energy, Btu
Teflon-wire insulation sleeving and solid	1.1	2,400
Aluminum (all forms)	0.8	20,500
Stainless steel	2.4	15,000
Inconel alloys	1.7	2,900

The design of the oxygen tank is such that once the upper and lower halves of the inner and outer shells are assembled and welded, the heater assembly must be inserted in the tank, moved to one side, and bolted in place. Then the quantity probe is inserted into the tank and the heater assembly wires (to the heaters, the thermostats, and the fan motors) must be pulled through the head of the quantity probe and the 32-inch coiled conduit in the dome. Thus, the design required during assembly a substantial amount of wire movement inside the tank, where movement cannot be readily observed, and where possible damage to wire insulation by scraping or flexing cannot be easily detected before the tank is capped off and welded closed.

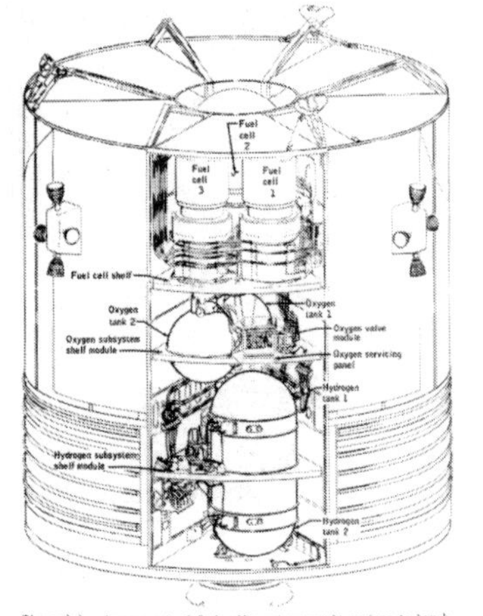

Figure 4-4.- Arrangement of fuel cells and cryogenic systems in bay 4.

Several minor manufacturing flaws were discovered in oxygen tank no. 2 in the course of testing. A porosity in a weld on the lower half of the outer shell necessitated grinding and re-welding. Rewelding was also required when it was determined that incorrect welding wire had been inadvertently used for a small weld on a vacuum pump mounted on the outside of the tank dome . The upper fan motor originally installed was noisy and drew excessive current. The tank was disassembled and the heater assembly, fans, and heaters were replaced with a new assembly and new fan. The tank was then assembled and sealed for the second time, and the space between the inner and outer shells was pumped down over a 28-day period to create the necessary vacuum.

TANK TESTS AT BEECH

Acceptance testing of oxygen tank no. 2 at Beech included extensive dielectric, insulation, and functional tests of heaters, fans, and vacuum pumps. The tank was then leak tested at 500 psi and proof tested at 1335 psi with helium.

After the helium proof test, the tank was filled with liquid oxygen and pressurized to a proof pressure of 1335 Psi by use of the tank heaters powered by 65 V ac. Extensive heat-leak tests were run at 900 psi for 25 to 30 hours over a range of ambient conditions and outflow rates. At the conclusion of the heat-leak tests, about 1 pounds of oxygen remained in the tank. About three-fourths of this was released by venting the tank at a controlled rate through the supply line to about 20 psi. The tank

Figure 4-5.- Fuel cells shelf.

was then emptied by applying warm gas at about 30 psi to the vent line to force the liquid oxygen (LOX) in the tank out the fill line (see fig. 4-2). No difficulties were recorded in this detanking operation.

The acceptance test indicated that the rate of heat leak into the tank was higher than permitted by the specifications. After some reworking the rate improved, but was still somewhat higher than specified. The tank was accepted with a formal waiver of this condition. Several other minor discrepancies were also accepted. These included oversized holes in the support for the electrical plug in the tank dome, and an oversized rivet hole in the heater assembly just above the lower fan. None of these items were serious, and the tank was accepted, filled with helium at 5 psi, and shipped to NR on May 3, 1967.

ASSEMBLY AND TEST AT NORTH AMERICAN ROCKWELL

The assembly of oxygen shelf serial number 0632AAG3277, with Beech oxygen tank serial number 10024XTA0009 as oxygen tank no. 1 and serial number 10024XTA0008 as oxygen tank no. 2, was completed on March 11, 1968. The shelf was to be installed in SM 106 for flight in the Apollo 10 mission.

Figure 4-6.- Oxygen tank shelf.

Beginning on April 27, the assembled oxygen shelf underwent standard proof-pressure leak, and functional checks. One valve on the shelf leaked and was repaired, but no anomalies were noted with regard to oxygen tank no. 2, and therefore no rework of oxygen tank no. 2 was required. None of the oxygen tank testing at NR requires use of LOX in the tanks.

On June 4, 1968, the shelf was installed in SM 106.

Between August 3 and August 8, 1968, testing of the shelf in the SM was conducted. No anomalies were noted.

Due to electromagnetic interference problem with the vac-ion pumps on cryogenic tank domes in earlier Apollo spacecraft, a modification was introduced and a decision was made to replace the complete oxygen shelf in SM 106. An oxygen shelf with approved modifications was prepared for installation in SM 106. On October 21, 1968, the oxygen shelf was removed from SM 106 for the required modification and installation in a later spacecraft.

Figure 4-7.- Hydrogen tank shelf.

The oxygen shelf was removed in the manner shown in figure 4-9. After various lines and wires were disconnected and bolts which hold the shelf in the SM were removed, a fixture suspended from a crane was placed under the shelf and used to lift the shelf and extract it from bay 4. One shelf bolt was mistakenly left in place during the initial attempt to remove the shelf; and as a consequence, after the front of the shelf was raised about 2 inches, the fixture broke, allowing the shelf to drop back into place. Photographs of the underside of the fuel cell shelf in SM 106 indicate that the closeout cap on the dome of oxygen tank no. 2 may have struck the underside of that shelf during this incident. At the time, however, it was believed that the oxygen shelf had simply dropped back into place and an analysis was performed to calculate the forces resulting from a drop of 2 inches. It now seems likely that the shelf was first accelerated upward and then dropped.

The remaining bolt was then removed, the incident recorded, and the oxygen shelf was removed without further difficulty. Following removal, the oxygen shelf was re-tested to check shelf integrity, including proof-pressure tests, leak tests, and functional tests of pressure transducers and switches, thermal switches, and vac-ion pumps. No cryogenic testing was conducted. Visual inspection revealed no problem. These tests would have disclosed external leakage or serious internal malfunctions of most types, but would not disclose fill line leakage within oxygen tank no. 2. Further calculations and tests conducted during this investigation, however, have indicated that the forces experienced by the shelf were probably close to those originally calculated assuming a 2-inch drop only. The probability of tank damage from this incident, therefore, is now considered to be rather low, although it is possible that a loosely fitting fill tube could have been displaced by the event.

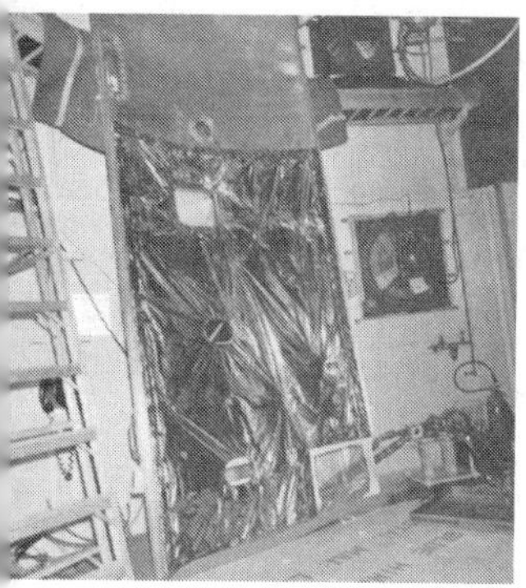
Figure 4-8.- Inside view of panel covering bay 4.

The shelf passed these tests and was installed in SM 109 on November 22, 1968. The shelf tests accomplished earlier in SM 106 were repeated in SM 109 in late December and early January, with no significant problem, and SM 109 was shipped to Kennedy Space Center (KSC) in June of 1969 for further testing, assembly on the launch vehicle, and launch.

TESTING AT KSC

At the Kennedy Space Center the CM and the SM were mated, checked, assembled on the Saturn V launch vehicle, and the total vehicle was moved to the launch pad.

The countdown demonstration test (CDDT) began on March 16. 1970. Up to this point, nothing unusual about oxygen tank no. 2 had been noted during the extensive testing at KSC. The oxygen tanks were evacuated to 5mm Hg followed by an oxygen pressure of about 80 psi. After the cooling of the fuel cells, cryogenic oxygen loading and tank pressurization to 331 psi were ompleted without abnormalities. At the time during CDDT when the oxygen tanks are normally partially emptied to about 50 percent f capacity, oxygen tank no. 1 behaved normally, but oxygen tank no. 2 only went down to 92 percent of its capacity. The normal rocedure during CDDT to reduce the quantity in the tank is to apply gaseous oxygen at 80 psi through the vent line and to open the line. When this procedure failed, it was decided to proceed with the CDDT until completion and then look at the oxygen detanking roblem in detail. An Interim Discrepancy Report was written and transferred to a Ground Support Equipment (GSE) Discrepancy eport, since a GSE filter was suspected.

n Friday, March 27, 1970, detanking operations were resumed, after discussions of the problem had been held with KSC, MSC, NR, d Beech personnel participating, either personally or by telephone. As a first step, oxygen tank no. 2, which had self-pressurized to 178 and was about 83 percent full, was vented through its fill line. The quantity decreased to 65 percent. Further discussions between

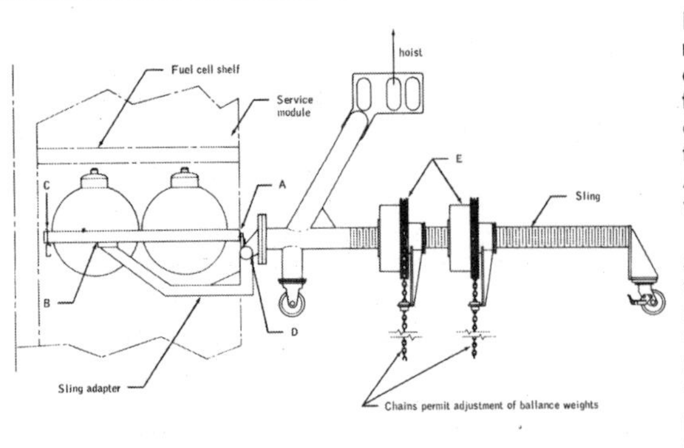

Figure 4-9.- Hoist and sling arrangement - oxygen shelf.

KSC, MSC, NR and Beech personnel considered that the problem might be due to a leak in the path between the fill line and the quantity probe due to loose fit in the sleeves and tube. Referring to figure 4-2, it will be noted that such a leak would allow the gaseous oxygen (GOX) being supplied to the vent line to leak directly to the fill line without forcing any significant amount of LOX out of the tank. At this point, a discrepancy report against the spacecraft system was written.

A "normal" detanking procedure was then conducted on both oxygen tanks, pressurizing through the vent line and opening the fill lines. Tank no. 1 emptied in a few minutes. Tank no. 2 did not. Additional attempts were made with higher pressures without effect, and a decision was made to try to "boil off" the remaining oxygen in tank no. 2 by use of the tank heaters. The heaters were energized with the 65 V dc. GSE power supply, and, about 1½ hours later, the fans were turned on to add more heat and mixing. After 6 hours of heater operation the quantity had only decreased to 35 percent, and it was decided to attempt a pressure cycling technique. With the heaters and fans still energized, the tank was pressurized to about 300 psi, held for a few minutes, and then vented through the fill line. The first cycle produced a 7-percent quantity decrease, and the process was continued, with the tank emptied after five pressure/vent cycles. The fans and heaters were turned off after about 8 hours of heater operation.

Suspecting the loosely fitting fill line connection to the quantity probe inner cylinder, KSC personnel consulted with cognizant personnel at MSC and at NR and decided to test whether the oxygen tank no. 2 could be filled without problems. It was decided that if the tank could be filled, the leak in the fill line would not be a problem in flight, since it was felt that even a loose tube resulting in an electrical short between the capacitance plates of the quantity gage would result in an energy level too low to cause any other damage.

Replacement of the oxygen shelf in the CM would have been difficult and would have taken at least 45 hours. In addition, shelf replacement would have had the potential of damaging or degrading other elements of the SM in the course of replacement activity. Therefore, the decision was made to test the ability to fill oxygen tank no. 2 on March 30, 1970, twelve days prior to the scheduled Saturday, April 11, launch, so as to be in a position to decide on shelf replacement well before the launch date.

Accordingly, flow tests with GOX were run on oxygen tank no. 2 and on oxygen tank no. 1 for comparison. No problems were encountered, and the flow rates in the two tanks were similar. In addition, Beech was asked to test the electrical energy level reached in the event of a short circuit between plates of the quantity probe capacitance gage. This test showed that very low energy levels would result. On the filling test, oxygen tanks no. 1 and no. 2 were filled with LOX to about 20 percent of capacity on March 30 with no difficulty. Tank no. 1 emptied in the normal manner, but emptying oxygen tank no. 2 again required pressure cycling with the heaters turned on.

As the launch date approached, the oxygen tank no. 2 detanking problem was considered by the Apollo organization. At this point, the "shelf drop" incident on October 21, 1968, at NR was not considered and it was felt that the apparently normal detanking which had occurred in 1967 at Beech was not pertinent because it was believed that a different procedure was used by Beech. In fact, however the last portion of the procedure was quite similar, although a slightly lower GOX pressure was utilized.

Throughout these considerations, which involved technical and management personnel of KSC, MC, NR, Beech, and NASA Headquarters, emphasis was directed toward the possibility and consequences of a loose fill tube; very little attention was paid to the extended operation of heaters and fans except to note that they apparently operated during and after the detanking sequences.

Many of the principals in the discussions were not aware of the extended heater operations. Those that did know the details of the procedure did not consider the possibility of damage due to excessive heat within the tank, and therefore did not advise management officials of any possible consequences of the unusually long heater operations.

As noted earlier in this chapter, and shown in figure 4-2, each heater is protected with a thermostatic switch, mounted on the heater tube, which is intended to open the heater circuit when it senses a temperature of 80° F. In tests conducted at MSC since the accident however, it was found that the switches failed to open when the heaters were powered from a 65 V dc supply similar to the power used at KSC during the detanking sequence. Subsequent investigations have shown that the thermostatic switches used, while rated as satisfactory for the 28 V dc spacecraft power supply, could not open properly at 65 V dc. Qualification and test procedures for the heater assemblies and switches do not at any time test the capability of the switches to open while under full current conditions. A review of the voltage recordings made during the detanking at KSC indicates that, in fact, the switches did not open when the temperature indication from within the tank rose past 80°F. Further tests have shown that the temperatures on the heater tube may have reached as much as 1000° F during the detanking. This temperature will cause serious damage to adjacent Teflon insulation, and such damage almost certainly occurred.

None of the above, however, was known at the time and, after extensive consideration was given to all possibilities of damage from a loose fill tube, it was decided to leave the oxygen shelf and oxygen tank no. 2 in the SM and to proceed with preparations for the launch of Apollo 13.

The manufacture and test history of oxygen tank no. 2 is discussed in more detail in Appendix C to this report.

PART 3. THE APOLLO 13 FLIGHT

The Apollo 13 mission was designed to perform the third manned lunar landing. The selected site was in the hilly uplands of the Fra Mauro formation. A package of five scientific experiments was planned for emplacement on the lunar surface near the lunar module (LM) landing point: (1) a lunar passive seismometer to measure and relay meteoroid impact and moonquakes and to serve as the second point in a seismic net begun with the Apollo 12 seismometer; (2) a heat flow device for measuring the heat flux from the lunar interior to the surface and surface material conductivity to a depth of 3 meters; (3) a charged-particle lunar environment experiment for measuring solar wind proton and electron effects on the lunar environment; (4) a cold cathode gage for measuring density and temperature variations in the lunar atmosphere; and (5) a dust detector experiment.

Additionally, the Apollo 13 landing crew was to gather the third set of selenological samples of the lunar surface for return to earth for extensive scientific analysis. Candidate future landing sites were scheduled to be photographed from lunar orbit with a high-resolution topographic camera carried aboard the command module.

During the week prior to launch, backup Lunar Module Pilot Charles M. Duke, Jr., contracted rubella. Blood tests were performed to determine prime crew immunity, since Duke had been in close contact with the prime crew. These tests determined that prime Commander James A. Lovell and prime Lunar Module Pilot Fred Haise were immune to rubella, but that prime Command Module Pilot Thomas K. Mattingly III did not have immunity. Consequently, following 2 days of intensive simulator training at the Kennedy Space Center, backup Command Module Pilot John L. Swigert, Jr., was substituted in the prime crew to replace Mattingly. Swigert had trained for several months with the backup crew, and this additional work in the simulators was aimed toward integrating him into the prime crew so that the new combination of crewmen could function as a team during the mission.

Launch was on time at 2:13 p.m., e.s.t., on April 11, 1970, from the KSC Launch Complex 39A. The spacecraft was inserted into a 100-nautical mile circular earth orbit. The only significant launch phase anomaly was premature shutdown of the center engine of the S-II second stage. As a result, the remaining four S-II engines burned 34 seconds longer than planned and the S-IVB third stage burned a few seconds longer than planned. At orbital insertion, the velocity was within 1.2 feet per second of the planned velocity. Moreover, an adequate propellant margin was maintained in the S-IVB for the translunar injection burn.

Orbital insertion was at 00:12:39 ground elapsed time (g.e.t.). The initial one and one-half earth orbits before translunar injection (TLI) were spent in spacecraft system checkout and included television transmissions as Apollo 13 passed over the Merritt Island Launch Area, Florida, tracking station.

The S-IVB restarted at 02:35:46 g.e.t. for the translunar injection burn , with shutdown coming some 5 minutes 51 seconds later. Accuracy of the Saturn V instrument unit guidance for the TLI burn was such that a planned midcourse correction maneuver at 11:41:23 g.e.t. was not necessary. After TLI, Apollo 13 was calculated to be on a free-return trajectory with a predicted closest approach to the lunar surface of 210 nautical miles.

The CSM was separated from the S-IVB about 3 hours after launch, and after a brief period of station-keeping, the crew maneuvered the CSM to dock with the LM vehicle in the LM adapter atop the S-IVB stage. The S-IVB stage was separated from the docked CSM and LM shortly after 4 hours into the Mission.

n manned lunar missions prior to Apollo 13, the spent S-IVB third stages were accelerated into solar orbit by a "slingshot" maneuver n which residual liquid oxygen was dumped through the J-2 engine to provide propulsive energy. On Apollo 13, the plan was to impact the S-IVB stage on the lunar surface in proximity to the seismometer emplaced in the Ocean of Storms by the crew of Apollo 12.

wo hours after TLI, the S-IVB attitude thrusters were ground commanded on to adjust the stage's trajectory toward the designated mpact at latitude 3°S. by longitude 30° W. Actual impact was at latitude 2.4° S. by longitude 27.9° W. — 74 nautical miles from the Apollo 2 seismometer and well within the desired range. Impact was at 77:56:40 g.e.t. Seismic signals relayed by the Apollo 12 seismometer s the 30,700-pound stage hit the Moon lasted almost 4 hours and provided lunar scientists with additional data on the structure of the Moon.

As in previous lunar missions, the Apollo 13 spacecraft was set up in the passive thermal control (PTC) mode which calls for a continuous roll rate of three longitudinal axis revolutions each hour. During crew rest periods and at other times in translunar and transearth coast when a stable attitude is not required, the spacecraft is placed in PTC to stabilize the thermal response by spacecraft structures and system.

At 30:40:49 g.e.t., a midcourse correction maneuver was made using the service module propulsion system. The crew preparations for the burn and the burn itself were monitored by the Mission Control Center (MMC) at MSC by telemetered data and by television from the spacecraft. This midcourse correction maneuver was a 23.2 feet per second hybrid transfer burn which took Apollo 13 off a free-return trajectory and placed it on a non-free-return trajectory. A similar trajectory had been flown on Apollo 12. The objective of having a free-return trajectory is to control the arrival time at the Mom to insure the proper lighting conditions at the landing site. Apollo 8, 10, and 11 flew a pure free return trajectory until lunar orbit insertion. The Apollo 13 hybrid transfer maneuver lowered the predicted closest approach, or pericynthion, altitude at the Moon from 210 to 64 nautical miles.

From launch through the first 46 hours of the mission, the performance of oxygen tank no. 2 was normal, so far as telemetered data and crew observations indicate. At 46:40:02, the crew turned on the fans in oxygen tank no. 2 as a routine operation. Within 3 seconds, the oxygen tank no. 2 quantity indication changed from a normal reading of about 82 percent full to an obviously incorrect reading "off-scale high," of over 100 percent. Analysis of the electrical wiring of the quantity gage shows that this erroneous reading could be caused by either a short circuit or an open circuit in the gage wiring or a short circuit between the gage plates. Subsequent events indicated that a short was the more likely failure mode.

At 47:54:50 and at 51:07:44, the oxygen tank no. 2 fans were turned on again, with no apparent adverse effects. The quantity gage continued to read off-scale high.

Following a rest period, the Apollo 13 crew began preparations for activating and powering up the LM for checkout. At 53:27 g.e.t., the Commander (CMR) and Lunar Module Pilot (LMP) were cleared to enter the LM to commence inflight inspection of the LM. Ground tests before launch had indicated the possibility of a high heat-leak rate in the LM descent stage supercritical helium tank. Crew verification of actual pressures found the helium pressure to be within normal limits. Supercritical helium is stored in the LM for pressurizing propellant tanks.

The LM was powered down and preparations were underway to close the LM hatch and run through the pre-sleep checklist when the accident in oxygen tank no. 2 occurred.

At 55:52:30 g.e.t., a master alarm on the CM caution and warning system alerted the crew to a low pressure indication in the cryogenic hydrogen tank no. 1. This tank had reached the low end of its normal operating pressure range several times previously during the flight. At 55:52:58, flight controllers in the MCC requested the crew to turn on the cryogenic system fans and heaters.

The Command Module Pilot (CMP) acknowledged the fan cycle request at 55:53:06 g.e.t., and data indicate that current was applied to the oxygen tank no. 2 fan motors at 55:53:20.

About 1-1/2 minutes later. at 55:54:53.555, telemetry from the spacecraft was lost almost totally for 1.8 seconds. During the period of data loss, the caution and warning system alerted the crew to a low voltage condition on dc main bus B. At about the same time, the crew heard a loud "bang" and realized that a problem existed in the spacecraft.

The events between fan turn on at 55:53:20 and the time when the problem was evident to the crew and Mission Control are covered in some detail in Part 4 of this chapter, "Summary Analysis of the Accident." It is now clear that oxygen tank no. 2 or its associated tubing lost pressure integrity because of combustion within the tank, and that effects of oxygen escaping from the tank caused the removal of the panel covering bay 4 and a relatively slow leak in oxygen tank no. 1 or its lines or valves. Photos of the SM taken by the crew later in the mission show the panel missing, the fuel cells on the shelf show the oxygen shelf tilted, and the high-gain antenna damaged.

The resultant loss of oxygen made the fuel cells inoperative, leaving the CM with batteries normally used only during reentry as the sole power source and with only that oxygen contained in a surge tank and repressurization packages (used to repressurize the CM after cabin venting). The LM, therefore, became the only source of sufficient electrical power and oxygen to permit safe return of the crew to Earth.

The various telemetered parameters of primary interest are shown in figure 4-10 and listed in table 4-11.

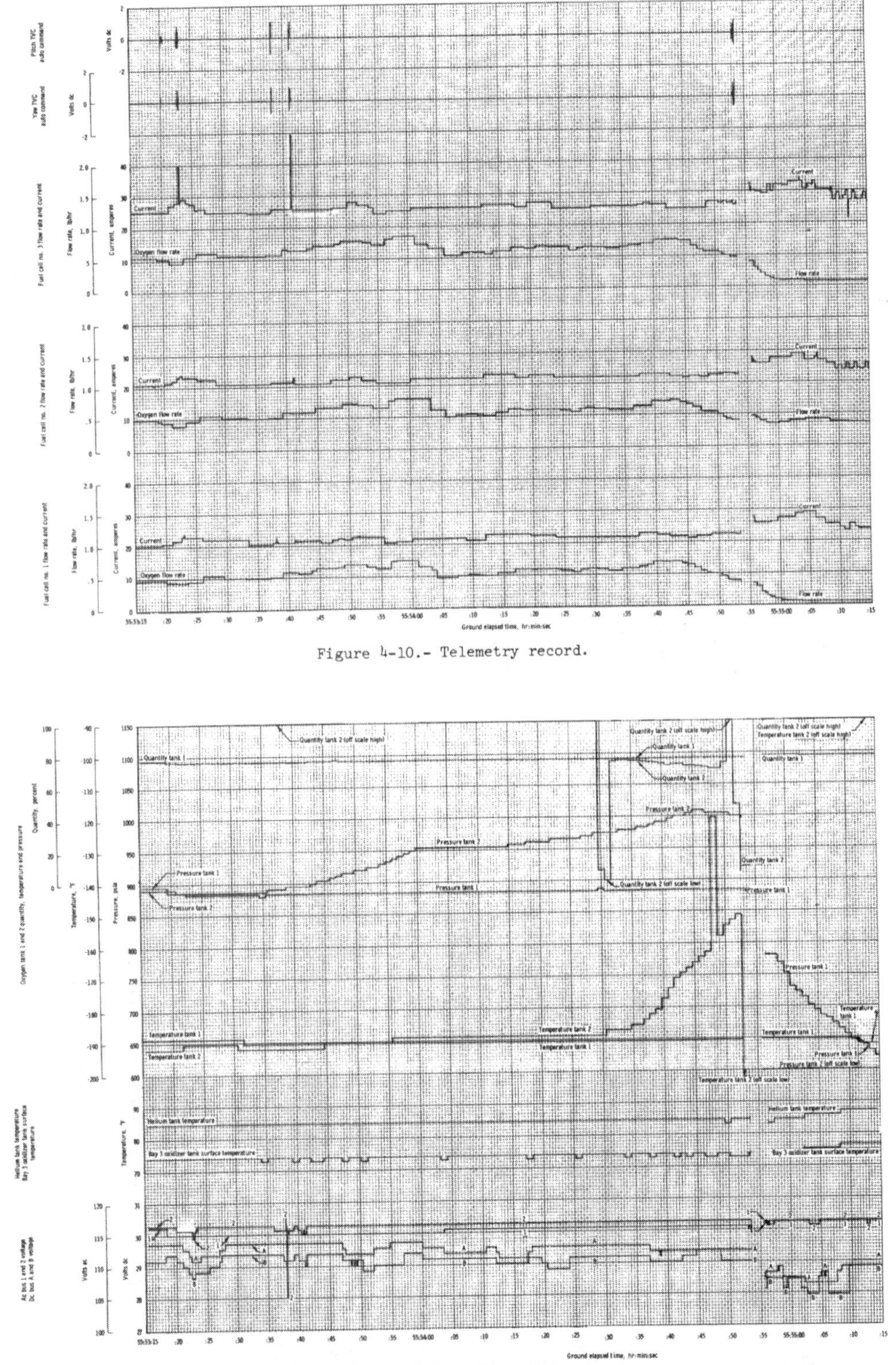

Figure 4-10.- Telemetry record.

Figure 4-10.- Concluded.

TABLE 4-II.- DETAILED CHRONOLOGY FROM
2.5 MINUTES BEFORE THE ACCIDENT TO 5 MINUTES AFTER THE ACCIDENT

Time, g.e.t.	Event
	Events During 52 Seconds Prior to First Observed Abnormality
55:52:31	Master caution and warning triggered by low hydrogen pressure in tank no. 1. Alarm is turned off after 4 seconds.
55:52:58	Ground requests tank stir.
55:53:06	Crew acknowledges tank stir.
55:53:18	Oxygen tank no. 1 fans on.
55:53:19	Oxygen tank no. 1 pressure decreases 8 Psi.
55:53:20	Oxygen tank no. 2 fans turned on.
55:53:20	Stabilization control system electrical disturbance indicates a power transient.
55:53:21	Oxygen tank no. 2 pressure decreases 4 psi.
	Abnormal Events During 90 Seconds Preceding the Accident
55:53:22.718	Stabilization control system electrical disturbance indicates a power transient.
55:53:22.757	1.2-volt decrease in ac bus 2 voltage.
55:53:22.772	11.1 amp rise in fuel cell 3 current for one sample.
55:53:36	Oxygen tank no. 2 pressure begins rise lasting for 24 Seconds
55:53:38.057	11-volt decrease in ac bus 2 voltage for one sample.
55:53:38.085	Stabilization control system electrical disturbance indicates a power transient.
55:53:41.172	22.9 amp rise in fuel cell 3 current for one sample.
55:53:41.192	Stabilization control system electrical disturbance indicates a power transient.
55:54:00	Oxygen tank no. 2 pressure rise ends at a pressure of 953.8 psia.
55:54:15	Oxygen tank no. 2 pressure begins to rise.
55:54:30	Oxygen tank no. 2 quantity drops from full scale for 2 seconds and then reads 75.3%.
55:54:31	Oxygen tank no. 2 temperature begins to rise rapidly.
55:54:43	Flow rate of oxygen to all three fuel cells begins to decrease.
55:54:45	Oxygen tank no. 2 pressure reaches maximum value of 1008.3 psia.
55:54:48	Oxygen tank no. 2 temperature rises 40° F for one sample (invalid reading).
55:54:51	Oxygen tank no. 2 quantity jumps to off-scale high and then begins to drop until the time of telemetry loss, indicating failed sensor.
55:54:52	Oxygen tank no. 2 temperature reads -151.3° F.
55:54:52.703	Oxygen tank no. 2 temperature suddenly goes off scale low, indicating failed sensor.
55:54:52.763	Last telemetered pressure from oxygen tank no. 2 before telemetry loss is 995.7 psia.
55:54:53.182	Sudden accelerometer activity on X, Y, and Z axes.
55:54:53.220	Stabilization control system body rate changes begin.
55:54:53.323	Oxygen tank no. 1 pressure drops 4.2 psi.
55:54:53.5	2.8-amp rise in total fuel cell current.
5:54:53.542	X, Y, and Z accelerations in CM indicate 1.17g, 0.65g and 0.65g, respectively.
	1.8-Second Data Loss
55:54:53.555	Loss of telemetry begins.
5:54:53.555+	Master caution and warning triggered by dc main bus B under-voltage Alarm is turned off in 6 seconds. All indications are that the cryogenic oxygen tank no. 2 lost pressure in this time period and the panel separated.
5:54:54.741	Nitrogen pressure in fuel cell 1 is off-scale low indicating failed sensor.
5:54:55.35	Recovery of telemetry data.
	Events During 5 Minutes Following the Accident
5:54:56	Service propulsion system engine valve body temperature begins a rise of 1.65°F in 7 seconds.
5:54:56	Dc main bus A decreases 0.9 volt to 28.5 volts and dc main bus B decreases 0.9 volt to 29.0 volts.
5:54:56	Total fuel cell current is 15 amps higher than the final value before telemetry loss. High current continues for 19 seconds.
5:54:56	Oxygen tank no. 2 temperature reads off-scale high after telemetry recovery, probably indicating failed sensors.
5:54:56	Oxygen tank no. 2 pressure reads off-scale low following telemetry recovery, indicating a broken supply line, a tank pressure below 19 psi, or a failed sensor.
5:54:56	Oxygen tank no. 1 pressure reads 781.9 psia and begins to drop steadily.
5:54:57	Oxygen tank no. 2 quantity reads off-scale high following telemetry recovery indicating failed sensor.
5:54:59	The reaction control system helium tank C temperature begins a 1.66° F increase in 36 seconds.
5:55:01	Oxygen flow rates to fuel cells 1 and 3 approached zero after decreasing for 7 seconds.
5:55:02	The surface temperature of the service module oxidizer tank in bay 3 begins a 3.8° F increase in a15-second period.
5:55:02	The service propulsion system helium tank temperature begins a 3.8° F increase in a 32-second period.
5:55:09	Dc main bus A voltage recovers to 29.0 volts; dc main bus B recovers to 28.8 volts.
5:55:20	Crew reports, "I believe we've had a problem here."
5:55:35	Crew reports, "We've had a main B bus undervolt
5:55:49	Oxygen tank no. 2 temperature begins steady drop lasting 59 seconds, probably indicating failed sensor.
5:56:10	Crew reports, "Okay right now, Houston. The voltage is looking good, and we had a pretty large bang associated with the caution and warning there. And as I recall, main B was the one that had had an amp spike on it once before."

55:56:38 Oxygen tank no. 2 quantity becomes erratic for 69 seconds before assuming an off-scale-low state, indicating failed sensor.
55:57:04 Crew reports, "That jolt most have rocked the sensor on — see now — oxygen quantity 2. It was oscillating down around 20 to 60 percent. Now it's full-scale high again."
55:57:39 Master caution and warning triggered by dc main bus B under-voltage Alarm is turned off in 6 seconds.
55:57:40 Dc main bus B drops below 26.25 volts and continues to fall rapidly.
55:57:44 Ac bus 2 fails within 2 seconds
55:57:45 Fuel cell 3 fails.
55:57:59 Fuel cell 1 current begins to decrease.
55:58:02 Master caution and warning caused by ac bus 2 being reset. Alarm is turned off after 2 seconds.
55:58:06 Master caution and warning triggered by dc main bus A under-voltage Alarm is turned off in 13 seconds.
55:58:07 Dc main bus A drops below 26.25 volts and in the next few seconds levels off at 25.5 volts.
55:58:07 Crew reports, "ac 2 is showing zip."
55:58:25 Crew reports, "Yes, we got a main bus A undervolt now, too, showing. It's reading about 25½. Main B is reading zip right now."
56:00:06 Master caution and warning triggered by high hydrogen flow rate to fuel cell 2. Alarm is turned off in 2 seconds.

PART 4. SUMMARY ANALYSIS OF THE ACCIDENT

Combustion in oxygen tank no. 2 led to failure of that tank, damage to oxygen tank no. 1 or its lines or valves adjacent to tank no. 2 removal of the bay 4 panel and, through the resultant loss of all three fuel cells, to the decision to abort the Apollo 13 mission. In the attempt to determine the cause of ignition in oxygen tank no. 2, the course of propagation of the combustion, the mode of tank failure and the way in which subsequent damage occurred, the Board has carefully sifted through all available evidence and examined the results of special tests and analyses conducted by the Apollo organization and by or for the Board after the accident. (For more information on details of mission events, design, manufacture and test of the system, and special tests and analyses conducted in this investigation refer to Appendices B, C, D, E, and F of this report.)

Although tests and analyses are continuing, sufficient information is now available to provide a reasonably clear picture of the nature of the accident and the events which led up to it. It is now apparent that the extended heater operation at KSC damaged the insulation on wiring in the tank and thus made the wiring susceptible to the electrical short circuit which probably initiated combustion within the tank. While the exact point of initiation of combustion may never be known with certainty, the nature of the occurrence is sufficiently understood to permit taking corrective steps to prevent its recurrence.

The Board has identified the most probable failure mode.

The following discussion treats the accident in its key phases: initiation, propagation of combustion, loss of oxygen task no. 2 system integrity, and loss of oxygen tank no. 1 system integrity.

INITIATION

Key Data

55:53:20* Oxygen tank no. 2 fans turned on.
55:53:22.757 1.2-volt decrease in ac bus 2 voltage.
55:53:22.772 11.1-ampere "spike" recorded in fuel cell 3 current followed by drop in current and rise in voltage typical of removal of power from one fan motor — indicating opening of motor circuit.
55:53:36 Oxygen tank no. 2 pressure begins to rise.

*In evaluating telemetry data, consideration must be given to the fact that the Apollo pulse code modulation (PCM) system samples dat in time and quantitizes in amplitude. For further information, reference my be made to Part B7 of Appendix B.

The evidence points strongly to an electrical short circuit with arcing as the initiating event. About 2.7 seconds after the fans were turne on in the SM oxygen tanks, an 11.1-ampere current spike and simultaneously a voltage-drop spike were recorded in the spacecraf electrical system. Immediately thereafter, current drawn from the fuel cells decreased by an amount consistent with the loss of powe to one fan. No other changes in spacecraft power were being made at the time. No power was on the heaters in the tanks at the tim and the quantity gage and temperature sensor are very low power devices. The next anomalous event recorded was the beginning of pressure rise in oxygen tank no. 2, 13 seconds later. Such a time lag is possible with low level combustion at the time. These facts poir to the likelihood that an electrical short circuit with arcing occurred in the fan motor or its leads to initiate the accident sequence. Th energy available from the short circuit was probably 10 to 20 joules. Tests conducted during this investigation have shown that thi energy is more than adequate to ignite Teflon of the type contained within the task. (The quantity gage in oxygen tank no. 2 had faile at 46:40 g.e.t. There is no evidence tying the quantity gage failure directly to accident initiation, particularly in view of the very low energ available from the gage.)

This likelihood of electrical initiation is enhanced by the high probability that the electrical wires within the tank were damaged durir the abnormal detanking operation at KSC prior to launch.

Furthermore, there is no evidence pointing to any other mechanism of initiation.

PROPAGATION OF COMBUSTION

Key Data

55:53:36 Oxygen tank no. 2 pressure begins rise (same event noted previously).
55:53:38.057 11-volt decrease recorded in ac bus 2 voltage.
55:53:41.172 22.9-ampere "spike" recorded in fuel cell 3 current, followed by drop in current and rise in voltage typical of one fan motor — indicating opening of another motor circuit.
55:54:00 Oxygen tank no. 2 pressure levels off at 954 psia.
55:54:15 Oxygen tank no. 2 pressure begins to rise again.

55:54:30	Oxygen tank no. 2 quantity gage reading drops from full scale (to which it had failed at 46:40 g.e.t.) to zero and then read 75-percent full. This behavior indicates the gage short circuit may have corrected itself.
55:54:31	Oxygen tank no. 2 temperature begins to rise rapidly.
55:54:45	Oxygen tank no. 2 pressure reading reaches maximum recorded value of 1008 psia.
55:54:52.763	Oxygen tank no. 2 pressure reading had dropped to 996 psia.

The available evidence points to a combustion process as the cause of the pressure and temperature increases recorded in oxygen tank no. 2. The pressure reading for oxygen tank no. 2 began to increase about 13 seconds after the first electrical spike, and about 55 seconds later the temperature began to increase. The temperature sensor reads local temperature, which need not represent bulk fluid temperature. Since the rate of pressure rise in the tank indicates a relatively slow propagation of burning, it is likely that the region immediately around the temperature sensor did not become heated until this time.

There are materials within the tank that can, if ignited in the presence of supercritical oxygen, react chemically with the oxygen in exothermic chemical reactions. The most readily reactive is Teflon used for electrical insulation in the tank. Also potentially reactive are metals, particularly aluminum. There is more than sufficient Teflon in the tank, if reacted with oxygen, to account for the pressure and temperature increases recorded. Furthermore, the pressure rise took place over a period of more than 69 seconds, a relatively long period, and one which would be more likely characteristic of Teflon combustion than metal-oxygen reactions.

While the data available on the combustion of Teflon in supercritical oxygen in zero-g are extremely limited, those which are available indicate that the rate of combustion is generally consistent with these observations. The cause of the 15-second period of relatively constant pressure first indicated at 55:53:59.763 has not been precisely determined; it is believed to be associated with a change in reaction rate as combustion proceeded through various Teflon elements.

While there is enough electrical power in the tank to cause ignition in the event of a short circuit or abnormal heating in defective wire, there is not sufficient electric power to account for all of the energy required to produce the observed pressure rise.

LOSS OF OXYGEN TANK NO. 2 SYSTEM INTEGRITY

Key Data

55:54:52	Last valid temperature indication (-151° F) from oxygen tank no. 2.
55:54:52.763	Last pressure reading from oxygen tank no. 2 before loss of data — 996 psia.
55:54:53.182	Sudden accelerometer activity on X, Y, and Z axes.
55:54:53.220	Stabilization control system body rate changes begin.
55:54:53.555*	Loss of telemetry data begins,
55:54:55.35	Recovery of telemetry data.
55:54:56	Various temperature indications in SM begin slight rises.
55:54:56	Oxygen tank no. 2 temperature reads off-scale high.
55:54:56	Oxygen tank no. 2 pressure reads off-scale low.

*Several bits of data have been obtained from this "loss of telemetry" data" period.

After the relatively slow propagation process described above took place, there was a relatively abrupt loss of oxygen tank no. 2 integrity. About 69 seconds after the pressure began to rise, it reached the peak recorded, 1008 psia, the pressure at which the cryogenic oxygen tank relief valve is designed to be fully open. Pressure began a decrease for 8 seconds, dropping to 996 psia before readings were lost. Virtually all signals from the spacecraft were lost about 1.85 seconds after the last presumably valid reading from within the tank, a temperature reading, and 0.8 second after the last presumably valid pressure reading (which may or may not reflect the pressure within the tank itself since the pressure transducer is about 20 feet of tubing length distant). Abnormal spacecraft accelerations were recorded approximately 0.42 second after the last pressure reading and approximately 0.38 second before the loss of signal. These facts all point to a relatively sudden loss of integrity. At about this time, several solenoid valves, including the oxygen valves feeding two of the three fuel cells, were shocked to the closed position. The "bang" reported by the crew also probably occurred in this time period. Telemetry signals from Apollo 13 were lost for a period of 1.8 seconds. When signal was reacquired, all instrument indicators from oxygen tank no, 2 were off-scale, high or low. Temperatures recorded by sensors in several different locations in the SM showed slight increases in the several seconds following reacquisition of signal. Photographs taken later by the Apollo 13 crew as the SM was jettisoned show that the bay 4 panel was ejected, undoubtedly during this event.

Figure 4-11.- Closeup view of oxygen tank shelf.

Data are not adequate to determine precisely the way in which the oxygen tank no. 2 system lost its integrity. However, available information analyses, and tests performed during this investigation indicate that most probably the combustion within the pressure vessel ultimately led to localized heating and failure at the pressure vessel closure. It is at this point, the upper end of the quantity probe, that the 1/2-inch Inconel conduit is located, through which the Teflon-insulated wires enter the pressure vessel. It is likely that the combustion progressed along the wire insulation and reached this location where all of the wires come together. This, possibly augmented by ignition of the metal in the upper end of the probe, led to weakening and failure of the closure or the conduit, or both.

Failure at this point would lead immediately to pressurization of the tank dome , which is equipped with a rupture disc rated at about 75 psi. Rupture of this disc or of the entire dome would then release oxygen, accompanied by combustion products, into bay 4. The accelerations recorded were probably caused by this release.

Release of the oxygen then began to pressurize the oxygen shelf space of bay 4. If the hole formed in the pressure vessel were large enough and formed rapidly enough, the escaping oxygen alone would be adequate to blow off the bay 4 panel. However, it is also quite possible that the escape of oxygen was accompanied by combustion of Mylar and Kapton (used extensively as thermal insulation in the oxygen shelf compartment figure 4-11 , and in the tank dome) which would augment the pressure caused by the oxygen itself. The slight temperature increases recorded at various SM locations indicate that combustion external to the tank probably took place. Further testing may shed additional light on the exact mechanism of panel ejection. The ejected panel then struck the high-gain antenna, disrupting communications from the spacecraft for the 1.8 seconds.

LOSS OF OXYGEN TANK NO. 1 INTEGRITY

Key Data

55:54:53.323	Oxygen tank no. 1 pressure drops 4 psia (from 883 psia to 879 psia).
55:54:53.555 to 55:54:55.35	Loss of telemetry data.
55:54:56	Oxygen tank no. 1 pressure reads 782 psia and drops steadily. Pressure drops over a period of 130 minutes to the point at which it was insufficient to sustain operation of fuel cell no. 2.

There is no clear evidence of abnormal behavior associated with oxygen tank no. 1 prior to loss of signal, although the one data bit (4 psi) drop in pressure in the last tank no. 1 pressure reading prior to loss of signal may indicate that a problem was beginning. Immediately after signal strength was regained, data show that tank no. 1 system had lost its integrity. Pressure decreases were recorded over a period of approximately 130 minutes, indicating that a relatively slow leak had developed in the tank no. 1 system. Analysis has indicated that the leak rate is less than that which would result from a completely ruptured line, but could be consistent with a partial line rupture or a leaking check or relief valve.

Since there is no evidence that there was any anomalous condition arising within oxygen tank no. 1, it is presumed that the loss of oxygen tank no. 1 integrity resulted from the oxygen tank no. 2 system failure. The relatively sudden, and possibly violent, event associated with loss of integrity of the oxygen tank no. 2 system could have ruptured a line to oxygen task no. 1, or have caused a valve to leak because of mechanical shock.

PART 5. APOLLO 13 RECOVERY

UNDERSTANDING THE PROBLEM

In the period immediately following the caution and warning alarm for main bus B under-voltage and the associated "bang" reported by the crew, the cause of the difficulty and the degree of its seriousness were not apparent.

The 1.8-second loss of telemetered data was accompanied by the switching of the CSM high-gain antenna mounted an the SM adjacent to bay 4 from narrow beam width to wide beam width. The high-gain antenna does this automatically 200 milliseconds after its directional lock on the ground signal has been lost.

A confusing factor was the repeated firings of various SM attitude control thrusters during the period after data loss. In all probability these thrusters were being fired to overcome the effects that oxygen venting and panel blow-off were having on spacecraft attitude, but it was believed for a time that perhaps the thrusters were malfunctioning.

The failure of oxygen tank no. 2 and consequent removal of the bay 4 panel produced a shock which closed valves in the oxygen supply lines to fuel cells 1 and 3. These fuel cells ceased to provide power in about 3 minutes, when the supply of oxygen between the closed valves and the cells was depleted. Fuel cell 2 continued to power ac bus 1 through dc main bus A, but the failure of fuel cell 3 left dc main bus B and ac bus 2 unpowered (see fig. 4-12). The oxygen tank no. 2 temperature and quantity gages were connected to ac bus 2 at the time of the accident. Thus, these parameters could not be read once fuel cell 3 failed at 55:57:44 until power was applied to ac bus 2 from main bus A.

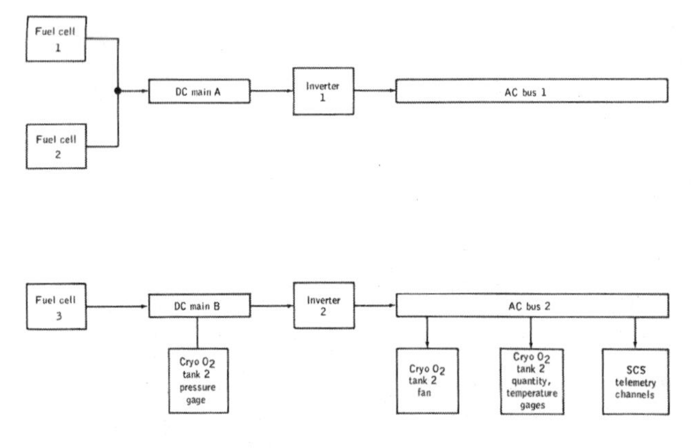

Figure 4-12.- Electrical configuration at 55:54:53 g.e.t.

The crew was not alerted to closure of the oxygen feed valves to fuel cells 1 and 3 because the valve position indicators in the CM were arranged to give warning only if both the oxygen and hydrogen valves closed. The hydrogen valves remained open. The crew had not been alerted to the oxygen tank no. 2 pressure rise or to its subsequent drop because a hydrogen tank low pressure warning had blocked the cryogenic subsystem portion of the caution and warning system several minutes before the accident.

When the crew heard the bang and got the master alarm for low dc main bus B voltage, the Commander was in the lower equipment bay of the command module, stowing a television camera which had just been in use.

The Lunar Module Pilot was in the tunnel between the CSM and the LM, returning to the CSM. The Command Module Pilot was in the left-hand couch, monitoring spacecraft performance. Because of the master alarm indicating low voltage, the CMP moved across to the right-hand couch where CSM voltages can be observed. He reported that voltages were "looking good" at 55:56:10. At this time, main bus B had recovered and fuel cell 3 did not fail for another 1½ minutes. He also reported fluctuations in the oxygen tank no. 2 quantity, followed by a return to the off-scale high position. (See fig. 4-13 for CM panel arrangements.

When fuel cells 1 and 3 electrical output readings went to zero, the ground controllers could not be certain that the cells had not somehow been disconnected from their respective busses and were not otherwise all right. Attention continued to be focused on electrical problems.

Five minutes after the accident, controllers asked the crew to connect fuel cell 3 to dc main bus B in order to be sure that the configuration was known. When it was realized that fuel cells 1 and 3 were not functioning, the crew was directed to perform an emergency powerdown to lower the load on the remaining fuel cell. Observing the rapid decay in oxygen tank no. 1 pressure, controllers asked the crew to switch power to the oxygen tank no. 2 instrumentation. When this was done, and it was realized that oxygen tank no. 2 had failed, the extreme seriousness of the situation became clear.

During the succeeding period, efforts were made to save the remaining oxygen in the oxygen tank no. 1. Several attempts were made

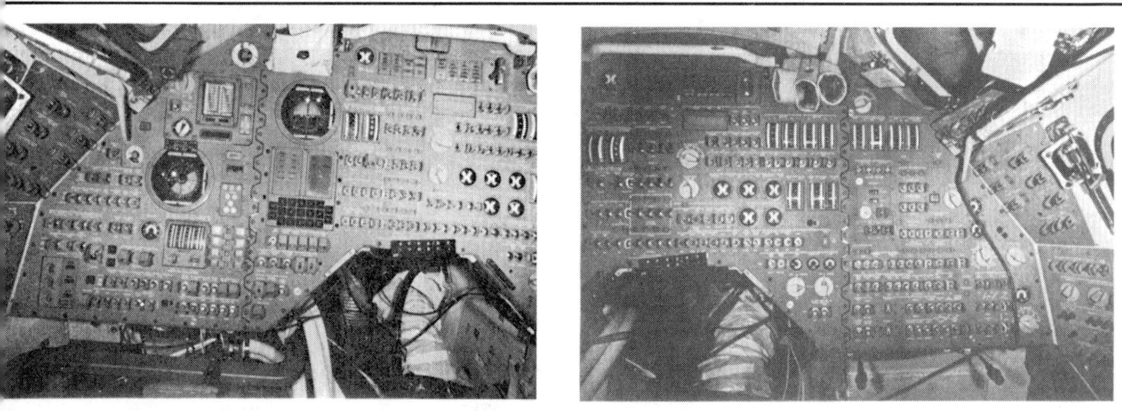

Figure 4-13.- Main display panel (left half).

Figure 4-13.- Main display panel (right half).

ut had no effect. The pressure continued to decrease.

: was obvious by about 1½ hours after the accident that the oxygen tank no. 1 leak could not be stopped and that shortly it would be ecessary to use the LM as a "lifeboat" for the remainder of the Mission.

y 58:40 g.e.t., the LM had been activated, the inertial guidance reference transferred from the CSM guidance system to the LM guidance ystem, and the CSM systems were turned off.

RETURN TO EARTH

'he remainder of the mission was characterized by two main activities — planning and conducting the necessary propulsion maneuvers ɔ return the spacecraft to Earth, and managing the use of consumables in such a way that the LM, which is designed for a basic mission vith two crewmen for a relatively short duration, could support three men and serve as the actual control vehicle for the time required.

)ne significant anomaly was noted during the remainder of the mission. At about 97 hours 14 minutes into the mission, the LMP eported hearing a "thump" and observing venting from the LM. Subsequent data review shows that the LM electrical power system xperienced a brief but major abnormal current flow at that time. There is no evidence that this anomaly was related to the accident. nalysis by the Apollo organization is continuing.

number of propulsion options were developed and considered. It was necessary to return the spacecraft to a free-return trajectory nd to make any required midcourse corrections. Normally, the service propulsion system (SPS) in the SM would be used for such naneuvers. However, because of the high electrical power requirements for using that engine, and in view of its uncertain condition and ne uncertain nature of the structure of the SM after the accident, it was decided to we the LM descent engine if possible.

he minimum practical return time was 133 hours g.e.t. to the Atlantic Ocean, and the maximum was 152 hours g.e.t. to the Indian)cean. Recovery forces were deployed in the Pacific. The return path selected was for splashdown in the Pacific Ocean at 142:40 g.e.t. his required a minimum of two burns of the LM descent engine. A third burn was subsequently made to correct the normal maneuver xecution variations in the first two burns. One small velocity adjustment was also made with reaction control system thrusters. All urns were satisfactory. Figures 4-14 and 4-15 depict the flight plan followed from the time of the accident to splashdown.

'he most critical consumables were water, used to cool the CSM and LM systems during use; CSM and LM battery power, the CSM atteries being for use during reentry and the LM batteries being needed for the rest of the mission; LM oxygen for breathing; and lithium ydroxide (LiOH) filter canisters used to remove carbon dioxide from the spacecraft cabin atmosphere. These consumables, and in articular the water and LiOH canisters, appeared to be extremely marginal in quantity shortly after the accident, but once the LM was owered down to conserve electric power and to generate less heat and thus use less water, the situation improved greatly. Engineers t MSC developed a method which allowed the crew to use materials on board to fashion a device allowing use of the CM LiOH anisters in the LM cabin atmosphere cleaning system (see fig. 4-16). At splashdown, many hours of each consumable remained available see figs. 4-17 through 4-19 and table 4-III).

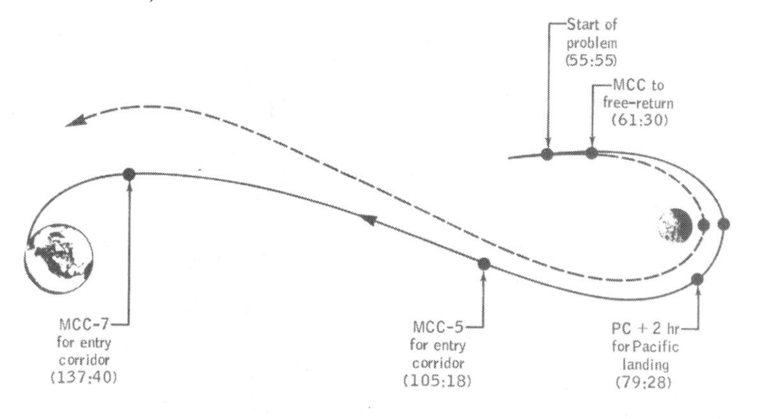

Figure 4-14.- Translunar trajectory phase.

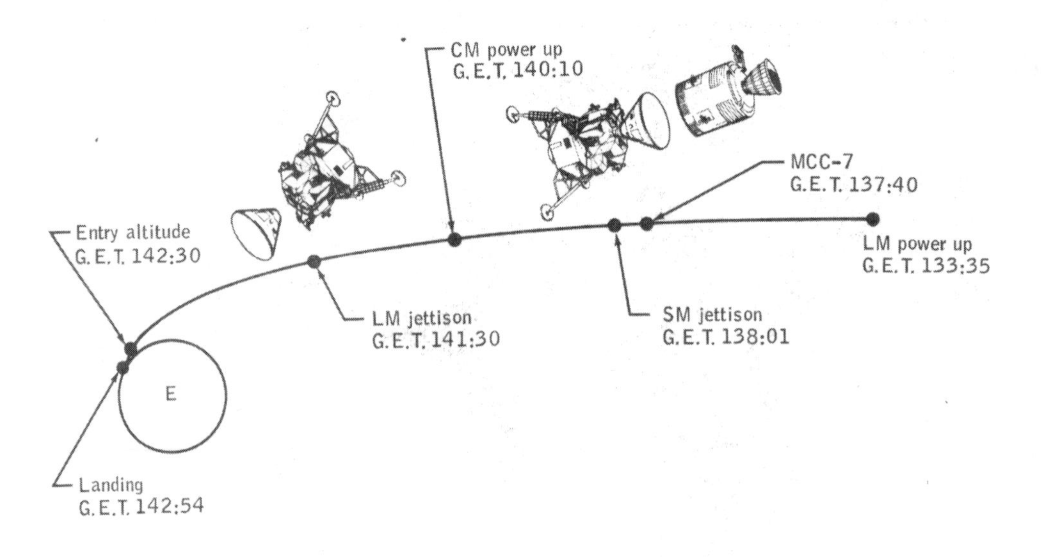

Figure 4-15.- Final trajectory phase.

Figure 4-16.- Lithium hydroxide canister modification.

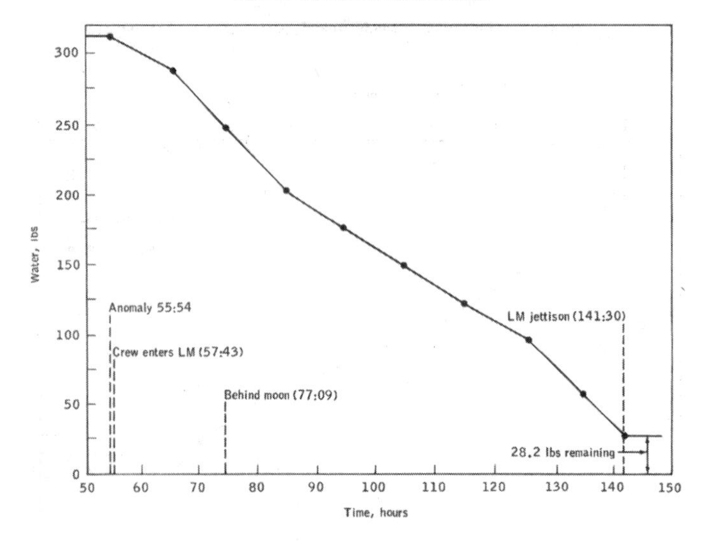

Figure 4-17.- Usable remaining water.

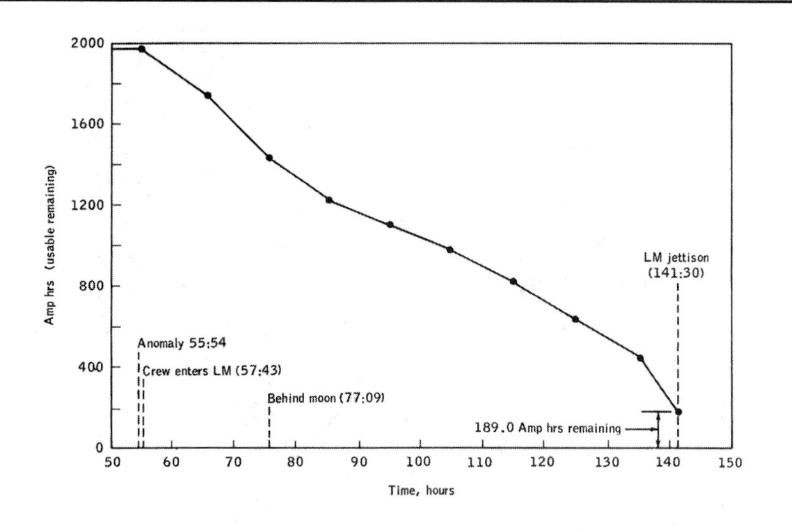

Figure 4-18.- Electrical power system consumables status.

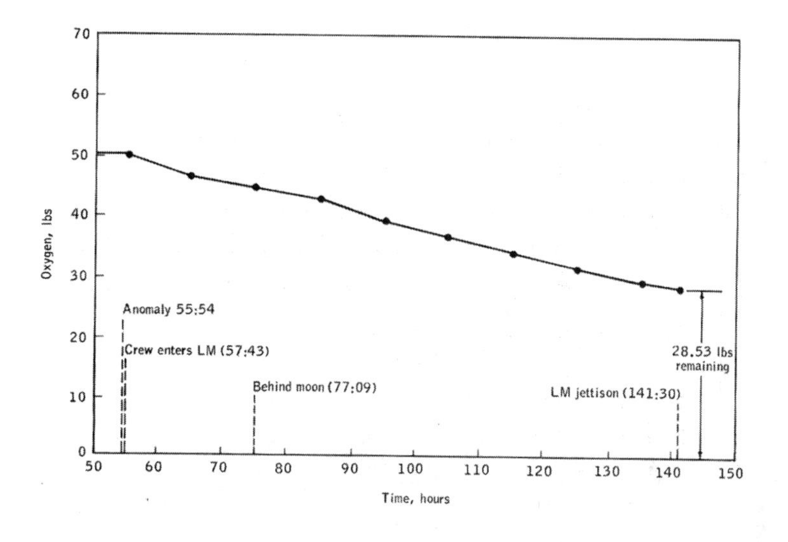

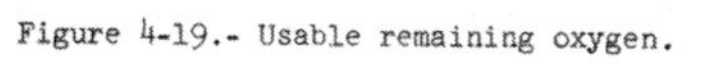
Figure 4-19.- Usable remaining oxygen.

TABLE 4-III.- CABIN ATMOSPHERE CARBON DIOXIDE REMOVAL BY LITHIUM HYDROXIDE

Required	85 hours
Available in LM	53 hours
Available in CM	182 hours

. more detailed recounting of the events during the Apollo 13 unch countdown and mission will be found in Appendix B to this report.

CHAPTER 5

FINDINGS, DETERMINATIONS, AND RECOMMENDATIONS

PART 1. INTRODUCTION

The following findings, determinations, and recommendations are the product of about 7 weeks of concentrated review of the Apoll 13 accident by the Apollo 13 Review Board. They are based on that review, on the accident Investigation by the Manned Spacecra Center (MSC) and its contractors, and on an extensive series of special tests and analyses performed by or for the Board and its panel

Sufficient work has been done to identify and understand the nature of the malfunction and the direction which the corrective action must take. All indications are that an electrically initiated fire in oxygen tank no. 2 in the service module (SM) was the cause of th accident. Accordingly the Board has concentrated on this tank; on its design, manufacture, test, handling, checkout, use, failure mode, an eventual effects on the rest of the spacecraft. The accident is generally understood, and the most probable cause has been identifie However, at the time of this report, some details of the accident are not completely clear.

Further tests and analyses, which will be carried out under the overall direction of MSC, will continue to generate new informatio relative to this accident. It is possible that this evidence may lead to conclusions differing in detail from those which can be drawn nov However, it is most unlikely that fundamentally different results will be obtained.

Recommendations are provided as to the general direction which the corrective actions should take. Significant modifications should b made to the SM oxygen storage tanks and related equipment. The modified hardware should go through a rigorous requalification te program. This is the responsibility of the Apollo organization in the months ahead.

In reaching its findings, determinations, and recommendations, it was necessary for the Board to review critically the equipment and th organizational elements responsible for it. It was found that the accident was not the result of a chance malfunction in a statistical sens but rather resulted from an unusual combination of mistakes, coupled with a somewhat deficient and unforgiving design. In brief, this what happened:

a). After assembly and acceptance testing, the oxygen tank no. 2 which flew on Apollo 13 was shipped from Beech Aircra Corporation to North American Rockwell (NR) in apparently satisfactory condition.

b). It is now known, however, that the tank contained two protective thermostatic switches on the heater assembly, whic were inadequate and would subsequently fail during ground test operations at Kennedy Space Center (KSC).

c). In addition, it is probable that the tank contained a loosely fitting fill tube assembly. This assembly was probably displace during subsequent handling, which included an incident at the prime contractor's plant in which the tank was jarred.

d). In itself, the displaced fill tube assembly was not particularly serious, but it led to the use of improvised detankir procedures at KSC which almost certainly set the stage for the accident.

e). Although Beech did not encounter any problem in detanking during acceptance tests, it was not possible to detank oxyge tank no. 2 using normal procedures at KSC. Tests and analyses indicate that this was due to gas leakage through the displace fill tube assembly.

f). The special detanking procedures at KSC subjected the tank to extended periods of heater operation and pressure cyclin These procedures had not been used before, and the tank had not been qualified by test for the conditions experience However, the procedures did not violate the specifications which governed the operation of the heaters at KSC.

g). In reviewing these procedures before the flight, officials of NASA, NR, and Beech did not recognize the possibility damage due to overheating. Many of these officials were not aware of the extended heater operation. In any event, adequa thermostatic switches might have been expected to protect the tank.

h). A number of factors contributed to the presence of inadequate thermostatic switches in the heater assembly. The origin 1962 specifications from NR to Beech Aircraft Corporation for the tank and heater assembly specified the use of 28 V c power, which is used in the spacecraft. In 1965, NR issued a revised specification which stated that the heaters should use 65 V dc power supply for tank pressurization; this was the power supply used at KSC to reduce pressurization time. Beec ordered switches for the Block II tanks but did not change the switch specifications to be compatible with 65 V dc.

i). The thermostatic switch discrepancy was not detected by NASA, NR, or Beech in their review of documentation, nor d tests identify the incompatibility of the switches with the ground support equipment (GSE) at KSC, since neither qualificatic nor acceptance testing required switch cycling under load as should have been done. It was a serious oversight in which a parties shared.

j). The thermostatic switches could accommodate the 65 V dc during tank pressurization because they normally remaine cool and closed. However, they could not open without damage with 65 V dc power applied. They were never required t do so until the special detanking During this procedure, as the switches started to open when they reached their upp temperature limit, they were welded permanently closed by the resulting arc and were rendered inoperative as protecti thermostats.

k). Failure of the thermostatic switches to open could have been detected at KSC if switch operation had been checked b observing heater current readings on the oxygen tank heater control panel. Although it was not recognized at that time, th tank temperature readings indicated that the heaters had reached their temperature limit and switch opening should ha been expected.

l). As shown by subsequent tests, failure of the thermostatic switches probably permitted the temperature of the heater tub assembly to reach about 1000° F in spots during the continuous 8-hour period of heater operation. Such heating has bee shown by tests to severely damage the Teflon insulation on the fan motor wires in the vicinity of the heater assembly. Fro that time on, including pad occupancy, the oxygen tank no. 2 was in a hazardous condition when filled with oxygen ar electrically powered.

m). It was not until nearly 56 hours into the mission, however, that the fan motor wiring, possibly moved by the fan stirring, short circuited and ignited its insulation by means of an electric arc. The resulting combustion in the oxygen tank probably overheated and failed the wiring conduit where it enters the tank, and possibly a portion of the tank itself.

n). The rapid expulsion of high-pressure oxygen which followed, possibly augmented by combustion of insulation in the space surrounding the tank, blew off the outer panel to bay 4 of the SM, caused a leak in the high-pressure system of oxygen tank no. 1, damaged the high-gain antenna, caused other miscellaneous damage, and aborted the mission.

'he accident is judged to have been nearly catastrophic. Only outstanding performance on the part of the crew, Mission Control, and ther members of the team which supported the operations successfully returned the crew to Earth.

investigating the accident to Apollo 13, the Board has also attempted to identify those additional technical and management lessons which can be applied to help assure the success of future space flight missions; several recommendations of this nature are included.

he Board recognizes that the contents of its report are largely of a critical nature . The report highlights in detail faults or deficiencies equipment and procedures that the Board has identified. This is the nature of a review board report.

is important, however, to view the criticisms in this report in a broader context. The Apollo spacecraft system is not without hortcomings, but it is the only system of its type ever built and successfully demonstrated. It has flown to the Moon five times and nded twice. The tank which failed, the design of which is criticized in this report, is one of a series which had thousands of hours of uccessful operation in space prior to Apollo 13.

Vhile the team of designers, engineers, and technicians that build and operate the Apollo spacecraft also has shortcomings, the ccomplishments speak for themselves. By hardheaded self-criticism and continued dedication, this team can maintain this nation's reeminence in space.

PART 2. ASSESSMENT OF ACCIDENT

FAILURE OF OXYGEN TANK NO. 2

. Findings

a). The Apollo 13 mission was aborted as the direct result of the rapid loss of oxygen from oxygen tank no. 2 in the SM, followed by a gradual loss of oxygen from tank no. 1. and a resulting loss of power from the oxygen-fed fuel cells.

b). There is no evidence of any forces external to oxygen tank no. 2 during the flight which might have caused its failure.

c). Oxygen tank no. 2 contained materials, including Teflon and aluminum, which if ignited will burn in supercritical oxygen.

d). Oxygen tank no. 2 contained potential ignition sources: electrical wiring, unsealed electric motors, and rotating aluminum fans.

e). During the special detanking of oxygen tank no. 2 following the countdown demonstration test (CDDT) at KSC, the thermostatic switches on the heaters were required to open while powered by 65 V dc in order to protect the heaters from overheating. The switches were only rated at 30 V dc and have been shown to weld closed at the higher voltage.

f). Data indicate that in flight the tank heaters located in oxygen tanks no. 1 and no. 2 operated normally prior to the accident, and they were not on at the time of the accident.

g). The electrical circuit for the quantity probe would generate only about 7 millijoules in the event of a short circuit and the temperature sensor wires less than 3 millijoules per second.

h). Telemetry data immediately prior to the accident indicate electrical disturbances of a character which would be caused by short circuits accompanied by electrical arcs in the fan motor or its leads in oxygen tank no. 2.

i). The pressure and temperature within oxygen tank no. 2 rose abnormally during the 1½ minutes immediately prior to the accident.

Determinations

(1) The cause of the failure of oxygen tank no. 2 was combustion within the tank.

(2) Analysis showed that the electrical energy flowing into the tank could not account for the observed increases in pressure and temperature.

(3) The heater, temperature sensor, and quantity probe did not initiate the accident sequence.

(4) The cause of the combustion was most probably the ignition of Teflon wire insulation on the fan motor wires caused by electric arcs in this wiring.

(5) The protective thermostatic switches on the heaters in oxygen tank no. 2 failed closed during the initial portion of the first special detanking operation. This subjected the wiring in the vicinity of the heaters to very high temperatures which have been subsequently shown to severely degrade Teflon insulation.

(6) The telemetered data indicated electrical arcs of sufficient energy to ignite the Teflon insulation, as verified by subsequent tests. These tests also verified that the 1-ampere fuses on the fan motors would pass sufficient energy to ignite the insulation by the mechanism of an electric arc.

(7) The combustion of Teflon wire insulation alone could release sufficient heat to account for the observed increases in tank pressure and local temperature, and could locally overheat and fail the tank or its associated tubing. The possibility of such failure at the top of the tank was demonstrated by subsequent tests.

(8) The rate of flame propagation along Teflon-insulated wires as measured in subsequent tests is consistent with the indicated rates of pressure rise within the tank.

SECONDARY EFFECTS OF TANK FAILURE

2. Findings

a). Failure of the tank was accompanied by several events including:

A "bang" as heard by the crew.

Spacecraft motion as felt by the crew and as measured by the attitude control system and the accelerometers in th command module (CM).

Momentary loss of telemetry.

Closing of several valves by shock loading.

Loss of integrity of the oxygen tank no. I system.

Slight temperature increases in bay 4 and adjacent sectors of the SM.

Loss of the panel covering bay 4 of the SM, as observed and photographed by the crew.

Displacement of the fuel cells as photographed by the crew.

Damage to the high-gain antenna as photographed by the crew.

b).The panel covering of bay 4 could be blown off by pressurization of the bay. About 25 psi of uniform pressure in bay 4 required to blow off the panel.

c).The various bays and sectors of the SM are interconnected with open passages so that all would be pressurized if any on were supplied with a pressurant at a relatively slow rate.

d).The CM attachments would be failed by an average pressure of about 10 psi on the CM heat shield and this would separat the CM from the SM.

Determinations

(1) Failure of the oxygen tank no. 2 caused a rapid local pressurization of bay 4 of the SM by the high-pressure oxyge that escaped from the tank This pressure pulse may have blown off the panel covering bay 4. This possibility wa substantiated by a series of special tests.

(2) The pressure pulse from a tank failure might have been augmented by combustion of Mylar or Kapton insulatio or both when subjected to a stream of oxygen and hot particles emerging from the top of the tank, a demonstrated in subsequent tests.

(3) Combustion or vaporization of the Mylar or Kapton might account for the discoloration of the SM engine nozz as observed and photographed by the crew.

(4) Photographs Of the SM by the crew did not establish the condition of the oxygen tank no. 2.

(5) The high-gain antenna damage probably resulted from striking by the panel, or a portion thereof, as it left the S

(6) The loss of pressure on oxygen tank no. I and the subsequent loss of power resulted from the tank no. 2 failur

(7) Telemetry, although good, is insufficient to pin down the exact nature, sequence, and location of each event of th accident in detail.

(8) The telemetry data, crew testimony, photographs, and special tests and analyses already completed are sufficie to understand the problem and to proceed with corrective actions.

OXYGEN TANK NO. 2 DESIGN

3. Findings

a). The cryogenic oxygen storage tanks contained a combination of oxidizer, combustible material, and potenti ignition sources.

b). Supercritical oxygen was used to minimize the weight, volume, and fluid-handling problems of the oxygen supp system.

c). The heaters, fans and tank instrumentation are used in the measurement and management of the oxygen supp

Determinations

(1) The storage of supercritical oxygen was appropriate for the Apollo system.

(2) Heaters are required to maintain tank pressure " the oxygen supply is used,

(3) Fans were used to prevent excessive pressure drops due to stratification, to mix the oxygen to improve accura of quantity measurements, and to insure adequate heater input at low densities and high oxygen utilization rate The need for oxygen stirring on future flights requires further investigation.

(4) The amount of material in the tank which could be ignited and burned in the given environment could have be reduced significantly.

(5) The potential ignition sources constituted an undue hazard when considered in the light of the particular tank design with its assembly difficulties.

(6) NASA, the prime contractor, and the supplier of the tank were not fully aware of the extent of this hazard.

(7) Examination of the high-pressure oxygen system in the service module following the Apollo 204 fire, which directed attention to the danger of fire in a pure oxygen environment, failed to recognize the deficiencies of the tank.

PREFLIGHT DAMAGE TO TANK WIRING

4. Findings

a). The oxygen tank no. 2 heater assembly contained two thermostatic switches designed to protect the heaters from overheating.

b). The thermostatic switches were designed to open and interrupt the heater current at 80° ± 10° F.

c). The heaters are operated an 28 V dc in flight and at NR.

d). The heaters are operated on 65 V dc at Beech Aircraft Corporation and 65 V dc at the Kennedy Space Center. These higher voltages are used to accelerate tank pressurization.

e). The thermostatic switches were rated at 7 amps at 30 V dc. While they would carry this current at 65 V dc in a closed position, they would fail if they started to open to interrupt this load.

f). Neither qualification nor acceptance testing of the heater assemblies or the tanks required thermostatic switch opening to be checked at 65 V dc. The only test of switch opening was a continuity check at Beech in which the switch was cycled open and closed in an oven.

g). The thermostatic switches had never operated in flight because this would only happen if the oxygen supply in a tank were depleted to nearly zero.

h). The thermostatic switches had never operated on the ground under load because the heaters had only been used with a relatively full tank which kept the switches cool and closed.

i). During the CDDT, the oxygen tank no. 2 would not detank in a normal manner. On March 27 and 28, a special detanking procedure was followed which subjected the heater to about 8 hours of continuous operation until the tanks were nearly depleted of oxygen.

j). A second special detanking of shorter duration followed on March 30, 1970.

k). The oxygen tanks had not been qualification tested for the conditions encountered in this procedure. However, specified allowable heater voltages and currents were not exceeded.

l). The recorded internal tank temperature went off-scale high early in the special detanking. The thermostatic switches would normally open at this point but the electrical records show no thermostatic switch operation. These indications were not detected at the time.

m). The oxygen tank heater controls at KSC contained ammeters which would have indicated thermostatic switch operation.

Determinations

(1) During the special detanking of March 27 and 28 at KSC, when the heaters in oxygen tank no. 2 were left on for an extended period, the thermostatic switches started to open while powered by 65 V dc and were probably welded shut.

(2) Failure of the thermostatic switches to open could have been detected at KSC if switch operation had been checked by observing heater current readings on the oxygen tank heater control panel. Although it was not recognized at the time, the tank temperature readings indicated that the heaters had reached their temperature limit and switch opening should have been expected.

(3) The fact that the switches were not rated to open at 65 V dc was not detected by NASA, NR, or Beech in their reviews of documentation or in qualification and acceptance testing.

(4) The failed switches resulted in severe overheating. Subsequent tests showed that heater assembly temperatures could have reached about 1000° F.

(5) The high temperatures severely damaged the Teflon insulation on the wiring in the vicinity of the heater assembly and set the stage for subsequent short circuiting. As shown in subsequent tests, this damage could range from cracking to total oxidation and disappearance of the insulation.

(6) During and following the special detanking, the oxygen tank no. 2 was in a hazardous condition whenever it contained oxygen and was electrically energized.

PART 3. SUPPORTING CONSIDERATIONS

DESIGN, MANUFACTURING, AND TEST

5. Finding

The pressure vessel of the supercritical oxygen tank is constructed of Inconel 718, and is moderately stressed at norma operating pressure.

Determination

From a structural viewpoint, the supercritical oxygen pressure vessel is quite adequately designed, employing a tough materia well chosen for this application. The stress analysis and the results of the qualification burst test program confirm the ability of the tank to exhibit adequate performance in its intended application.

6. Findings

a). The oxygen tank design includes two unsealed electric fan motors immersed in supercritical oxygen.

b). Fan motors of this design have a test history of failure during acceptance test which includes phase-to-phase and phase-to-ground faults.

c). The fan motor stator windings are constructed with Teflon coated, ceramic-insulated number 36 AWG wire. Full phase to-phase and phase - to ground insulation is not used in the motor design.

d). The motor case is largely aluminum.

Determinations

(1) The stator winding insulation is brittle and easily fractured during manufacture of the stator coils.

(2) The use of these motors in supercritical oxygen was a questionable practice.

7. Findings

a). The cryogenic oxygen storage tanks contained materials that could be ignited and which will burn under the conditions prevailing within the tank, including Teflon, aluminum, solder, and Drilube 822.

b). The tank contained electrical wiring exposed to the super critical oxygen. The wiring was insulated with Teflon.

c). Some wiring was in close proximity to heater elements and to the rotating fan.

d). The design was such that the assembly of the equipment was essentially "blind" and not amenable to inspection after completion.

e). Teflon insulation of the electrical wiring inside the cryogenic oxygen storage tanks of the SM was exposed to relatively sharp metal edges of tank inner parts during manufacturing assembly operations.

f). Portions of this wiring remained unsupported in the tank on completion of assembly.

Determinations

(1) The tank contained a hazardous combination of materials and potential ignition sources.

(2) Scraping of the electrical wiring insulation against metal inner parts of the tank constituted a substantial cumulative hazard during assembly, handling, test, checkout, and operational use.

(3) "Cold flow" of the Teflon insulation, when pressed against metal corners within the tank for an extended period of time, could result in an eventual degradation of insulation protection.

(4) The externally applied electrical tests (500-volt Hi-pot) could not reveal the extent of such possible insulation damage but could only indicate that the relative positions of the wires at the time of the tests were such that the separation or insulation would withstand the 500-volt potential without electrical breakdown.

(5) The design was such that it was difficult to insure against these hazards.

(6) There is no evidence that the wiring was damaged during manufacturing.

9. Findings

a). Dimensioning of the short Teflon and Inconel tube segments of the cryogenic oxygen storage tank fill line was such that looseness to the point of incomplete connection was possible in the event of worst-case tolerance buildup.

b). The insertion of these segments into the top of the tank quantity probe assembly at the point of its final closure and welding was difficult to achieve.

c). Probing with a hand tool was used in manufacturing to compensate for limited visibility of the tube segment positions.

Determination

It was possible for a tank to have been assembled with a set of relatively loose fill tube parts that could go undetected in final inspection and be subsequently displaced.

10. Findings

a). The Apollo spacecraft system contains numerous pressure vessels, many of which carry oxidants, plus related valves and other plumbing.

b). Investigation of potential hazards associated with these other systems was not complete at the time of the report, but is being pursued by the Manned Spacecraft Center.

c). One piece of equipment, the fuel cell oxygen supply valve module, has been identified as containing a similar combination of high-pressure oxygen, Teflon, and electrical wiring as in the oxygen tank no. 2. The wiring is unfused and is routed through a 10-amp circuit breaker.

Determination

The fuel cell oxygen supply valve module has been identified as potentially hazardous.

11. Findings

a). In the normal sequence of cryogenic oxygen storage tank integration and checkout, each tank undergoes shipping, assembly into an oxygen shelf for a service module, factory transportation to facilitate shelf assembly test, and then integration of shelf assembly to the SM.

b). The SM undergoes factory transportation, air shipment to KSC, and subsequent ground transportation and handling.

Determination

There were environments during the normal sequence of operations subsequent to the final acceptance tests at Beech that could cause a loose-fitting set of fill tube parts to become displaced.

12. Findings

a). At North American Rockwell, Downey, California, in the attempt to remove the oxygen shelf assembly from SM 106, a bolt restraining the inner edge of the shelf was not removed.

b). Attempts to lift the shelf with the bolt in place broke the lifting fixture, thereby jarring the oxygen tanks and valves.

c). The oxygen shelf assembly incorporating SIN XTA0008 in the tank no. 2 position, which had been shaken during removal from SM 106, was installed in SM 109 one month later.

d). An analysis, shelf inspection, and a partial retest emphasizing electrical continuity of internal wiring were accomplished before reinstallation.

Determinations

(1) Displacement of fill tube parts could have occurred, during the "shelf drop" incident at the prime contractor's plant, without detection.

(2) Other damage to the tank may have occurred from the jolt, but special tests and analyses indicate that this is unlikely

(3) The "shelf drop" incident was not brought to the attention of project officials during subsequent detanking difficulties at KSC.

13. Finding

Detanking expulsion of liquid oxygen out the fill line of the oxygen tank by warm gas pressure applied through the vent line, was a regular activity at Beech Aircraft, Boulder, Colorado, in emptying a portion of the oxygen used in end-item acceptance tests.

Determination

The latter stages of the detanking operation on oxygen tank no. 2 conducted at Beech on February 3, 1967, were similar to the standard procedure followed at KSC during the CDDT.

14. Findings

a). The attempt to detank the cryogenic oxygen tanks at KSC after the CDDT by the standard procedures on March 23, 1970, was unsuccessful with regard to tank no. 2.

b). A special detanking procedure was used to empty oxygen tank no. 2 after CDDT. This procedure involved continuous protracted heating with repeated cycles of pressurization to
about 300 psi with gas followed by venting.

c). It was employed both after CDDT and after a special test to verify that the tank could be filled.

d). There is no indication from the heater voltage recording that the thermostatic switches functioned and cycled the heaters off and on during these special detanking procedures.

e). At the completion of detanking following CDDT, the switches are only checked to see that they remain closed at -75° F as the tank is warmed up. They are not checked to verify that they will open at +80° F.

f). Tests subsequent to the flight showed that the current associated with the KSC 65 V dc ground powering of the heaters would cause the thermostatic switch contacts to weld closed if they attempted to interrupt this current.

g). A second test showed that without functioning thermostatic switches, temperatures in the 800° to 1000° F range would exist at locations on the heater tube assembly that were in close proximity with the motor wires. These temperatures are high enough to damage Teflon and melt solder.

Determinations

(1) Oxygen tank no. 2 (XTA 0008) did not detank after CDDT in a manner comparable to its performance the last time it had contained liquid oxygen, i.e., in acceptance test at Beech.

(2) Such evidence indicates that the tank had undergone some change of internal configuration during the intervening events of the previous 3 years.

(3) The tank conditions during the special detanking procedures were outside all prior testing of Apollo CSM cryogenic oxygen storage tanks. Heater assembly temperatures measured in subsequent tests exceeded 1000° F.

(4) Severe damage to the insulation of electrical wiring internal to the tank as determined from subsequent tests, resulted from the special procedure .

(5) Damage to the insulation, particularly on the long unsupported lengths of wiring, may also have occurred due to boiling associated with this procedure.

(6) MSC, KSC, and NR personnel did not know that the thermostatic switches were not rated to open with 65 V dc GSE power applied.

15. Findings

a). The change in detanking procedures on the cryogenic oxygen tank was made in accordance with the existing change control system during final launch preparations for Apollo 13.

b). Launch operations personnel who made the change did not have a detailed understanding of the tank internal components, or the tank history. They made appropriate contacts before making the change.

c). Communications, primarily by telephone, among MSC, KSC, NR, and Beech personnel during final launch preparations regarding the cryogenic oxygen system included incomplete and inaccurate information.

d). The MSC Test Specification Criteria Document (TSCD) which was used by KSC in preparing detailed tank test procedures states the tank allowable heater voltage and current as 65 to 85 V dc and 9 to 17 amperes with no restrictions on time.

Determinations

(1) NR and MSC personnel who prepared the TSCD did not know that the tank heater thermostatic switches would not protect the tank.

(2) Launch operations personnel assumed the tank was protected from overheating by the switches.

(3) Launch operations personnel at KSC stayed within the specified tank heater voltage and current limits during the detanking at KSC.

16. Findings

a). After receipt of the Block II oxygen tank specifications from NR, which required the tank heater assembly to operate with 65 V dc GSE power only during tank pressurization Beech Aircraft did not require their Block I thermostatic switch supplier to make a change in the switch to operate at the higher voltage.

b). NR did not review the tank or heater to assure compatibility between the switch and the GSE.

c). MSC did not review the tank or heater to assure compatibility between the switch and the GSE.

d). No tests were specified by MSC, NR, or Beech to check this switch under load.

Determinations

(1) NR and Beech specifications governing the powering and the thermostatic switch protection of the heater assemblies were inadequate.

(2) The specifications governing the testing of the heater assemblies were inadequate.

17. Finding

The hazard associated with the long heater cycle during detanking was not given consideration in the decision to fly oxygen tank no. 2.

Determinations

(1) MSC, KSC, and NR personnel did not know that the tank heater thermostatic switches did not protect the tank from overheating.

(2) If the long period of continuous heater operation with failed thermostatic switches had been known, the tank would have been replaced.

18. Findings

a). Management controls requiring detailed reviews and approvals of design, manufacturing processes, assembly procedures, test procedures, hardware acceptance, safety, reliability, and flight readiness are in effect for all Apollo hardware and operations.

b). When the Apollo 13 cryogenic oxygen system was originally designed, the management controls were not defined in as great detail as they are now.

Determination

From review of documents and interviews, it appears that the management controls existing at that time were adhered to in the case of the cryogenic oxygen system incorporated in Apollo 13.

19. Finding

The only oxygen tank no. 2 anomaly during the final countdown was a small leak through the vent quick disconnect, which was corrected.

Determination

No indications of a potential inflight malfunction of the oxygen tank no. 2 were present during the launch countdown.

MISSION EVENTS THROUGH ACCIDENT

20. Findings

a). The center engine of the S-II stage of the Saturn V launch vehicle prematurely shut down at 132 seconds due to large 16 hertz oscillations in thrust chamber pressure .

b). Data indicated less than 0.1g vibration in the CM.

Determinations

(1) Investigation of this S-II anomaly was not within the purview of the Board except insofar as it relates to the Apollo 13 accident.

(2) The resulting Oscillations or vibration of the space vehicle probably did not affect the oxygen tank

21. Findings

a). Fuel cell current increased between 46:40:05 and 46:40:08 indicating that oxygen tank no. 1 and tank no. 2 fans were turned on during this interval.

b). The oxygen tank no. 2 quantity indicated off-scale high at 46:40:08

Determinations

(1) The oxygen tank no. 2 quantity probe short circuited at 46:40:08.

(2) The short circuit could have been caused by either a completely loose fill tube part or a solder splash being carried by the moving fluid into contact with both elements of the probe capacitor.

22. Findings

a). The crew acknowledged Mission Control's request to turn on the tank fans at 55:53:06.

b). Spacecraft current increased by 1 ampere at 55:53:19.

c). The oxygen tank no. 1 pressure decreased 8 psi at 55:53:19 due to normal destratification.

Determination

The fans in oxygen tank no. 1 were turned on and began rotating at 55:55:19.

23. Findings

a). Spacecraft current increased by 1½ amperes and ac bus 2 voltage decreased 0.6 volt at 55:53:20.

b). Stabilization and Control System (SCS) gimbal command telemetry channels, which are sensitive indicators of electrical transients associated with switching on or off of certain spacecraft electrical loads, showed a negative initial transient during oxygen tank no. 2 fan turn-on cycles and a positive initial transient during oxygen tank no. 2 fan turnoff cycles during the Apollo 13 mission. A negative initial transient was measured in the SCS at 55:53:20.

c). The oxygen tank no. 2 pressure decreased about 4 psi when the fans were turned on at 55:53:21.

Determinations

(1) The fans in oxygen tank no. 2 were turned on at 55:53:20.

(2) It cannot be determined whether or not they were rotating because the pressure decrease was too small to conclusively show destratification. It is likely that they were .

24. Finding

An 11.1-amp spike in fuel cell 3 current and a momentary 1.2-volt decrease were measured in ac bus 2 at 55:53:23.

Determinations

(1) A short circuit occurred in the circuits of the fans in oxygen tank no. 2 which resulted in either blown fuses or opened wiring, and one fan ceased to function.

(2) The short circuit probably dissipated an energy in excess of 10 joules which, as shown in subsequent tests, is more than sufficient to ignite Teflon wire insulation by means of an electric arc.

25. Findings

a). A momentary 11-volt decrease in ac bus 2 voltage was measured at 55:53:38.

b). A 22.9-amp spike in fuel cell 3 current was measured at 55:53:41.

c). After the electrical transients, CM current and ac bus 2 voltage returned to the values indicated prior to the turn-on of the fans in oxygen tank no. 2.

Determination

Two short circuits occurred in the oxygen tank no. 2 fan circuits between 55:53:38 and 55:53:41 which resulted in either blown fuses or opened wiring, and the second fan ceased to function.

26. Finding

Oxygen tank no. 2 telemetry showed a pressure rise from 887 to 954 psia between 55:53:36 and 55:54:00. It then remained nearly constant for about 15 seconds and then rose again from 954 to 1008 psia, beginning at 55:54:15 and ending at 55:54:45

Determinations

(1) An abnormal pressure rise occurred in oxygen tank no. 2.

(2) Since no other known energy source in the tank could produce this pressure buildup, it is concluded to have resulted from combustion initiated by the first short circuit which started a wire insulation fire in the tank.

27. Findings

a). The pressure relief valve was designed to be fully open at about 1000 psi.

b). Oxygen tank no. 2 telemetry showed a pressure drop from 1008 psia at 55:54:45 to 996 psia at 55:54:55, at which time telemetry data were lost.

Determination

This drop resulted from the normal operation of the pressure relief valve as verified in subsequent tests.

28. Findings

a). At 55:54:29, when the pressure in oxygen tank no. 2 exceeded the master caution and warning trip level of 975 psia, the CM master alarm was inhibited by the fact that a warning of low hydrogen pressure was already in effect, and neither the crew nor Mission Control was alerted to the pressure rise.

b). The master caution and warning system logic for the cryogenic system is such that an out-of-tolerance condition of one measurement which triggers a master alarm prevents another master alarm from being generated when any other parameter in the same system becomes out-of-tolerance.

c). The low-pressure trip level of the master caution and warning system for the cryogenic storage system is only 1 psi below the specified lower limit of the pressure switch which controls the tank heaters. A small imbalance in hydrogen tank pressures or a shift in transducer or switch calibration can cause the master caution and warning to be triggered preceding each heater cycle. This occurred several times on Apollo 13.

d). A limit sense light indicating abnormal oxygen tank no. 2 pressure should have come on in Mission Control about 30 seconds before oxygen tank no. 2 failed. There is no way to ascertain that the light did, in fact, come on. If it did come on, Mission Control did not observe it.

Determinations

(1) If the pressure switch setting and master caution and warning trip levels were separated by a greater pressure differential, there would be less likelihood of unnecessary master alarms.

(2) With the present master caution and warning system, a spacecraft problem can go unnoticed because of the presence of a previous out-of-tolerance condition in the same subsystem.

(3) Although a master alarm at 55:54:29 or observance of a limit sense light in Mission Control could have alerted the crew or Mission Control in sufficient time to detect the pressure rise in oxygen tank no. 2, no action could have been taken at that time to prevent the tank failure. However, the information could have been helpful to Mission Control and the crew in diagnosis of spacecraft malfunctions.

(4) The limit sense system in Mission Control can be modified to constitute a more positive backup warning system.

29. Finding

Oxygen tank no. 2 telemetry showed a temperature rise of 38° F beginning at 55:54:31 sensed by a single sensor which measured local temperature. This sensor indicated off-scale low at 55:54:55.

Determinations

(1) An abnormal and sudden temperature rise occurred in oxygen tank no. 2 at approximately 55:54:31.

(2) The temperature was a local value which rose when combustion had progressed to the vicinity of the sensor.

(3) The temperature sensor failed at 55:54:55.

30. Finding

Oxygen tank no. 2 telemetry indicated the following changes: (1) quantity decreased from off-scale high to off-scale low in 2 seconds at 55:54:30, (2) quantity increased to 75.3 percent at 55:54:32, and (3) quantity was off-scale high at 55:54:51 and later became erratic.

Determinations

(1) Oxygen tank no. 2 quantity data between 55:54:32 and 55:54:50 may represent valid measurements.

(2) Immediately preceding and following this time period, the indications were caused by electrical faults.

31. Findings

a). At about 55:54:53, or about half a second before telemetry loss, the body-mounted linear accelerometers in the command module, which are sampled at 100 times per second, began indicating spacecraft motions. These disturbances were erratic, but reached peak values of 1.17g, 0.65g, and 0.65g in the X, Y, and Z directions, respectively, about 15 milliseconds before data loss.

b). The body-mounted roll, pitch, and yaw rate gyros showed low level activity for 1/4 second beginning at 55:54:53.220.

c). The integrating accelerometers indicated that a velocity increment of approximately 0.5 fps was imparted to the spacecraft between 55:54:53 and 55:54:55.

d). Doppler tracking data measured an incremental velocity component of 0.26 fps along a line from the Earth to the spacecraft at approximately 55:54:55.

e). The crew heard a loud "bang" at about this time.

f). Telemetry data were lost between approximately 55:54:53 and 55:54:55 and the spacecraft switched from the narrow-beam antenna to the wide-beam antenna.

g). Crew observations and photographs showed the bay 4 panel to be missing and the high-gain antenna to be damaged.

Determinations

(1) The spacecraft was subjected to abnormal forces at approximately 55:54:53. These disturbances were reactions resulting from failure and venting of the oxygen tank no. 2 system and subsequent separation and ejection of the bay 4 panel.

(2) The high-gain antenna was damaged either by the panel or a section thereof from bay 4 at the time of panel separation.

32. Finding

Temperature sensors in bay 3, bay 4, and the central column of the SM indicated abnormal increases following reacquisition of data at 55:54:55.

Determination

Heating took place in the SM at approximately the time of panel separation.

33. Findings

a). The telemetered nitrogen pressure in fuel cell 1 was off-scale low at reacquisition of data at 55:54:55.

b). Fuel cell 1 continued to operate for about 3 minutes past this time.

c). The wiring to the nitrogen sensor passes along the top of the shelf which supports the fuel cells immediately above the oxygen tanks.

Determinations

(1) The nitrogen pressure sensor in fuel cell 1 or its wiring failed at the time of the accident.

(2) The failure was probably caused by physical damage to the sensor wiring or shock.

(3) This is the only known instrumentation failure outside the oxygen system at that time.

34. Finding

Oxygen tank no. 1 pressure decreased rapidly from 879 psia to 782 psia at approximately 55:54:54 and then began to decrease more slowly at 55:54:56.

Determination

A leak caused loss of oxygen from tank no. 1 beginning at approximately 55:54:54.

35. Findings

a). Oxygen flow rates to fuel cells 1 and 3 decreased in a 5-second period beginning at 55:54:55, but sufficient volume existed in lines feeding the fuel cells to allow them to operate about 5 minutes after the oxygen supply valves were cut off.

b). The crew reported at 55:57:44 that five valves in the reaction control system (RCS) were closed. The shock required to close the oxygen supply valves is of the same order of magnitude as the shock required to close the RCS valves.

c). Fuel cells 1 and 3 failed at about 55:58.

Determination

The oxygen supply valves to fuel cells 1 and 3, and the five RCS valves, were probably closed by the shock of tank failure or panel ejection or both.

MISSION EVENTS AFTER ACCIDENT

36. Findings

a). Since data presented to flight controllers in Mission Control are updated only once per second, the 1.8-second loss of data which occurred in Mission Control was not directly noticed. However, the Guidance Officer did note and report a "hardware restart" of the spacecraft computer. This was quickly followed by the crew's report of a problem.

b). Immediately after the crew's report of a "bang" and a main bus B undervolt all fuel cell output currents and all bus voltages were normal, and the cryogenic oxygen tank indications were as follows:

Oxygen tank no. 1:
Pressure: Several hundred psi below normal
Quantity: Normal
Temperature : Normal

Oxygen tank no. 2:
Pressure : Off-scale low
Quantity: Off-scale high
Temperature: Off-scale high

c). The nitrogen pressure in fuel cell 1 indicated zero, which was incompatible with the hydrogen and oxygen pressures in this fuel cell, which were normal. The nitrogen pressure is used to regulate the oxygen and hydrogen pressure, and hydrogen and oxygen pressures in the fuel cell would follow the nitrogen pressure.

d). Neither the crew nor Mission Control was aware at the time that oxygen tank no. 2 pressure had risen abnormally just before the data loss.

e). The flight controllers believed that a probable cause of these indications could have been a cryogenic storage system instrumentation failure, and began pursuing this line of investigation.

Determination

Under these conditions it was reasonable to suspect a cryogenic storage system instrumentation problem, and to attempt to verify the readings before taking any action. The fact that the oxygen tank no. 2 quantity measurement was known to have failed several hours earlier also contributed to the doubt about the creditability of the telemetered data.

37. Findings

a). During the 3 minutes following data loss, neither the flight controllers nor the crew noticed the oxygen flows to fuel cells 1 and 3 were less than 0.1 lb./hr. These were unusually low readings for the current being drawn.

b). Fuel cells 1 and 5 failed at about 5 minutes after the data loss.

c). After the fuel cell failures, which resulted in dc main bus B failure and the under-voltage condition on dc main bus A, Mission Control diverted its prime concern from what was initially believed to be a cryogenic system instrumentation problem to the electrical power system.

d). Near-zero oxygen flow to fuel cells 1 and 3 was noted after the main bus B failure, but this was consistent with no power output from the fuel cells.

e). The flight controllers believed that the fuel cells could have been disconnected from the busses and directed the crew to connect fuel cell 1 to dc main bus A and fuel cell 3 to dc main bus B.

f). The crew reported the fuel cells were configured as directed and that the talkback indicators confirmed this.

Determinations

(1) Under these conditions it was logical for the flight controllers to attempt to regain power to the busses since the fuel cells might have been disconnected as a result of a short circuit in the electrical system. Telemetry does not indicate whether or not fuel cells are connected to busses, and the available data would not distinguish between a disconnected fuel cell and a failed one.
(2) If the crew had been aware of the reactant valve closure, they could have opened them before the fuel cells were starved of oxygen. This would have simplified subsequent actions.

38. Finding

The fuel cell reactant valve talkback indicators in the spacecraft do not indicate closed unless both the hydrogen and oxygen valves are closed.

Determinations

(1) If these talkbacks were designed so that either a hydrogen or oxygen valve closure would indicate "barberpole" the Apollo 13 crew could possibly have acted in time to delay the failure of fuel cells 1 and 3, although they would nevertheless have failed when oxygen tank no. 1 ceased to supply oxygen.

(2) The ultimate outcome would not have been changed, but had the fuel cells not failed, Mission Control and the crew would not have had to contend with the failure of dc main bus B and ac bus 2 or attitude control problems while trying to evaluate the situation.

Reaction Control System

39. Findings

a). The crew reported the talkback indicators for the helium isolation valves in the SM RCS quads B and D indicated closed shortly after the dc main bus B failure. The secondary fuel pressurization valves for quads A and C also were reported closed.

b). The SM RCS quad D propellant tank pressures decreased until shortly after the crew was requested to confirm that the helium isolation valves were opened by the crew.

c). During the 1½ hour period following the accident, Mission Control noted that SM RCS quad C propellant was not being used, although numerous firing signals were being sent to it.

d). Both the valve solenoids and the onboard indications of valve position of the propellant isolation valves for quad C are powered by dc main bus B.

e). During the 1½ hour period immediately following the accident, Mission Control advised the crew which SM RCS thrusters to power and which ones to unpower.

Determinations

(1) The following valves were closed by shock at the time of the accident: Helium isolation valves in quads B and D, Secondary fuel pressurization valves in quads A and C

(2) The propellant isolation valves in quad C probably were closed by the same shock.

(3) Mission Control correctly determined the status of the RCS system and Properly advised the crew on how to regain automatic attitude control.

Management of Electrical System

40. Findings

a). After fuel cell 1 failed, the total dc main bus A load was placed on fuel cell 2 and the voltage dropped to approximately 25 volts, causing a caution and warning indication and a master alarm.

b). After determining the fuel cell 2 could not supply enough power to dc main bus A to maintain adequate voltage, the crew connected entry battery A to this bus as an emergency measure to increase the bus voltage to its normal operating value.

c). Mission Control directed the crew to reduce the electrical load on dc main bus A by following the emergency powerdown checklist contained in the onboard Flight Data File.

d). When the power requirements were sufficiently reduced so that the one remaining fuel cell could maintain adequate bus voltage, Mission Control directed the crew to take the entry battery off line.

e). Mission Control then directed the crew to charge this battery in order to get as much energy back into it as possible, before the inevitable loss of the one functioning fuel cell.

Determinations

(1) Emergency use of the entry battery helped prevent potential loss of dc main bus A, which could have led to loss of communications between spacecraft and ground and other vital CM functions.

(2) Available emergency powerdown lists facilitated rapid reduction of loads on the fuel cell and batteries.

Attempts to Restore Oxygen Pressure

41. Findings

a). After determining that the CM problems were not due to instrumentation malfunctions, and after temporarily securing a stable electrical system configuration, Mission Control sought to improve oxygen pressures by energizing the fan and heater circuits in both oxygen tanks.
b). When these procedures failed to arrest the oxygen loss, Mission Control directed the crew to shut down fuel cells 1 and 3 by closing the hydrogen and oxygen flow valves.

Determinations

(1) Under more normal conditions oxygen pressure might have been increased by turning on heaters and fans in the oxygen tanks; no other known actions had such a possibility.

(2) There was a possibility that oxygen was leaking downstream of the valves; had this been true, closing of the valves might have preserved the remaining oxygen in oxygen tank no. 1.

Lunar Module Activation

42. Findings

a). With imminent loss of oxygen from oxygen tanks no. 1 and no. 2, and failing electrical power in the CM, it was necessary to use the lunar module (LM) as a "lifeboat" for the return to Earth.

b). Mission Control and the crew delayed LM activation until about 15 minutes before the SM oxygen supply was depleted.

c). There were three different LM activation checklists contained in the Flight Data File for normal and contingency situations; however, none of these was appropriate for the existing situation. It was necessary to activate the LM as rapidly as possible to conserve LM consumables and CM reentry batteries to the maximum extent possible.

d). Mission Control modified the normal LM activation checklist and referred the crew to specific pages and instructions. This bypassed unnecessary steps and reduced the activation time to less than an hour.

e). The LM inertial platform was aligned during an onboard checklist procedure which manually transferred the CM alignment to the LM

Determinations

(1) Initiation of LM activation was not undertaken sooner because the crew was properly more concerned with attempts to conserve remaining SM oxygen.

(2) Mission Control was able to make workable on-the-spot modifications to the checklists which sufficiently shortened the time normally required for powering up the LM.

43. Findings

a). During the LM power-up and the CSM powerdown there was a brief time interval during which Mission Control gave the crew directions which resulted in neither module having an active attitude control system.

b). This caused some concern in Mission Control because of the possibility of the spacecraft drifting into inertial platform gimbal lock condition.

c). The Command Module Pilot (CMP) stated that he was not concerned because he could have quickly reestablished direct manual attitude control if it became necessary.

Determination

This situation was not hazardous to the crew because had gimbal lock actually occurred, sufficient time was available to reestablish an attitude reference.

44. Findings

a). LM flight controllers were on duty in Mission Control at the time of the accident in support of the scheduled crew entry into the LM.

b). If the accident had occurred at some other time during the translunar coast phase, LM system specialists would not have been on duty, and it would have taken at least 30 minutes to get a fully manned team in Mission Control.

Determination

Although LM flight controllers were not required until more than an hour after the accident, it was beneficial for them to be present as the problem developed.

LM Consumables Management

45. Findings

a). The LM was designed to support two men on a 2-day expedition to the lunar surface. Mission Control made major revisions in the use rate of water, oxygen, and electrical power to sustain three men for the 4-day return trip to the Earth.

b). An emergency powerdown checklist was available in the Flight Data File on board the LM. Minor revisions were made to the list to reduce electrical energy requirements to about 20 percent of normal operational values with a corresponding reduction in usage of coolant loop water.

c). Mission Control determined that this maximum powerdown could be delayed until after 80 hours ground elapsed time, allowing the LM primary guidance and navigation system to be kept powered up for the second abort maneuver.

d). Mission Control developed contingency plans for further reduction of LM power for use in case an LM battery problem developed. Procedures for use of CM water in the LM also were developed for use if needed.
e). Toward the end of the mission, sufficient consumable margins existed to allow usage rates to be increased above earlier planned levels. This was done.

f). When the LM was jettisoned at 141:30 the approximate remaining margins were:

Electrical power	4½ hours
Water	5½ hours
Oxygen	124 hours

Determinations

(1) Earlier contingency plans and available checklists were adequate to extend life support capability of the LM well beyond its normal intended capability.

(2) Mission Control maintained the flexibility of being able to further increase the LM consumables margins.

Modification of LM Carbon Dioxide Removal System

46. Findings

a). The lithium hydroxide (LiOH) cartridges, which remove water and carbon dioxide from the LM cabin atmosphere, would have become ineffective due to saturation at about 100 hour.

b). Mission rules set maximum allowable carbon dioxide partial pressure at 7.5mm Hg. LiOH cartridges are normally changed before cabin atmosphere carbon dioxide partial pressure reaches this value.

c). Manned Spacecraft Center engineers devised and checked out a procedure for using the CM LiOH canisters to achieve carbon dioxide removal. Instructions were given on how to build a modified cartridge container using materials in the spacecraft.

d). The crew made the modification at 95 hours, and carbon dioxide partial pressure in the LM dropped rapidly from 7.5mm Hg to 0.1mm Hg.

e). Mission Control gave the crew further instructions for attaching additional cartridges in series with the first modification. After this addition, the carbon dioxide partial pressure remained below 2mm Hg for the remainder of the Earth return trip.

Determination

The Manned Spacecraft Center succeeded in improvising and checking out a modification to the filter system which maintained carbon dioxide concentration well within safe tolerances.

LM Anomaly

47. Findings

a). During the time interval between 97:15:53 and 97:13:55, LM descent battery current measurements on telemetry showed a rapid increase from values of no more than 3 amperes per
battery to values in excess of 30 amperes per battery. The exact value in one battery cannot be determined because the measurement for battery 2 was off-scale high at 60 amperes.

b). At about that time the Lunar Module Pilot (LMP) heard a "thump" from the vicinity of the LM descent stage.

c). When the LMP looked out the LM right-hand window, he observed a venting of small particles from the general area where the LM descent batteries 1 and 2 are located. This venting continued for a few minutes.

d). Prior to 97:13 the battery load-sharing among the four batteries had been equal, but immediately after the battery currents returned to nominal, batteries 1 and 2 supplied 9 of the 11 amperes total. By 97:23 the load-sharing had turned to equal.

e). There was no electrical interface between the LM and the CSM at this time.

f). An MSC investigation of the anomaly is in progress.

Determinations

(1) An anomalous incident occurred in the LM electrical system at about 97:13:53 which appeared to be a short circuit.

(2) The thump and the venting were related to this anomaly.

(3) The apparent short circuit cleared itself.

(4) This anomaly was not directly related to the CSM or to the accident.

(5) This anomaly represents a potentially serious electrical problem.

CM Battery Recharging

48. Findings

a). About one half of the electrical capacity of reentry battery A (20 of 40 amp-hours) was used during emergency conditions following the accident. A small part of the capacity of reentry battery B was used in checking out dc main bus B at 95 hours. The reduced charge remaining in the batteries limited the amount of time the CM could operate after separation from the LM.

b). Extrapolation of LM electrical power use rates indicated a capacity in excess of that required for LM operation for the remainder of the flight.

c). Mission Control worked out a procedure for using LM battery power to recharge CM batteries A and B. This procedure used the electrical umbilical between the LM and the CM which normally carried electrical energy from the CM to the LM. The procedure was nonstandard and was not included in checklists.

d). The procedure was initiated at 112 hours and CM batteries A and B were fully recharged by 128 hours.

Determination

Although there is always some risk involved in using new, untested procedures, analysis in advance of use indicated no hazards were involved. The procedure worked very well to provide an extra margin of safety for the reentry operation.

Trajectory Changes For Safe Return to Earth

49. Findings

a). After the accident, it became apparent that the lunar landing could not be accomplished and that the spacecraft trajectory must be altered for a return to Earth.

b). At the time of the accident, the spacecraft trajectory was one which would have returned it to the vicinity of the Earth, but it would have been left in orbit about the Earth rather than reentering for a safe splashdown.

c). To return the spacecraft to Earth, the following midcourse corrections were made:

A 38-fps correction at 61:30, using the LM descent propulsion system (DPS), required to return the spacecraft to the Earth.

An 81-fps burn at 79:28, after swinging past the Moon, using the DPS engine, to shift the landing point from the Indian Ocean to the Pacific and to shorten the return trip by 9 hours.

A 7.8-fps burn at 105:18 using the DPS engine to lower Earth perigee from 87 miles to 21 miles.

A 3.2-fps correction at 137:40 using LM RCS thrusters, to assure that the CM would reenter the Earth's atmosphere at the center of its corridor.

d). All course corrections were executed with expected accuracy and the CM reentered the Earth's atmosphere at 142:40 to return the crew safely at 142:54, near the prime recovery ship.

e). Without the CM guidance and navigation system, the crew could not navigate or compute return-to-Earth maneuver target parameters

Determinations

(1) This series of course corrections was logical and had the best chance of success because, as compared to other options, it avoided use of the damaged SM; it put the spacecraft on a trajectory, within a few hours after the accident, which had the best chance for a safe return to Earth; it placed splashdown where the best recovery forces were located; it shortened the flight time to increase safety margins in the use of electrical power and water; it conserved fuel for other course corrections which might have become necessary; and it kept open an option to further reduce the flight time.

(2) Mission Control trajectory planning and maneuver targeting were essential for the safe return of the crew.

Entry Procedures and Checklists

50. Findings

a). Preparation for reentry required nonstandard procedures because of the lack of SM oxygen and electrical power supplies.

b). The SM RCS engines normally provide separation between the SM and the CM by continuing to fire after separation.

c). Apollo 13 SM RCS engines could not continue to fire after separation because of the earlier failure of the fuel cells.

d). The CM guidance and navigation system was powered down due to the accident. The LM guidance and navigation system had also been powered down to conserve electrical energy and water. A spacecraft inertial attitude reference had to be established prior to reentry.

e). The reentry preparation time had to be extended in order to accomplish the additional steps required by the unusual situation.

f). In order to conserve the CM batteries, LM jettison was delayed as long as practical. The LM batteries were used to supply part of the power necessary for CM activation.

g). The procedures for accomplishing the final course correction and the reentry preparation were developed by operations support personnel under the direction of Mission Control.

h). An initial set of procedures was defined within 12 hours after the accident. These were refined and modified during the following 2 days, and evaluated in simulators at MSC and KSC by members of the backup crew.

i). The procedures were read to the crew about 24 hours prior to reentry, allowing the crew time to study and rehearse them.

j). Trajectory evaluations of contingency conditions for LM and SM separation were conducted and documented prior to the mission by mission-planning personnel at MSC.

k). Most of the steps taken were extracted from other procedures which had been developed, tested, and simulated earlier.

Determinations

(1) The procedures developed worked well and generated no new hazards beyond those unavoidably inherent in using procedures which have not been carefully developed, simulated, and practiced over a long, training period.
(2) It is not practical to develop, simulate, and practice procedures for use in every possible contingency.

51. Findings

a). During the reentry preparations, after SM jettison, there was a half-hour period of very poor communications with the CM due to the spacecraft being in a poor attitude with the LM present.

b). This condition was not recognized by the crew or by Mission Control.

Determination

Some of the reentry preparations were unnecessarily prolonged by the poor communications, but since the reentry preparation timeline was not crowded, the delay was more of a nuisance than additional hazard to the crew.

52. Findings

a). The crew maneuvered the spacecraft to the wrong LM roll attitude in preparation for LM jettison. This attitude put the CM very close to gimbal lock which, had it occurred, would have lost the inertial attitude reference essential for automatic guidance system control of reentry.

b). If gimbal lock had occurred, a less accurate but adequate attitude reference could have been reestablished prior to reentry.

Determination

The most significant consequence of losing the attitude reference in this situation would have been the subsequent impact on the remaining reentry preparation timeline. In taking the time to reestablish this reference, less time would have been available to accomplish the rest of the necessary procedures. The occurrence of gimbal lock in itself would not have significantly increased the crew hazard.

PART 4. RECOMMENDATIONS

1. The cryogenic oxygen storage system in the service module should be modified to:

a). Remove from contact with the oxygen all wiring, and the unsealed motors, which can potentially short circuit and ignite adjacent material, or otherwise insure against a catastrophic electrically induced fire in the tank

b). Minimize the use of Teflon, aluminum , and other relatively combustible materials in the presence of the oxygen and potential ignition sources.

2. The modified cryogenic oxygen storage system should be subjected to a rigorous requalification program, including careful attention to potential operational problems.

3. The warning systems on board the Apollo spacecraft and in the Mission Control Center should be carefully reviewed and modified when appropriate, with specific attention to the following:

a). Increasing the differential between master alarm trip levels and expected normal operating ranges to avoid unnecessary alarms.

b). Changing the caution and warning system logic to prevent an out-of-limits alarm from blocking another alarm when a second quantity in the same subsystem goes out of limits.

c). Establishing a second level of limit sensing in Mission Control on critical quantities with a visual or audible alarm which cannot be easily overlooked.

d). Providing independent talkback indicators for each of the six fuel cell reactant valves plus a master alarm when any valve closes.

4. Consumables and emergency equipment in the LM and the CM should be viewed to determine whether steps should be taken to enhance their potential for use in a "lifeboat" mode.

5. The Manned Spacecraft Center should complete the special tests and analyses now underway in order to understand more completely the details of the Apollo 13 accident. In addition, the lunar module power system anomalies should receive careful attention. Other NASA Centers should continue their support to MSC in the areas of analysis and test.

6. Whenever significant anomalies occur in critical subsystems during final preparation for launch, standard procedures should require a presentation of all prior anomalies on that particular piece of equipment, including those which have previously been corrected or explained. Furthermore, critical decisions involving the flightworthiness of subsystems should require the presence and full participation of an expert who is intimately familiar with the details of that subsystem.

7. NASA should conduct a thorough reexamination of all of its spacecraft, launch vehicle, and ground systems which contain high-density oxygen, or other strong oxidizers, to identify and evaluate potential combustion hazards in the light of information developed in this investigation.

8. NASA should conduct additional research on materials compatibility, ignition, and combustion in strong oxidizers at various g levels; and on the characteristics of supercritical fluids. Where appropriate, new NASA design standards should be developed.

9. The Manned Spacecraft Center should reassess all Apollo spacecraft subsystems, and the engineering organizations responsible for them at MSC and at its prime contractors, to insure adequate understanding and control of the engineering and manufacturing details of these subsystems at the subcontractor and vendor level. Where necessary, organizational elements should be strengthened and in-depth reviews conducted on selected subsystems with emphasis on soundness of design, quality of manufacturing, adequacy of test, and operational experience.

The NASA Mission Reports

The NASA Mission Report books come with a bonus Windows CDROM featuring related movies and pictures. Forthcoming books in the NASA Mission Reports Series:-

Apollo 8 - 1-896522-50-5 10" x 7" 192 pg (8 color) $14.95 AVAILABLE NOW!
Apollo 9 - 1-896522-51-3 10" x 7" 240 pg (8 color) $14.95 AVAILABLE NOW!
Friendship 7 - 1-896522-60-2 10" x 7" 216 pg (8 color) $14.95 AVAILABLE NOW!
Apollo 10 - 1-896522-52-1 10" x 7" 184 pg (8 color) $14.95 AVAILABLE NOW!
Apollo 11 Volume 1 - 1-896522-53-X 10" x 7" 248 pg (8 color) $16.95 AVAILABLE NOW!
Apollo 11 Volume 2 - 1-896522-49-1 10" x 7" 168 pg $13.95 AVAILABLE NOW!
Apollo 12 - 1-896522-54-8 10" x 7" 248 pg (8 color) $16.95 AVAILABLE NOW!
Gemini 6 - 1-896522-61-0 10" x 7" 200 pg (8 color) $16.95 AVAILABLE NOW!
Apollo 13 1-896522-55-6 10" x 7" 256 (8 color) $16.95 AVAILABLE NOW!
Apollo 14 November 2000 1-896522-56-4 10" x 7" TBA (w. color) (approx $16.95)
Apollo 15 TBA 1-896522-57-2 10" x 7" TBA (w. color) (approx $16.95)

The series will continue on through Gemini, Mercury, Skylab, STS and unmanned missions.
All prices are in US dollars

Available from Apogee Books, Box 62034, Burlington, Ontario, L7R 4K2, Canada.
http://www.cgpublishing.com

Also available on-line at all good booksellers.

"A unique reference providing details available heretofore only to researchers with access to the archives."...
Library Journal - Feb '99

"(the series) will serve as an invaluable reference tool for aficionados of human spaceflight."...
Astronomy - Jul '99

"The package is guaranteed to put space enthusiasts into orbit."..
Today's Librarian - Jul '99

"..budding Tsiolkovskys, Goddards, and von Brauns will devour each title."...
Booklist - Jul '99

"Highly recommended for space buffs who want detailed information on these flights."... **Choice - Nov '99**

"This series is highly recommended ...A bargain at twice the price!"...
Ad Astra - National Space Society - Jan '00

"...the most ambitious look at our space program to date."...
Playboy - Mar '00

The Watch

You've probably heard about Hollywood's latest version of Doomsday.

It involves either a very large Asteroid or Comet headed on a collision course for Planet Earth. The prospects for humanity are grim indeed, as our very civilization is threatened. Of course along comes either Bruce Willis or Robert Duvall to save the day and everything ends up with a happy ending.

The fact is that the scenarios of a large object possibly hitting the Earth is all too real! Unfortunately we don't have any secret spaceships to deal with this threat. More to the point, we have very little idea of where these objects actually are. Current projections based on empirical data suggest that there are in excess of 2,000 Near Earth Objects (NEO's) currently crossing Earth's orbit of a magnitude and size capable of ending life on Earth as we know it. We know where about 10% of them are, or about 200. We know that these 200 pose no threat to us at present. Of the other 1800, one could hit at any time and we would be defenseless against it.

We must find these objects as a matter of urgency as our very survival as a species depends on it.

The Watch has been formed as a project of The Space Frontier Foundation to raise funds as quickly as possible, to enable our most brilliant astronomers to access the necessary machinery and observatories as quickly as possible. This will enable a global search of the heavens, from the USA to Europe, India, Chile, Australia, Canada and wherever else we need observing stations. The rate of discovery has to increase 10 fold in order to catalog most of these killer objects within the next ten years. The difference that this information makes, just might give us a long-term shot at survival on this small planet. At some point in the near future we might be able to deflect these projectiles from impacting our homeworld, but we can't even think about this in the first place without knowing where they are.

You might agree that this is a possible scenario, but that the chances are that it won't happen in our lifetime, that these events only happen every 10 million years or so. This is just not the case! There are substantial impacts each and every year. Besides when mentioning "Chances" all this indicates is that we just "Don't Know!" Substantial impacts have occurred in this century in Siberia and the Amazon, we have no idea how many have occurred at ocean impacts, and we have had no idea (until late this century) how many tsunamis in the past have been caused by impacts.

This is not panic! It is being prudent! We must find these objects as quickly and efficiently as possible. If we don't, the downside is unimaginable.

Remember as a species all of our eggs are in one basket called Earth.

The Watch as mentioned is a project of The Space Frontier Foundation, a not for profit organisation of private individuals dedicated to opening the Space Frontier to human settlement as rapidly as possible. Our goals include the protection of the Earth's fragile biosphere and creating a freer and more prosperous life for each generation by utilizing the unlimited energy and material resources of space. Our purpose is to unleash the power of free enterprise and lead a united humanity permanently into the Solar System.

The Watch organisation is comprised of an executive staff consisting of Richard Godwin as Executive Director and Rick Tumlinson President of the SFF overseeing the project. There is also The Watch Advisory Council which consists of some of the world's most renowned Planetary Scientists and Astronomers, who provide an active input into the workings of The Watch, and who will advise as to where raised funds can be most actively employed.

The Watch council consists of:

Dr. Richard Binzel, Professor at the Department of Earth, Atmospheric and Planetary Sciences at Massachusetts Institute of Technology (MIT) a leading authority on the Spectra and compositions of the near Earth and Belt asteroids.

Dr. Tom Gehrels: Professor of Planetary Sciences at the University of Arizona. Is the father and leader of the Spacewatch asteroid search programme, which pioneered the use of CCD imaging and real time computer analysis to increase discovery rates of NEO's

Dr. Eleanor Helin: renowned astronomer affiliated with Jet Propulsion Laboratory and for many years a pioneer in the discovery of NEO's.

Dr. John Lewis: University of Arizona, Professor of Planetary Sciences and Co. Director of the Space Engineering Research Centre, U of Arizona. Author of 150 research publications as well as the popular science books, "Rain of Iron & Ice" and "Mining the Sky."

Dr. Brian Marsden: Heads the Minor Planet Center in Massachusetts. A specialist in tracking and categorization of Asteroids and Comets. Member of IAU Nomenclature Committee for Minor planets.

Apollo 13 CD-ROM

The accompanying CD-ROM is designed to use your World Wide Web browser to be viewed. It is programmed to not leave any footprint on your computer (i.e. no drivers, no installation, no updating your Windows registry etc.)

On inserting the disc in your CD-ROM drive you may be prompted to locate the file "Autorun.exe". This file is located in the root directory of your CD-ROM drive. (i.e. it is usually drive "D" but may be "E", "F" etc.) Most computers will find the Autorun program unassisted.

Autorun will open your default Web browser and you can then navigate the contents of the disc just like any web page.

Included on the disc are hundreds of images as well as hours of video. All of the video is in MPEG1 format and should automatically launch your default media player software (such as Windows Media Player).

Included on this disc:

The NASA video documentary "*Houston We've Got A Problem.*"

The pre-explosion in-flight Television broadcast from Apollo 13.

An exclusive video interview from 1999 with Commander James Lovell.

Video of the launch of Apollo 13.

The Post-flight Apollo 13 Press Conference video including the question and answer session.

The complete run of 70mm Hasselblad photographs taken by the Apollo 13 crew.

An extensive NASA acronym database.

The 198 page technical document "Separation Procedures/ Alternate & Abort Missions for Apollo 13". (You will need the program Adobe Acrobat Reader to view this file.)